LYELL IN AMERICA

Charles Lyell, ca. 1842. Daguerreotype. Courtesy of the Department of Geology, King's College, University of London.

LYELL IN AMERICA

Transatlantic Geology, 1841–1853

Leonard G. Wilson

The Johns Hopkins University Press
Baltimore and London

Printed in the United States of America on acid-free paper
2 4 6 8 9 7 5 3 1

The Johns Hopkins University Press
2715 North Charles Street,
Baltimore, Maryland 21218-4363
www.press.jhu.edu

Library of Congress Cataloging-in-Publication Data will
be found at the end of this book.
A catalog record for this book is available from
the British Library.

ISBN 0-8018-5797-X

To Adelia

Contents

Illustrations

Preface

THE first draft of the present volume was written largely between 1970 and 1976, but, in consequence of other demands, its revision and completion have extended over a much longer time. From 1968 to 1972 my research on Lyell was supported by National Science Foundation Grant GS-1819, and from 1975 to 1979 by National Science Foundation Grant SOC 75-00406-Aol. During the academic year 1975–1976, the University of Minnesota gave me a sabbatical leave to work on Lyell.

Mrs. Helen Mammen, my secretary from 1968 to 1979, typed the first draft of the manuscript and additions to certain chapters. Mrs. Elizabeth Carveth, my secretary from 1987 to 1989, retyped the enlarged first draft on a word processor, and Mrs. Cynthia Godin, my secretary from 1989 to 1991, entered into the computer numerous and extensive revisions. Mrs. Leonor Soriano, my secretary since 1991, has done many further revisions required to complete the final manuscript. To each of these ladies, I am deeply indebted for faithful, accurate, and intelligent assistance.

In the course of research for this book, I have received generous help from many people, most especially from Lord Lyell of Kinnordy and his mother, the Lady Lyell, who gave me not only full access to the Lyell papers at Kinnordy House but also most generous hospitality during successive visits there. Mr. Malcolm Lyell of Baverstock, Wiltshire, kindly permitted me to have microfilms of family letters in his possession. I am similarly indebted to all the individuals, libraries, and archives listed under Manuscript Sources for access to letters and other materials relating to Lyell.

In 1980 Dr. James A. Secord read the first draft of the manuscript and offered valuable criticisms and suggestions. Ms. Deborah Klenotic, copyeditor for the Johns Hopkins University Press, in addition to general copyediting suggested the rearrangement of certain topics and improved transitions from one subject to another. I am especially indebted to Philip Heywood, cartographer at the Minnesota Geological Survey, who has drawn the maps of Lyell's American travels for the book.

My deepest debts are to my wife, Adelia, who has for more than twenty years sustained and encouraged me in research and writing. She has accompanied me on research trips to retrace in part the path of Lyell's travels in America: to Virginia, South Carolina, and Georgia in 1971; to New Harmony, Indiana, in 1982; to New Orleans and the Mississippi Delta in 1988; and to Montgomery, Claiborne, and Tuscaloosa, Alabama, in 1989. She has read the manuscript, edited it skillfully, and prepared the index.

LYELL IN AMERICA

Introduction

WHEN Charles Lyell sailed for America in 1841, he was already famous as the author of the *Principles of Geology,* published in 1830. In the *Principles,* Lyell set forth a radical new idea. The earth, he said, was always as it is now. In the distant past, geological changes proceeded slowly and imperceptibly. Rain and rivers wore down the land and waves washed away coastlines, just as they do today. At the same time, earthquakes and volcanic action slowly raised sedimentary strata from the sea bottom to form mountains. Lyell's theories shocked geologists, almost all of whom assumed that geological changes during the past were very different from those of the present. They envisioned a history of the earth punctuated by sudden, catastrophic elevations of mountain ranges and great floods that swept over the land.

In 1832 William Whewell coined the term *uniformitarianism* to describe Lyell's view of the pace of geological change as steady and uniform and the term *catastrophism* to describe the opposite point of view, then held by most geologists.[1] For catastrophist geologists, the steeply inclined, vertical, or even overturned rock strata in mountains suggested former violent convulsions of the earth's surface. In 1841 the Reverend William Conybeare argued that the Alps had risen to their present height in a single violent convulsion.[2] In contrast, Lyell explained the elevation and dislocation of stratified rocks in mountains as the result of ordinary volcanic processes acting slowly throughout long periods of time.

The *Principles of Geology* was a highly persuasive book. In clear, luminous prose, Lyell described geological changes now at work and presented evidence of calm, undisturbed geological conditions on the earth's surface in the past. The book was read widely. By 1841 it was in its sixth edition. It had won enthusiastic adherents, among whom was Charles Darwin. During the voyage of H.M.S. *Beagle* from 1831 to 1836, Darwin studied the geology of volcanic islands

1. Whewell, review of *Principles,* 126.
2. Rudwick, "Conybeare to Charles Lyell," 285.

and of South America in light of Lyell's *Principles*. In 1838 he reported to the Geological Society of London that the rise of the Andes Mountains in South America had occurred gradually, accompanied by earthquakes and volcanic action. Darwin observed the elevation of the coast of Chile by several feet after an 1834 earthquake.[3]

Lyell's view of earth history implied stable, but slowly changing, environments, maintained through the course of gradual geological change. Local changes in geography and climate contributed to the continual extinction of species. The extinction of a species altered the living conditions of surrounding species, producing further extinctions. New species arose in place of the old ones. Through the successive epochs of the Tertiary period, extinction occurred gradually and new species, usually closely related to their predecessors, replaced the extinct in similarly gradual fashion.

In contrast, catastrophist geologists assumed that almost all species became extinct abruptly at the close of each major geological period, destroyed by violent convulsions that shook the earth. After the convulsions, the Divine Creator brought into being a new assemblage of species. To catastrophist geologists, the fossils in successive geological formations were a record of successive Creations. In this way, catastrophist geologists sought to reconcile the emerging knowledge of the fossil record with the biblical account of Creation.

In 1796 the German mineralogist Abraham Gottlob Werner distinguished five classes of rocks. The oldest rocks, the Primitive, or Primary, were crystalline rocks, such as granite, found in the cores of mountains. Next came the Transition rocks, which were stratified rocks, sometimes crystalline. Above them were the Secondary stratified rocks, followed by Volcanic rocks and Alluvial, or water-washed, deposits. Other geologists described Alluvial strata, containing pebbles derived from Secondary rocks, as Tertiary. By 1820 geologists generally recognized four great classes of rocks: Primary, Transition, Secondary, and Tertiary.

Between 1820 and 1840 geology changed from a science based on the mineralogical characters of rocks to one that identified stratified rocks by the organic fossils found in them.[4] Before 1820, geologists traced the succession of stratified rocks by their order of superposition, assuming they had been deposited in horizontal layers

3. Darwin, "Volcanic Phenomena in South America."

4. For a detailed discussion of the early phases of the change as far as 1830, see Laudan, *Mineralogy to Geology*, esp. 138–79.

on the sea bottom. In England in the 1790s, William Smith noted that each stratified rock formation contained a characteristic group of fossils by which he could identify the formation wherever it appeared in the country. Smith studied primarily the Secondary formations from the Oolite of the Cotswold Hills (Jurassic) upward to the Chalk (Cretaceous). These formations outcrop in bands across England in a northeast-southwest direction. Smith's method of identifying formations profoundly influenced the work of early nineteenth-century geologists. Georges Cuvier and Alexandre Brongniart used fossils to distinguish the marine and freshwater Tertiary formations of the Paris Basin.[5] Brongniart also showed that wherever Chalk strata appeared in France or England, they contained the same characteristic fossils. Furthermore, Brongniart found that the fossils in a subdivision of the Chalk, the Gray Chalk (*Craie grise*), were the same as those in the hard black limestone that formed the peaks of mountains in the Savoy Alps. Similarly, fossils in the subdivision beneath the Gray Chalk, called the Chalky Chlorite (*Glauconie crayeuse*), were identical to those in the Greensand, a formation immediately beneath the Chalk in England. English geologists had considered the Greensand distinct from the Chalk, but Brongniart showed that the two were subdivisions of a single formation. Although it was the most conspicuous and familiar rock, the Chalk was only one among a series of limestones, sandstones, and shales constituting what Brongniart named the Cretaceous formation. Whatever their appearance or character, they invariably contained the characteristic Cretaceous fossils. Based on his study of the Cretaceous, Brongniart concluded that fossils offered the most reliable means for the identification of strata.[6]

Conybeare and his coauthor, William Phillips, echoed this opinion in their 1822 account of the geology of England and Wales. They wrote that each geological formation contained "an association of [fossil] species peculiar in many instances to itself, widely differing from those of other formations; and accompanying it throughout its whole course."[7] Conybeare and Phillips traced the succession of English formations from the Tertiary strata of the London and Hampshire Basins down through the series of Secondary formations to the Old Red Sandstone. They did not describe any fossils from the Old Red Sandstone, and they regarded the Transition strata beneath it as lacking fossils.

5. Cuvier and Brongniart, *Géographie minéralogique des environs de Paris.*

6. Alexandre Brongniart, "Terrains de craie."

7. Conybeare and Phillips, *Outlines,* x.

Nevertheless, in 1831 on the border with Wales, Roderick Murchison found a series of strata beneath the Old Red Sandstone containing an abundance of fossils. Earlier he had learned from Smith how to identify rocks by their fossils. He was also familiar with Lyell's use of fossils to classify Tertiary formations. In 1835 Murchison named the strata the *Silurian system* and later published his great monograph describing it.[8] Each subformation of the Silurian system possessed its own distinctive set of fossils. Murchison also showed that the Old Red Sandstone consisted of at least three formations, comparable in their total thickness to the Silurian system.[9] During the 1830s Old Red Sandstone strata yielded a growing number of fossil fishes. John Phillips described the fossils of the Mountain Limestone, a Carboniferous formation immediately above the Old Red Sandstone.[10] The fossils of the Old Red Sandstone were intermediate between those of the Mountain Limestone above and the Silurian strata beneath.

William Lonsdale, curator and librarian of the Geological Society of London, who was describing Murchison's fossil corals from both the Silurian and Old Red Sandstone formations, observed that the rich assemblage of fossils from the limestones of South Devon were, like those of the Old Red Sandstone, intermediate in character between fossils of the Silurian system and those of the overlying Mountain Limestone. The South Devon fossils corresponded to those of the Old Red Sandstone.[11] In 1839, after studying a much larger collection of fossils from the South Devon limestone, Lonsdale, John Phillips, and James Sowerby confirmed the intermediate character of the Devon fossils. The limestones of South Devon were, therefore, contemporary with the Old Red Sandstone. Both series of strata belonged to a distinct formation between the Silurian and Carboniferous that Murchison and Adam Sedgwick called the *Devonian.*[12] During the next two years, John Phillips prepared detailed descriptions of the Devonian fossils.[13]

The recognition of the Devonian formation in 1839 established

8. Murchison, "On the Silurian System"; Murchison, *Silurian System.*

9. Murchison, *Silurian System,* 169.

10. Phillips, *Mountain Limestone District.*

11. Lonsdale, "Limestones of South Devon"; Lonsdale, "Limestones of South Devonshire."

12. Sedgwick and Murchison, "Physical Structure of Devonshire." Murchison's and Sedgwick's conclusions were criticized by Henry De La Beche, who, reflecting the older mineralogical outlook, objected to the reliance on fossils for the determination of the relative ages of strata. See Rudwick, *Devonian Controversy.*

13. Phillips, *Palaeozoic Fossils.*

a succession of geological formations in Europe. They extended from the Silurian and Devonian up through the Secondary and Tertiary, with each formation defined by a characteristic set of fossils. Some fossil species survived through more than one geological period, but most disappeared to be replaced by other species. Even within the Silurian system, Murchison found a great change of species. "Scarcely a single species which existed when the Silurian aera commenced," he observed, "can be detected in the strata which mark its close."[14]

Even more exciting to geologists was the recognition, based on fossils, of European geological formations in distant parts of the world. In America in 1830, Samuel George Morton identified the New Jersey marl as a Cretaceous formation from its fossils. He showed that it was part of a band of Cretaceous strata extending from New Jersey southward along the Atlantic coastal plain as far as South Carolina.[15] In their geological survey of Virginia in 1834, the brothers William and Henry Rogers identified strata along the James River as belonging to Lyell's Eocene and Miocene formations.[16] From the Falkland Islands, Charles Darwin brought back in 1836 a collection of Silurian fossils. Silurian fossils also were brought to London from North America, South Africa, and India. A group of ammonite fossils from the coast of Africa were so similar to those from the English Lias that, when they were laid out at the Geological Society, many of the geologists thought they had been sent from Lyme Regis in Dorsetshire.[17]

In the summer of 1839, a few months after they announced the existence of a Devonian system, Sedgwick and Murchison traveled to Germany to examine the strata below the Coal along the Rhine Valley. They found the same succession of Silurian and Devonian formations as in England.[18] The French paleontologist Edouard de Verneuil accompanied them on their tour and afterward, with Vicomte d'Archiac, identified some twenty-seven hundred species of fossils from the Silurian, Devonian, and Carboniferous strata of Germany and Belgium.[19] The results of Murchison's and Sedg-

14. Murchison, *Silurian System,* 585.

15. Morton, "Organic Remains of the Ferruginous Sand Formation"; cf. Wilson, "Emergence of Geology," 424–29.

16. W. B. Rogers and H. D. Rogers, "Tertiary Formations of Virginia."

17. Murchison, *Silurian System,* 583.

18. Sedgwick and Murchison, "Palaeozoic Deposits of the North of Germany and Belgium."

19. Archiac de Saint Simon and Verneuil, "Fossils of the Older Deposits in the Rhenish Provinces."

wick's tour of the Rhine Valley led Murchison in 1841 to travel to Russia to seek further counterparts of the Silurian and Devonian.

In 1841 Lyell set out for America. The Lowell Institution at Boston had invited him to give a series of public lectures on geology. The honorarium from the lectures enabled him to travel through America and study its geology. He wanted to examine the Cretaceous and Tertiary formations of the Atlantic coastal plain of the United States and compare them with their European counterparts. The geologists of the New York and Pennsylvania surveys had studied the Silurian, Devonian, and Carboniferous rocks of these states, and Lyell wanted to examine them, too. He wished to compare American fossils with those of Europe and to learn whether the North American continent would confirm his belief in the uniformity of geological history.

CHAPTER 1

First Impressions of the New World: The Northeast

ON 20 JULY 1841, Charles and Mary Lyell sailed from Liverpool for Boston aboard the Cunard steamship *Acadia.* Before their departure, Lyell had been reading the recently published reports of the various American natural history surveys, each of which included geological information. "I would not," he told his father, "exchange a tour in Naples or Sicily for one of the Hudson or the St. Lawrence."[1] The ship steamed into Boston Harbor on Monday, 2 August, a clear hot summer's day. As they disembarked, Lyell—an erect, vigorous, sandy-haired gentleman of forty-four, observant but frequently lost in thought—and his dark-haired, vivacious lady must have made a striking couple.

The America they set foot on was a country in rapid transition. Now more than two hundred years old, the English settlements along the Atlantic seaboard had developed a culture that, although bearing the unmistakable stamp of British origin, had become distinctively American. In many of their beliefs and customs, the Americans were conservative. Among the green hills of New England and along the inlets of the Chesapeake Bay, they retained more of the habits of thought of their seventeenth-century British ancestors than did their British counterparts. Yet, except for those who settled in New England, the people who left England for America in the seventeenth century often came from the ragged fringes of British society. Once put down on the American coast, they became subject to the heat of its summers and the cold of its winters, its distances, and its loneliness. In their common struggle to live in a new country, Americans became singularly open in manner and spontaneous in action. They infused old attitudes and customs with a new spirit of freedom rooted in a belief in human equality.

The livelihood of the American coast was trade. Settlers who had not been traders before they arrived soon became traders by necessity. Living in a wilderness with no local markets, they had to take

1. Lyell, to his father, 16–17 July 1841, Kinnordy MSS.

the products of their land elsewhere to be sold. The tobacco of Virginia went to Bristol, Glasgow, and London. The wheat of New England and Pennsylvania went to the West Indies, where it was exchanged for sugar, molasses, and rum. Cotton from the South went to mills in New England or old England to be spun and woven into cloth. American timber went into American ships, which went everywhere.

After the Revolution, Americans had begun to move across the Appalachian Mountains to settle the wider lands of western New York, Ohio, and Kentucky. The interior settlements were handicapped by their great distance from markets and by the forests that enveloped them. Transportation was vital to their life, but at first their only connection with the eastern seaboard was by horseback and canoe. In such circumstances, the advent of the steamboat wrought a real revolution. It made the rivers of America great highways capable of conveying goods and people to distant places easily and quickly. The Hudson, Ohio, and Mississippi Rivers and the chain of Great Lakes began to bear an increasing load of commerce. In 1825 the completion of the Erie Canal offered barges a passage from the Hudson River to Lake Erie, from which point people or goods might be carried by lake steamboat or sailing ship to the whole western country. Within ten years of its completion, the Erie Canal was carrying so much traffic that the state of New York began to widen and deepen it. The great success of the Erie Canal led to the building of many other canals. By 1840 more than thirty-three hundred miles of canals connected the major rivers, lakes, and estuaries of the eastern United States. The river steamboat became a distinctive, dramatic feature of American life. On the western rivers the steamboats were driven by high-pressure engines to give them the power to force their way upriver against strong currents. Occasionally the boilers overstrained by the pressure blew up. The picture of a steamboat aflame at night on the black waters of some snag-filled western river was an all too familiar vignette of life in America. Yet the drive to overcome time and distance, and loneliness and poverty, brushed aside all such hazards.

Similarly, by 1841 railroads were beginning to extend their tentacles through the country, their services frequently linked with those of the steamboats on the rivers, estuaries, and sheltered passages of the eastern seaboard. A railroad had also been built much of the way across New York State, its path often closely following that of the Erie Canal.

To the British traveler of the early nineteenth century, America was both fascinating and appalling. Americans were at once like English people and different from them. Many differences seemed, and some perhaps were, merely the result of provincial uncouthness. Certainly that was what Charles Dickens and Frances Trollope thought. Unfortunately, there was no one himself quite so provincial as the ordinary British traveler in America. The Briton did not realize that with whatever zeal and willingness the American might ape British ways, the mass of the North American continent would defeat him. It would, in time, alter the shape and temper of his mind and his institutions, whatever he did.

By 1840 steam engines and boilers were improved enough to allow the establishment of regular trans-Atlantic steamship service, yet another innovation in travel of profound significance to America. The Lyells crossed from Liverpool to Boston in twelve and a half days. The average time for the crossing by the best sailing ships had been about thirty-eight days. On 29 July, Mary Lyell wrote to her father that they had had "a famous run of it having gone the whole way steadily from nine to eleven miles an hour."[2] They had enjoyed good weather but with frequent headwinds. Lyell marveled at the speed of modern communications. When they stopped at Halifax, Nova Scotia, on 31 July, they posted some letters to be taken by the steamship *Caledonia* back to England. "And in less than a month from the time of our quitting London," wrote Lyell, "our friends in remote parts of Great Britain (in Scotland & in Devonshire) were reading an account of the harbour of Halifax, of the Micmac Indians & other novelties seen on the shores of the New World."[3] Lyell noted that it was only with the help of railroads, recently established in Britain, that such rapid communication was achieved.

Among British travelers, Charles Lyell was perhaps uniquely prepared to view America in broader perspective. His Scottish birth and background, which had always prevented him from feeling fully at home in England, now came to his aid. At least to some small degree an outsider in England, he could view English ways and prejudices with a measure of detachment. A liberal in both politics and science, he had cultivated an open mind—and a calm and rational one. His numerous travels on the Continent had broadened him. He had seen the strength of the German universities in the development not only of the sciences, but of every kind of serious intel-

2. Mary Lyell to Leonard Horner, 29 July 1841, Kinnordy MSS.

3. Lyell, *Travels,* 1:4.

Mary Elizabeth Lyell, ca. 1842. Daguerreotype. Courtesy of the Department of Geology, King's College, University of London.

lectual discipline. The contrast with the miserable weakness of Oxford and Cambridge was all too apparent. In science and education Lyell was prepared to believe that British ways were not the only ones, nor necessarily the best.

Boston

Boston that Monday in August was in the full glow of a summer's day. "The heat here," Lyell wrote, "is intense, the harbour and city beautiful, the air clear and entirely free from smoke, so that the shipping may be seen far off at the end of many of the streets." The Lyells intended to stay in Boston about a week, before going on to look at the geology of New York and Pennsylvania of which he knew from the preliminary reports of their state geological surveys. He would return to Boston in the fall to deliver the Lowell Institute lectures. After the lectures, they would travel to the South for Lyell to study the geology of the Atlantic coastal plain. Shortly after they settled at the Tremont Hotel, "one of the best in the world,"[4] Lyell called on John Amory Lowell, the trustee of the Lowell Institute, whom he found "a plain, unpretending, sensible man and exceedingly kind."[5] They had introductions to the two brothers of Dr. Francis Boott of London, who were active in the cotton industry of Lowell, Massachusetts. Mr. and Mrs. Edward Brooks, among others, called on them. Professor Jared Sparks of Harvard College helped Lyell arrange for the publication of an American edition of his *Principles of Geology*. As a former publisher of the *North American Review* and editor of the writings of both George Washington and Benjamin Franklin, Sparks was well qualified to advise on questions of publication. He had recently visited England where he had heard Lyell deliver a paper at the Geological Society of London the previous November.

With its dignified brick houses and public buildings, its trees and squares, Boston reminded Lyell strongly of home. He wrote to his sister Eleanor that he was "surprised after going 3,000 miles to find the people, language, manners, buildings, churches, naturalists, preachers all so much more English, or less provincial than in numerous parts of Gt. Britain which are only as many hundred miles distant from our capital."[6]

In the surrounding country, Lyell noted "the entire distinctness

4. Ibid.

5. Mary Lyell to Mrs. Leonard Horner, 8 August 1841, Kinnordy MSS.

6. Lyell to Eleanor Lyell, 9 August 1841, Kinnordy MSS.

of the trees, shrubs, and plants from those of the other side of the Atlantic."[7] On a visit to the cliffs of the Nahant Peninsula with Cyrus Alger, the owner of an ironworks in South Boston and an amateur geologist, Lyell observed that in contrast to the plants, many shells on the beach at Nahant belonged to species familiar to him on European shores. On their return to Boston, the physician and naturalist Dr. Augustus Gould invited Lyell to examine his large collection of North American shells. Lyell found about 35 percent of the shells identical with those of Europe, while many of the rest were merely geographical varieties of European species. Such a close correspondence between America and Europe in the life of the sea stood in sharp contrast to the great distinctness of the plants and animals on land.

New Haven, Connecticut

On their departure from Boston, the Lyells traveled by train to Springfield, Massachusetts. There they boarded a steamboat to go down the Connecticut River to Hartford. In the August sunshine the riverbanks were aglow with yellow goldenrod. At Hartford they visited a quarry at Rocky Hill, a large mass of traprock, before boarding a train for New Haven.

When the train pulled into New Haven that evening, Professor Benjamin Silliman of Yale College, a tall, white-haired man of sixty-two, was waiting for them on the platform. As founder and editor of the *American Journal of Science* ("Silliman's *Journal*"), the leading scientific journal in the United States, Silliman had for many years corresponded widely with scientists in European countries and exchanged gifts of books and specimens. The *American Journal of Science* regularly reviewed new scientific books published in Europe, and European scientists looked to it for news of scientific discoveries in the United States. For Americans interested in science, usually amateurs who collected minerals or fossils, or observed the behavior of storms or tides, the quarterly appearance of Silliman's *Journal* was a regular source of stimulus and inspiration. It also offered them an avenue of publication. From western military posts and frontier settlements, from New Orleans, St. Louis, and Cincinnati, from plantations in the Low Country of the Carolinas and Georgia, from new colleges set down in the pine woods of Alabama and Mississippi, men sent items of scientific interest to Silliman in

7. Lyell, *Travels,* 1:4.

the full confidence that if their work possessed any merit he would publish it. Often these same men, living and working in remote corners of the country, had been students at Yale College, where they had attended Silliman's lectures on chemistry and mineralogy.[8]

Since 1830 Silliman had been exchanging letters regularly with Lyell's friend Gideon Mantell, a Sussex surgeon and geologist. A warm friendship had grown up between them based entirely on correspondence. When Silliman learned that Lyell was coming to America to deliver the Lowell lectures, he had been slightly hurt that neither John Amory Lowell nor Lyell had informed him. During the preceding winter, Silliman himself had delivered the first course of lectures sponsored by the Lowell Institute, lectures dealing with geology. Nevertheless, not a small-minded man, he had written immediately to Lyell to invite him to visit New Haven. At the same time, Silliman had written to Mantell to ask what Lyell was like and whether he would bring servants with him. In reply, Mantell had sent a sketch of Lyell based on their friendship of more than twenty years. Lyell's appearance Mantell described as "nothing remarkable, except a broad expanse of forehead," adding that "he is of the middle size, [with] a decided Scottish physiognomy, small eyes, fine chin, and a rather proud or reserved expression of countenance. He is very absent, and a slow but profound thinker."[9] Mantell was concerned that Lyell's reserved manner, which even in England was viewed sometimes as aloofness, might prove to be a liability in the free and democratic society of America. He reassured Silliman that Lyell was accustomed to a plain style of living and that he need not be concerned about being able to entertain him adequately.

As Silliman's carriage, its lamps lit, passed through the tree-lined streets of New Haven in the warm August darkness, Lyell asked Silliman whether he was not taking them to the Tontine Hotel. Silliman replied that he was taking them to his own house, where he urged them to "banish all reserve and be acquainted at once."[10] At the door of his house they were welcomed warmly by Harriet Silliman and her daughters.

New Haven was a small town, situated on an elevated plain between two small rivers at the head of a broad estuary. Large elms lined its streets, providing a pleasant setting for the buildings of Yale

8. Fulton and Thomson, *Silliman;* Wilson, ed., *Silliman and His Circle.*

9. Mantell to Silliman, 14 June 1841, Silliman Family Papers; see Fisher, *Silliman* 2:196–98.

10. Silliman to Mantell, 25 September 1841, Silliman Family Papers; see Fisher, *Silliman,* 2:198–99.

College, which faced the town green. It possessed handsome churches, of which the Lyells counted more than twenty.[11] Surrounding the town on three sides were four hills of red traprock that rose abruptly in precipices hundreds of feet high, like sentinels guarding the town. Two were named according to the direction from town in which they lay, East Rock and West Rock. To the north rose Mill Rock and Pine Rock. In 1806, after returning from a year in Great Britain, Silliman had identified the New Haven traprock as identical to the whinstone, or basalt, of Salisbury Craig at Edinburgh. During the 1830s, in the course of a geological survey of Connecticut, James G. Percival, a former student of Silliman's, observed the crescent shape of the Connecticut trap formations and demonstrated their origin as igneous intrusions, which had altered the sandstone strata above and below them.[12]

The next day, Silliman, his son, and Percival took Lyell to see the traprocks around New Haven, including the trap dike in the slates along Humphreysville Road, the columns and valley of West Rock, the trap dikes and sandstone quarries of East Haven, and the great columns of Mount Carmel. Whether it was the striking geology, the beauty of the scenery, Silliman's invitation to "banish all reserve," or perhaps all three, Lyell on this outing was not the stiff scientist Mantell had described. Silliman thought him "animated and interesting, often eloquent and full of geological zeal,"[13] shouting from one of the hills that it was a glorious country and most picturesque and beautiful. Though Lyell found much to ponder and now and then, with his hand on his forehead, would sink into deep thought, Silliman found him "as lively and agreeable as anyone could be." Silliman was charmed even more by Mary, whom he found "a beautiful and lovely woman of most accomplished manners."[14] That evening, to celebrate their visit, he opened a bottle of champagne.

Lyell continued to note many unfamiliar features of the American fauna and flora. In New Haven he saw his first hummingbird, hovering about the flower of a gladiolus in a garden. In the sleepy heat of August, most of the songbirds were silent, the chief sound being the whir of grasshoppers, but they did hear "a thrush with a red breast, which they call here the robin."[15] They met Harriet Sil-

11. Lyell, *Travels,* 1:12.

12. Percival, *Geology of Connecticut,* esp. 318–19, 429–31; cf. Lyell, *Travels,* 1:13.

13. Silliman to Mantell, 25 September 1841, Silliman Family Papers; see Fisher, *Silliman,* 2:199.

14. Silliman, "Personal Notices," vol. 4 [August 1841], Silliman Family Papers.

15. Mary Lyell to Mrs. Leonard Horner, 8 August 1841, Kinnordy MSS.

liman's uncle, Colonel John Trumbull, then eighty-six years old. Trumbull, the renowned painter of the American Revolution, had studied painting in London under Benjamin West and during the Revolutionary War had been an aide-de-camp to General Washington. Later, he had painted the large scenes of events of the Revolution in the Capitol at Washington.

One evening the Lyells accompanied the Sillimans to tea at the home of the mineralogist Charles Upham Shepard, who lectured on natural history at Yale. During the winters Shepard was accustomed to go to Charleston, South Carolina, to lecture on chemistry at the South Carolina Medical College. He had also helped with the geological survey of Connecticut and gave Lyell information about the mineralogy of the trap masses.[16]

New York and the State Survey

After three days in New Haven, the Lyells departed by steamboat and spent the better part of the day steaming down Long Island Sound to New York City. There they settled into the Astor House Hotel. In New York, Lyell met William Redfield, a businessman who since 1820 had been interested in steamboat navigation, first on the Connecticut River and then on the Hudson River. When frequent disastrous explosions discouraged steamboat traffic, Redfield came up with the idea of towing barges for passengers at the end of lines long enough to ensure safety in case a steamboat boiler exploded. This measure, used only until public confidence in steamboats returned, was typical of Redfield's resourcefulness. His scientific reputation was established by two significant papers published in Silliman's *Journal* in 1831 and 1834 on the behavior of storms along the Atlantic coast.[17] Redfield demonstrated that a storm was a "progressive whirlwind" rotating in a counterclockwise direction about a center that moved in the direction of the prevailing wind. In 1836 Redfield began to collect fossil fish from the sandstones of Connecticut and New Jersey.

Redfield took Lyell across the Hudson River to Newark, New Jersey, to see ripple marks and raindrop impressions in strata of the New Red Sandstone (Triassic); on 16 August he accompanied the Lyells on a steamboat trip up the Hudson to Albany, New York. The size and power of the great Hudson River boats, more than

16. Gloria Robinson, "Charles Upham Shepard," in Wilson, ed., *Silliman and His Circle,* 85–103.

17. Redfield, "Prevailing Storms of the Atlantic Coast"; Redfield, "Hurricanes and Storms of the West Indies."

three hundred feet long and carrying five hundred passengers, impressed Lyell. As they traveled upriver, he and Mary delighted in the extraordinary beauty of the Palisades and the Catskill Mountains.[18]

Lyell was intensely interested in the work of the geological survey of New York State, of which he had read preliminary reports. In Albany he met three of the four geologists who had worked on the survey: Ebenezer Emmons, James Hall, and William Mather. All were thoroughly familiar with his writings. When Governor Marcy was planning the New York survey in 1836, one of the books he read was Lyell's *Principles*, which became a model for the survey and strengthened its emphasis on scientific geology.[19] The oldest and best educated of the geologists to work on the New York survey was Lardner Vanuxem. He had studied in Paris under Alexandre Brongniart and had made geological surveys of North and South Carolina. When he made his initial reconnaissance of central and western New York in 1836, Vanuxem recognized that the horizontal strata north of the Appalachian Mountains dipped beneath the Pennsylvania Coal and were therefore lower and older than it. In the Wernerian classification then in use, they were considered Transition formations, that is, beds intermediate in age between Primary and Secondary rocks, the Secondary beginning with the Coal and extending upward to the Chalk. Previously, the New York strata had been considered Secondary. They were deeply intersected by numerous ravines and river valleys that provided rich sections for study.

The paleontologist to the New York survey was Timothy Abbot Conrad of Philadelphia. After publishing a short work on the living shells of the Atlantic coast in 1831, Conrad began to study the fossil shells of the Atlantic coastal plain. He recognized that these fossil shells corresponded to those of the London Clay in England, a Tertiary formation. Fossil shells from the Carolinas demonstrated to Conrad that the American Tertiary formation extended at least as far south as the Santee River in South Carolina. In 1832 he learned from fossil shells sent from Claiborne, Alabama, that the beds of the river bluff at Claiborne likewise corresponded in age to the London Clay. The many new species among the Claiborne shells made Conrad decide to investigate them. In March 1833 he spent several weeks collecting on the Claiborne bluff, which he found "a perfect Eldorado of fossils."[20] Soon after his return to Philadelphia

18. Lyell, *Travels*, 1:16.

19. Aldrich, "New York Survey," 92–94.

20. Conrad to Samuel George Morton, 20 April 1833, quoted in Wheeler, *Timothy Abbott Conrad*, 31.

in February 1834, Conrad read the third volume of Lyell's *Principles,* published at London while he was in Alabama, in which Lyell classified Tertiary formations as Eocene, Miocene, and Pliocene. He immediately adopted Lyell's term *Eocene* for the beds previously called the London Clay.

When the New York geologists began to study the strata of the state, they were confronted with the problem of naming formations in such a way as to avoid confusing them. They decided to name each formation after the locality where it was best developed and to identify it elsewhere by its characteristic assemblage of fossils. In 1837 Conrad noted that the fossils of the New York strata corresponded generally to those of Roderick Murchison's Silurian formation in England.[21] The following year he correlated three groups of New York strata with particular subdivisions of the Silurian system.[22] Meanwhile, in 1835 Adam Sedgwick had described the Transition strata in North Wales that he named Cambrian.[23] From Murchison's and Sedgwick's accounts, Conrad concluded that the New York formations corresponded to both the Cambrian and Silurian systems of England and Wales. The publication of Murchison's large work on the Silurian system in 1839 enabled Conrad to identify more precisely the New York formations that corresponded to Murchison's Silurian formations. He realized also that the series of New York strata was far more complete than Murchison's series in Wales, containing three times the number of formations, each clearly distinguishable by its fossil contents.[24]

Neither Conrad nor Vanuxem was at Albany, but James Hall offered to accompany Lyell across New York State to show him the geology of the country. At twenty-nine the youngest of the New York geologists, Hall had studied geology at the Rensselaer School in Troy, New York, under Amos Eaton and Ebenezer Emmons and had joined the state survey in 1836 as Emmons's assistant. In 1837, having been appointed geologist of the Fourth District, which comprised the western portion of the state, Hall made preliminary traverses of the Genesee River valley, and in 1838 began his detailed work there. The Genesee and its tributaries had cut deep gashes through the strata, exposing a complete series of beds from Lake Ontario to the Pennsylvania border.[25]

21. New York (State) Natural History Survey, *First Annual Report,* 176.
22. New York (State) Natural History Survey, *Second Annual Report,* 107–19.
23. Sedgwick, "Structure of the Cambrian Mountains."
24. New York (State) Natural History Survey, *Fourth Annual Report,* 200–201.
25. Aldrich, "New York Survey," 174–79, 232–34.

Western New York

All three New York geologists were willing to guide Lyell and Mary to the geology of their respective districts. "The only difficulty," wrote Mary, "is whom to choose. If we go on in this way I think it will be impossible for us to form an impartial judgment of America."[26] Lyell finally accepted Hall's offer to guide them over the geology of western New York, reasoning that it would enable him to see the broadest series of the New York strata. They planned to travel from Albany to Niagara Falls and back along a route that would allow them to see the entire succession of formations from the Potsdam Sandstone up to the Pennsylvania Coal.

Their trip across New York State was for all three travelers an exciting adventure. As they examined the beds of shales, sandstones, and limestones in the sides of ravines and river valleys, and beside pools and waterfalls, Lyell could not keep from bursting into exclamations and extemporaneous lectures on what the fossils told of the teeming life in the waters of ancient oceans. Some of the fossil shells, he told Hall, belonged to genera still living in modern seas. The vast scale of the New York formations, their nearly horizontal strata extending undisturbed and unaltered for hundreds of miles, portrayed vividly the tranquil conditions of the seas in which they had accumulated. The abundance of fossils showed how rich and varied had been the life of the ancient ocean. To Lyell, so familiar with the fossil shells of the Tertiary period and their modern counterparts, the luxuriant life of Paleozoic seas was striking proof of the continuity of life and uniformity of conditions throughout the history of the earth. As Lyell talked, Hall listened. Lyell opened for Hall new vistas of the Paleozoic seas that would influence his work for the rest of his life.

As they traveled together on trains, in hired wagons, and on swaying stagecoaches, scrambling along rock faces by day, and returning to simple country inns in the evening, Hall came to like and admire both Lyell and Mary. Always cheerful, they let no hardship bother them. They delighted in the beauty of the country and were fascinated by its natural history. In the field, Lyell's reserve melted away and Mary's laughter echoed through the stillness of August afternoons.

From Albany the Lyells and Hall traveled by train up the Mohawk River valley to Little Falls. As they rolled along in the warm open air,

26. Mary Lyell to Eleanor Lyell, 23 August 1841, Kinnordy MSS.

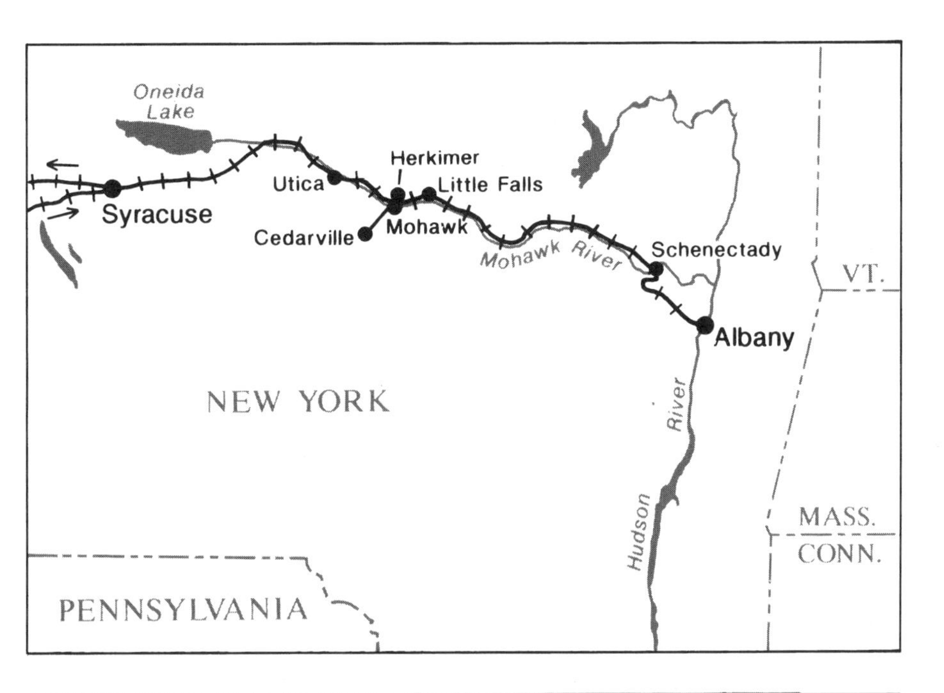

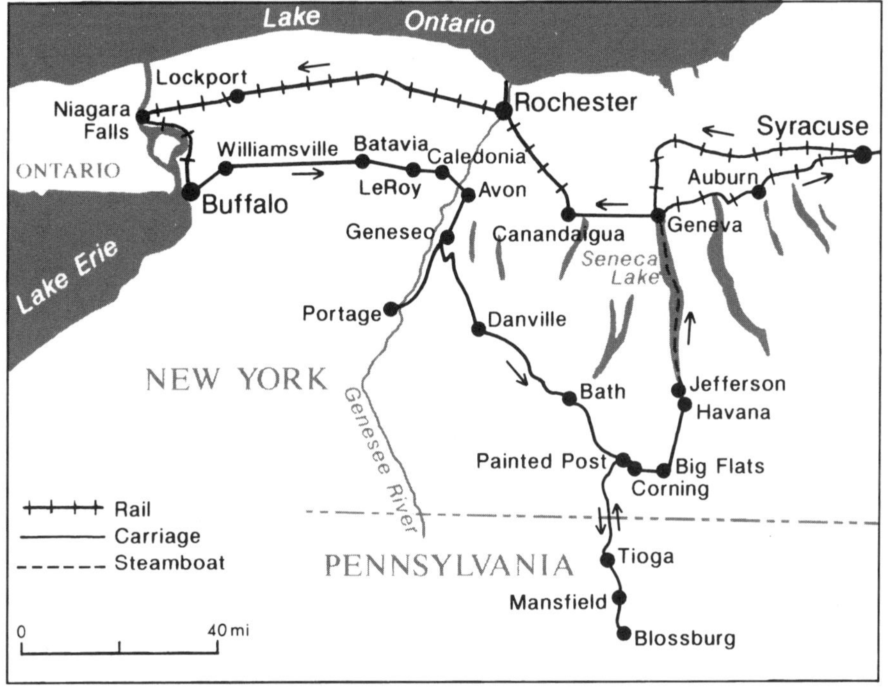

Lyell's Tour across New York State, August–September, 1841. Map by Philip Heywood.

sparks from the wood-burning locomotive sometimes flew into people's faces or burnt holes in their clothes. At several stations they stopped for refreshments.

At Little Falls, the Mohawk River had carved a deep gorge that revealed the line of contact of the nearly horizontal, unaltered Transition strata with the crystalline gneiss of the Adirondacks. The small flow of water over the falls, shrunken by summer drought, allowed Hall and Lyell to examine the rocks with ease. Here Lyell first saw the Potsdam Sandstone, the lowest rock in the New York Transition series, corresponding, he thought, to the lowest Silurian formation.[27] After staying overnight at Little Falls, they hired a carriage and horses to take them across the valley to visit several quarries and to spend the afternoon exploring the ravine of the Cedarville River, which cut through a series of strata, principally of the Onondaga Limestone. They then continued west from Utica to Syracuse and thence to Rochester, traveling most of the way by railroad, but for the twelve miles from Geneva to Canandaigua by stagecoach. "The stages are curious vehicles," wrote Mary, "exactly like some I have seen in country parts of France, with three seats holding three each with a strap across the middle. They go up and down like a boat enough to make one sea sick."[28] As they made their way across western New York, with its rolling hills, newly cleared fields, and pretty villages of white wooden houses set against groves of trees, the Lyells developed a new sense of the spaciousness, fertility, and abundance of America. At Geneva and Canandaigua they passed beautiful, long, narrow finger lakes, and everywhere they saw orchards of apple trees, loaded with ripening fruit, standing amid the stumps of the original forest only recently cleared away.

At Rochester, Hall and the Lyells went to see the upper falls on the Genesee River and explored the ravine below the lower falls. Here, Hall was in the midst of his home ground and could display the geology of the area with authority. Lyell was impressed by the great scale of the New York Silurian strata and their abundance of fossils, their nearly horizontal position revealing their succession in unmistakable order. As they went westward for hundreds of miles, some formations thinned out and were replaced by others.

The prosperity and rapid progress of the country were everywhere evident. At Rochester, where twenty-five years earlier the first

27. It was later found to correspond to Adam Sedgwick's Lower Cambrian.

28. Mary Lyell to Eleanor Lyell, 23 August 1841, Kinnordy MSS.

settler had built his log cabin, there were now twenty thousand people and streets lined with large houses. A few years earlier, the Lyells and Hall would have had to spend many days traveling by barge or canoe along the Erie Canal. Now they traveled at sixteen miles an hour on a railway, often through large swamps or thick forests with occasional clearings. Lyell felt the headlong energy of the people. The farms, villages, and towns all appeared prosperous. Everywhere new schoolhouses showed the eager desire for education, while the many churches reflected the active religious life of the people.[29]

Mary found the social customs more civilized than she had expected. "Though travellers have probably seen every one of the absurd or disagreeable customs they relate," she wrote, "they have either happened in remote half-civilized parts of the country, or in a class of society which could easily find its equivalent in our country." She saw less tobacco chewing and spitting than she had expected and less smoking than in Germany, and that almost entirely among persons who appeared to be farmers, laborers, or small shopkeepers.[30] In the small country inns, they frequently found that the sheets on their bed had not been changed for them, but when she asked for the bed to be made up with clean sheets, it was always done without question.

At Rochester, Lyell and Hall visited the site where the remains of a mastodon had been dug up. They also visited the shore of Lake Ontario and examined the Ridge Road, running parallel with the Lake Ontario shore for nearly a hundred miles. It marked the shoreline of an old proglacial lake and reminded Lyell of the eskers (former glacial river courses) he had seen in Sweden. On the beach of Lake Ontario, Lyell picked up freshwater shells and caught a small tortoise with a "gaily coloured shell," which he later let go. By the lakeshore and along the roads, Lyell saw many yellow butterflies, much like a British species, gathering for their migration southward. "Some times forty clustering on a small spot," noted Lyell, "resembled a plot of primroses, and as they rose altogether and flew off slowly on every side, it was like the play of a beautiful fountain."[31]

At Lockport the Lyells found their first letters from England. Lyell learned that John Phillips's book on the fossils of Devonian rocks had been published. He wrote immediately to order a copy for himself and to ask Phillips to send a copy to James Hall so that it

29. Lyell, *Travels*, 1:23.

30. Mary Lyell to Eleanor Lyell, 23 August 1841, Kinnordy MSS.

31. Lyell, *Travels*, 1:25.

might be cited in the final reports of the New York survey, to be published the following year.[32] Beside the series of locks where the Erie Canal climbed the Niagara escarpment at Lockport, Lyell collected fossils from each of the strata in the escarpment.[33]

In late August, the Lyells and Hall went on to Niagara Falls. "It is a sight never to be forgotten," wrote Mary Lyell, "but the first feeling was more the extreme beauty than the grandeur. Afterwards the immense body of water grew upon me."[34] Lyell recorded similar impressions of his first view of the falls:

> The sun was shining full upon them—no building in view—nothing but the green wood, the falling water and the white foam. At that moment they appeared to me more beautiful than I had expected and less grand; but after several days when I had enjoyed a nearer view of the two cataracts, had listened to their thundering sound, and gazed on them for hours from above and below, and had watched the river foaming over the rapids, then plunging headlong into the dark pool,—and when I had explored the delightful island which divides the falls, where the solitude of the ancient forest is still unbroken, I at last learned by degrees to comprehend the wonders of the scene, and to feel its full magnificence.[35]

For Lyell, Niagara was a geological as well as scenic wonder. He quickly decided that the great gorge, extending to Queenston seven miles below the falls, had been carved out by the action of the river over a long period of time. The falls were steadily retreating upriver toward Lake Erie, as the falling water eroded the soft shale at their base, undermining the harder limestone above. From time to time, great pieces of limestone broke off and fell into the river. As the falls retreated upriver, their height tended to decline with the downward dip of the strata forming them. Lyell reasoned that if the falls had gradually receded upriver from the Niagara escarpment at Queenston, the river above the falls must formerly have flowed at a higher level.

Before coming to Niagara, Lyell had learned that on Goat Island, in the midst of the falls, there were beds of sand and gravel containing river shells, which indicated that the Niagara River had for-

32. Lyell to Leonard Horner, 26 August 1841, Kinnordy MSS; Phillips, *Palaeozoic Fossils.*

33. Lyell, Notebook 91, 20–28, Kinnordy MSS.

34. Mary Lyell to Leonard Horner, 6 September 1841, Kinnordy MSS.

35. Lyell, *Travels,* 1:27–28.

Bird's Eye View of Niagara Falls. Fontispiece of Lyell's *Travels in North America* (1845).

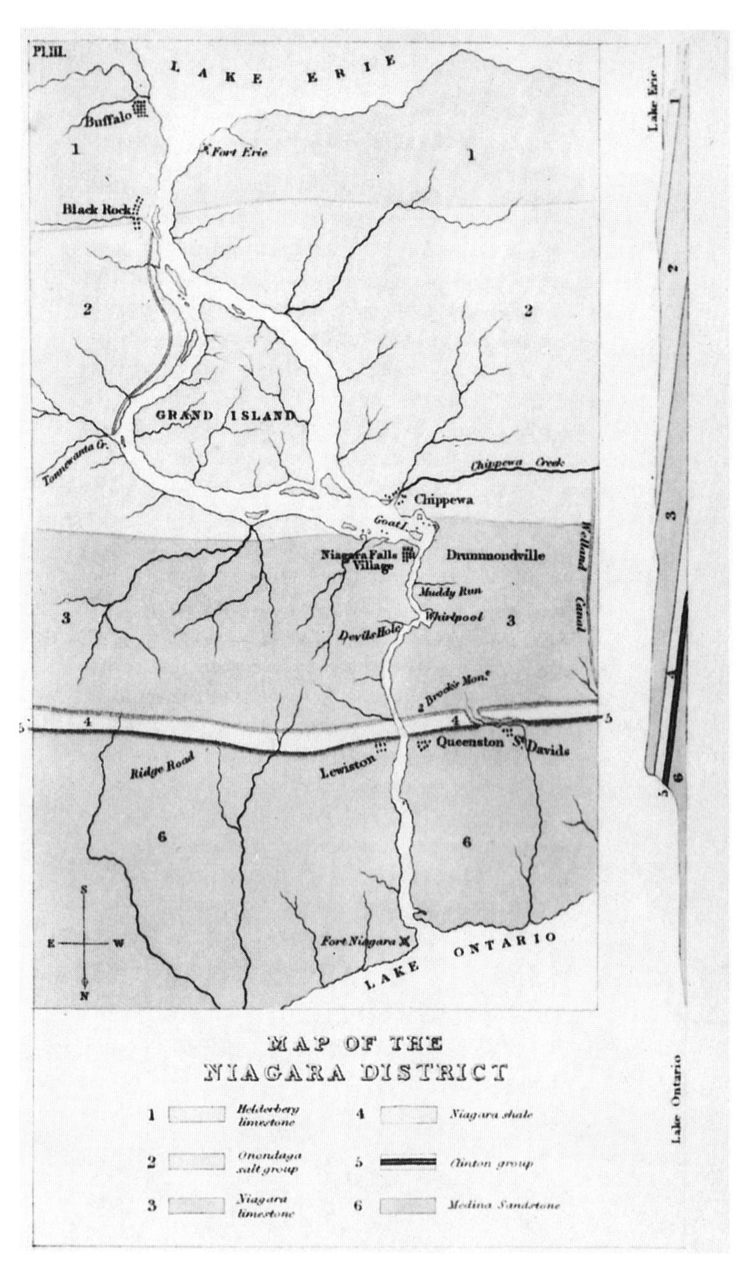

Map of the Niagara District. Lyell's *Travels in North America* (1845), Plate III.

merly flowed at a higher level than at present. He proposed to Hall that they should first examine the Goat Island beds and then see if they could trace them along the rim of the gorge below the falls. They immediately found the same deposits in two river terraces on the American side, cut through by streams at various levels above the falls. Lyell concluded that the original river had cut through deposits of glacial drift. He could trace the old riverbanks "facing each other on both sides of the ravine for many miles below the Falls."[36] He proved conclusively what had long been considered probable, namely, that the falls had originally broken over the Niagara escarpment at Queenston and had slowly receded upriver. He estimated that its recession was at the rate of about a foot a year, and therefore the falls were about thirty-five thousand years old. From the shells he had collected in marl beds below the Ridge Road near Rochester, Lyell knew that Lake Ontario had formerly stood at a higher level. He reasoned that when the level of Lake Ontario had been at the base of the Ridge Road, the falls of Niagara had fallen into a bay of the lake.[37]

For both Lyell and Hall, Niagara was an exciting and rewarding experience. Lyell was delighted to find the action of a river and its cataract during the immediate geological past demonstrated on such an immense scale. Hall, on his part, was seeing the leading geologist of Europe draw together diverse facts to reveal the history of Niagara with startling clarity. But Lyell saw even more at Niagara. He realized that when the geological history of the St. Lawrence River valley and Great Lakes basin was unraveled, it would tell much of the geological history of the whole continent. The Niagara gorge was very young, but the Silurian strata through which it was cut were very old.

> Many have been the successive revolutions in organic life, and many the vicissitudes in the physical geography of the globe, and often has sea been converted into land, and land into sea, since that rock was formed. The Alps, the Pyrenees, the Himalaya, have not only begun to exist as lofty mountain chains, but the solid materials of which they are composed have been slowly elaborated beneath the sea within the stupendous interval of ages here alluded to.[38]

36. Ibid., 39.
37. Lyell, Notebook 91, 14, Kinnordy MSS.
38. Lyell, *Travels,* 1:53.

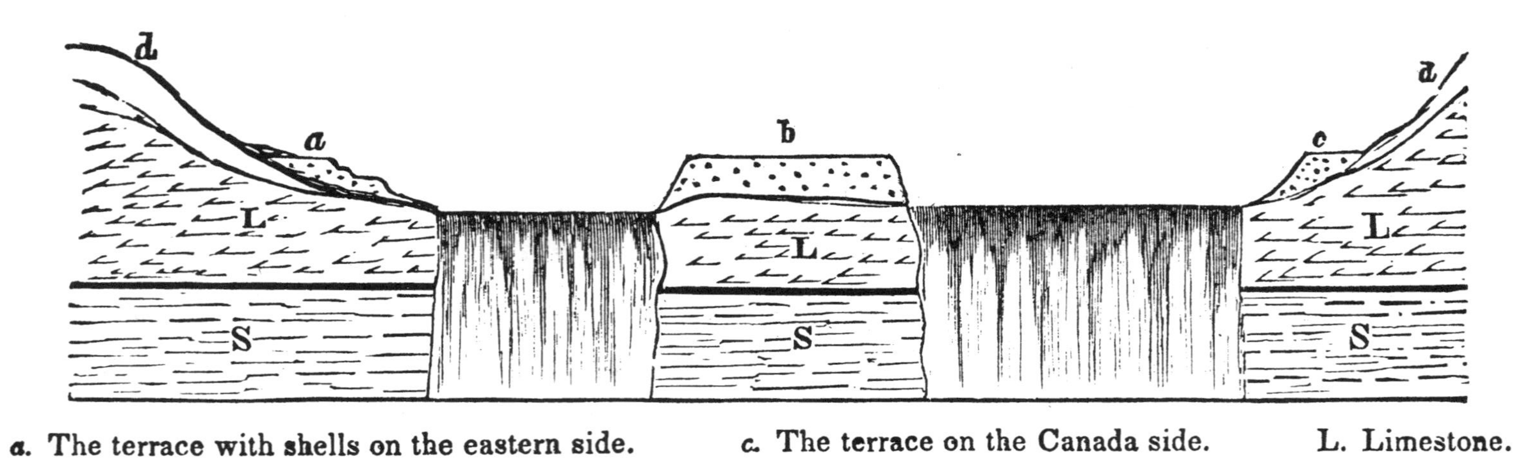

Section across Niagara Falls, Showing Terraces. From Hall's *Fourth Geological District* (1843), p. 397.

On 30 August, after spending three days studying Niagara, Hall and the Lyells went by railway to Buffalo, a large town on the shore of Lake Erie, where they stopped only a few hours. At Buffalo they turned eastward, traveling by stagecoach along a route well to the south of their former one so as to traverse a higher and younger series of strata. At Batavia they hired a carriage to drive to the village of Le Roy, where Lyell examined a falls on the Genesee River. They continued up the valley of the Genesee to Geneseo and thence to Portage. Lyell and Hall walked along the bed of the river to trace the succession of strata, while Mary went along the road in the carriage. They were to meet at the house of Edwin Johnson, who was supervising the building of a new canal.

When Mary drove up to Johnson's large but rustic log house at Portage, surrounded by a forest of tall pines and overlooking a waterfall on the Genesee, Johnson came out to say that they must stay that night at his house "as he knew Mr. Hall, & Mr. Lyell was public property." Johnson introduced her to his wife and three daughters and showed her his copy of Lyell's *Principles.* "Mr. Hall & Charles soon arrived," wrote Mary, "having scrambled up an extraordinary face of rock on hands and knees. We had tea & all sorts of good cakes & preserves & the young ladies played and sang to us."[39] They had a comfortable room, and the next morning the Johnsons were up at six o'clock to give them breakfast.

Lyell regarded the two falls of the Genesee River at Le Roy and Portage as Niagaras in miniature. Like Niagara, they were cutting their way back through the strata. Each had hollowed out a deep ravine or gorge "bearing the same proportion in volume to the body of water flowing through them which the great ravine of the Niagara does to that river."[40] At Geneseo on 3 September, Hall left them to return to Albany, where his wife was expecting a child. On his arrival at Albany, Hall immediately began preparations for a trigonometrical survey of the Niagara River to be included in his state survey report. In the survey Hall determined that the terraces on the two sides of the Niagara River at the falls were, as Lyell predicted, at exactly the same level.

Pennsylvania Coalfields

Meanwhile the Lyells set out for Blossburg, Pennsylvania, to trace the succession of strata upward to the Pennsylvania Coal. They trav-

39. Mary Lyell [to Leonora Horner, 6 September 1841], Journal, 1841–42, 29, Kinnordy MSS.
40. Lyell, *Travels,* 1:56.

Gorge of the Genesee at Portage. Sketch by Eben Horsford. From Hall's *Fourth Geological District* (1843), Plate XIX.

eled by stage to Bath, with a stagecoach driver who had never driven a vehicle before. "Let anyone who does not appreciate the value of railway travelling," wrote Mary, "try an American stage on such a road as we travelled, such jolts & blows & we were twelve hours going 46 miles & did not reach Bath till two in the morning."[41] From Bath they went in a light wagon alongside the Cohocton River to Painted Post, named for a post that marked a place in the forest where the Indians used to gather for hunting expeditions or war. From there they continued to Corning, a town established only four years before amid forested hills at the end of a new railway line. The houses stood among large pine stumps that had not yet been removed. At Corning, Lyell learned that the distinction between the appellations *man* and *gentleman*, of fundamental social significance in Britain, was meaningless in America. Upon asking the landlord of the inn to find his driver, he heard the landlord call out in the barroom, "Where is the gentleman that brought this man here?"[42] Lyell recognized that peculiar though American manners might seem to a Briton, they were infused with courtesy and goodwill.

Leaving most of their luggage at Corning to pick up on their return, the Lyells drove to Tioga, Pennsylvania, where Lyell hired a young man to drive him in a gig to study the geology of the country. He was amazed to learn that the lad, though not yet twenty, owned a farm and had been for several years editor of a newspaper, the *Tioga Democrat.* Meanwhile, Mary rode with "a respectable man, a settler from Vermont," to Mansfield, Pennsylvania, where Lyell joined her the following evening. The weather was beautiful, hot during the day but cool in the mornings and evenings. The American apples, which were beginning to ripen, they found excellent.[43] The prosperity and abundance of the country continued to impress them. Each village, so recently carved out of the forest, had several church steeples. The only drawback was the lack of reliable money, the currency consisting of notes issued by banks, many of which had failed. All bank notes were therefore suspect. "Two young ladies in the stage the other day had all their money refused," wrote Lyell, "because it was of Ohio banks, which led to the exclamation 'Well we are in a pretty fix.' We were recommended to have a monthly 'detector' with the list of banks which have failed in the four preceding weeks."[44]

41. Mary Lyell, Journal, 1841–42, 29, Kinnordy MSS.
42. Lyell, *Travels,* 1:60–61.
43. Mary Lyell and Lyell to Sophia Lyell, 7 September 1841, Kinnordy MSS.
44. Ibid.; Lyell, *Travels,* 1:61; Lyell, "Rocks of Pennsylvania," 555.

At Blossburg a portion of the Appalachian Coalfield overlay the uppermost of the New York formations. About seven years before, a German physician, Dr. Saynisch, had discovered coal at Blossburg and organized a company to mine it. The houses in the community were scattered among small clearings in the woods. Dr. and Mrs. Saynisch welcomed the Lyells to their house. The only luxuries were a rocking chair and a Viennese piano, on which Mrs. Saynisch played music from the opera *Korsna*. Her husband entertained them by recounting how he had shot a wolf from his bedroom window and how the year before a large panther had been killed in the woods nearby.

Saynisch took Lyell to visit the mine, which contained nine seams of bituminous coal, some of them as thick as six feet. Under each seam lay a bed of clay containing fossil *Stigmariae,* with leaves attached to the stems, which extended horizontally through the clay. Above the coal was a layer of white grit or sandstone with numerous impressions of fernlike leaves over its lower surface, resembling dried plants in an herbarium. All the features of the Blossburg Coal beds—the underclay containing *Stigmariae* and the overlying layer of sandstone containing impressions of leaves—were identical with those of coal beds in Britain. In the Pennsylvania Coal many fossil plants were identical with British fossil coal plants.

On their return into New York State, Lyell seized every opportunity along the road to observe the Chemung Shales in stream banks and railroad cuttings. As he and Mary took a steamboat the length of Lake Seneca to Geneva, he observed from the lake the terraces formed on the hills by successive strata of the Chemung formation. At Auburn on Sunday morning, 12 September, they rose at three o'clock to board a train to Albany, where that evening they found James Hall waiting for them at the door of their hotel.[45]

After weeks of staying in small country inns, where clean sheets were obtained only by special request and where they were awakened at five o'clock in the morning by a man ringing a bell in the passage to summon them to breakfast, the Lyells enjoyed the luxury of returning to a large hotel with clean, comfortable beds and a bell to ring for service. Nevertheless, reflecting on their trip through the frontier districts of western New York, Mary wrote to her sister:

> Oh! that some of the poor starving people of England could but migrate here where they want people & whoever is willing to work

45. Mary Lyell to Joanna Horner, 21 September 1841, Kinnordy MSS.

> will get good pay & abundance of the best food. Everybody eats alike at the inns & pays alike also, the pay high. Travelling altogether is as near as possible, at the same rate as in England. As to dress it is quite curious to see the newest Parisian fashions as far west as Niagara in the farmers' houses—gowns made with long peaks & drawn bodices. In short I have hardly seen a gown that could have been made 3 months ago.—The ladies dress extremely nicely—their hair is beautifully neat all which they must do themselves as I do not think I have seen more than two lady's maids all this time.

Their own clothes were nearly worn out. "The dust & the variety of vehicles will have reduced us to rags, I think, by the time we return to Boston."[46] At Albany, Lyell found a letter from Union College at Schenectady saying that the college had made him an honorary doctor of law.

Hall took the Lyells across the Hudson River to Troy to meet Amos Eaton, whom Mary described as "a dirty old geologist who was delighted to see Charles."[47] After climbing to the top of Mount Ida to see the view and observe the effects of an avalanche, which had occurred several years before, they dined at Hall's house in Troy. In the evening they returned across the river to Albany by steamboat.

The next day the Lyells accompanied Hall on a steamboat down the Hudson to Coxsackie, where a mastodon's tusk had been found in a freshwater marl similar to the marls with which Lyell was familiar in Scotland.[48] They proceeded into the Helderberg Mountains, an immense limestone escarpment formed of what were then considered Upper Silurian strata (actually Devonian). At a place called the Indian Ladder, the escarpment opened out into steep cliffs, which offered a wide view of the Mohawk and Hudson Valleys. In mid-September the sugar maples were showing the first changes of color and the panoramic view from the Indian Ladder was beginning to assume its autumn splendor.

At Schoharie, in a valley of the Helderberg Mountains, John Gebhard, a farmer, took them to his house to show them his large collection of Helderberg fossils. Gebhard showed Lyell a slab of what he called Tentaculite limestone, covered with awl-shaped shells. Lyell recognized them as the spines of a sea urchin and advised Gebhard to look for the body of the animal from which they came. Some

46. Ibid.
47. Mary Lyell, Journal, 1841–42, 43–44, Kinnordy MSS.
48. Ibid.

time later, Gebhard found the giant chambered crinoid bulb, *Camarocrinus,* which he called "Lyell's sea urchin."[49] At Schoharie Lyell attracted some attention. "While we were in Mr. Gebhard's collection," wrote Mary, "a number of different people came in & no one took the smallest notice of them. They just stood staring at Charles without speaking & then walked out. It was so very odd I could hardly help laughing."[50]

It was a beautiful day, and the Gebhards took them into the country to visit geological sites and picnic. Just as they were sitting down to their picnic lunch, a large hawk swooped down from a tree to seize a chipmunk so close to them that Hall tried to catch both. The chipmunk escaped and the hawk lost its dinner. "Whether it was that we were just sitting down to our own picnic or not I cannot say," Lyell commented, "but the three geologists, the Gebhards father & son & Mr. Hall, sympathized only with the bird."[51]

On their return to Albany, Lyell called on Governor William Henry Seward to discuss the state geological survey. Lyell urged Seward to publish its final report in the best way possible and with illustrations. John Adams Dix, who was responsible for the original planning of the survey, called on Lyell at his hotel to discuss the publication of the survey results.[52] Lyell told Dix how well he thought the survey had worked, "especially in creating a set of practical & scientific hammer bearers."[53]

On 22 September, the Lyells returned to New York City, whence they set out for Philadelphia. As they crossed New Jersey by train, they compared the comfortable farms with the backwoods country through which they had recently traveled. The houses looked older and the fields contained no stumps. At the Delaware River, they boarded a steamboat to go downriver to Philadelphia.

Philadelphia, with its clean streets and well-built brick houses, they found the handsomest city they had seen in America. At the United States Hotel on Chestnut Street, they were quite comfortable, except for being awakened during the night by the uproar of volunteer fire companies rushing to put out fires that were often false alarms.

49. Clarke, *James Hall,* 41.

50. Mary Lyell, Journal, 1841–42, 45, Kinnordy MSS.

51. Lyell to his father, 21 September 1841, Kinnordy MSS.

52. John Adams Dix (1798–1879) had been appointed adjutant general of New York in 1830 and from 1839 was secretary of state for New York. In 1841 he had just founded at Albany a literary and scientific journal, *Northern Light.*

53. Lyell to Leonard Horner, 21 September 1841, Kinnordy MSS.

In 1841, Philadelphia was the scientific capital of America. It was the former home of Benjamin Franklin, whose experiments on electricity had established him as an international scientific figure. In 1792 the Reverend Joseph Priestley, the discoverer of oxygen, had come to Philadelphia as a refugee from political persecution in England. The city was the home of the American Philosophical Society, the American counterpart of the Royal Society of London, and of the University of Pennsylvania Medical School, founded in 1765. Of more immediate bearing on geology was the founding in 1812 of the Academy of Natural Sciences of Philadelphia. Just two years before the Lyells' visit, the academy had moved into a splendid new building. It was a gift from William Maclure, entrepreneur and geologist, who in 1809 made the first geological survey of a broad area of the North American continent. In 1817 Maclure established the academy's journal, which published valuable papers on American geology and natural history.

Among the most active members of the Academy of Natural Sciences in 1841 was Dr. Samuel George Morton, a busy practicing physician, who, after studying medicine at the University of Pennsylvania, had studied in Edinburgh and Paris and traveled extensively through Europe.[54] Shortly after Morton had settled in practice at Philadelphia in 1826, he began to collect fossils from the marl beds in New Jersey and from the excavations being made for the Delaware and Chesapeake Canal. He was soon struck by the similarity between the fossils he found and those being described for the Greensand and Chalk formations in England and France, respectively; the species were almost all different, but the genera and larger groups were the same. The marl beds of New Jersey and Delaware must therefore correspond in age to the Greensand and Chalk. They were Cretaceous and, since the New Jersey strata were composed largely of green sands and marls, they must correspond to the Greensand of Europe.[55] Morton thus made one of the earliest attempts to correlate precisely, on the basis of fossils, a group of American strata with their European equivalents. Lyell was anxious to see the marl beds Morton described.

At Philadelphia Lyell met for the first time the eccentric bachelor Timothy Abbot Conrad, the paleontologist to the New York sur-

54. Meigs, *Samuel George Morton.*

55. Morton, "Organic Remains of the Ferruginous Sand Formation"; in 1834 Morton had published a revised and extended version of this paper as a separate volume, *Organic Remains of the Cretaceous Group.* See Wilson, "Geology as a Science."

vey. "I have been much pleased with Conrad," Lyell wrote to Hall, "& am sure that were he less isolated & could have more frequent intercourse with congenial souls, he would no longer see difficulties or dwell so much on his constitutional difficulties."[56] On 30 September Lyell and Conrad went to Bristol, Pennsylvania, to call on Lardner Vanuxem, who was then writing his portion of the New York survey report. At Bristol, they crossed the Delaware River into New Jersey to examine the Cretaceous marl beds. From some sixty species of fossils that he collected, Lyell concluded that the New Jersey marls corresponded more closely to the Chalk itself than to the Greensand lying beneath it.[57]

Conrad gave Lyell details of the Cretaceous and Tertiary strata in the southern United States. At Claiborne, Alabama, and other localities, Conrad thought there was a formation intermediate between the Cretaceous and Eocene formations that he called Cretaceo-Eocene. He told Lyell that Dr. Edmund Ravenel of Charleston had reported that a white limestone near Monk's Corner, South Carolina, contained casts of Cretaceous fossils. Conrad also gave Lyell lists of fossil shells collected from Miocene beds in the eastern United States. This was just the kind of information that Lyell had hoped to obtain in America. To her sister Susan, Mary wrote:

> Charles rejoices in having come to America. Not a day passes that he does not learn something by observation or through others. He is certainly stirring up the geologists here very much also. One said the other day, "If I could see you once a week for the rest of my life I think I should do something." Another said, "If I had but known you ten years ago, you would have found a very different collection here & I should have done a great deal more."[58]

Henry Darwin Rogers offered to act as Lyell's guide to the anthracite coalfields of eastern Pennsylvania. Rogers, whom Lyell had met at London in 1832, was professor of geology and mineralogy at the University of Pennsylvania and director of the Pennsylvania Geological Survey, begun like that of New York in 1836.[59] On 2 October the Lyells went with Rogers and William McIlvaine, his assistant, by rail along the valley of the Schuylkill River to Reading,[60] where the Schuylkill cut through the easternmost ridge of the Ap-

56. Lyell to Hall, 1 October 1841, Hall MSS.
57. Lyell, *Travels,* 1:78.
58. Mary Lyell to Susan Horner, 2 October 1841, Kinnordy MSS.
59. Gerstner, *Rogers.*
60. William McIlvaine (1789–1854) was born at Bristol, Pennsylvania, and educated at the Col-

palachians to reveal its structure in cross section. In their geological survey of Pennsylvania, Henry Rogers and his brother, William Barton Rogers, had worked out the remarkable structure of the Appalachian chain, determining that it consisted of a series of parallel ridges, hundreds of miles long, each part of a single simple fold in the strata. The discovery of these folds had a direct bearing on the question of mountain building and on the forces that brought about elevation. "Here the strata were folded," wrote Lyell, "as if they had been subjected to a great lateral pressure when in a soft and yielding state."[61] They had been forced into a series of wrinkles and furrows and thereby elevated into ridges sometimes two thousand feet above the intervening valleys, but the elevation and folding had been accomplished without any sign of igneous activity.

From Reading the Lyells, Henry Rogers, and McIlvaine drove in a hired carriage up through a pass in the Kittatinny Mountains. They traveled slowly, stopping frequently to collect fossils along the road. At the inn at Mount Carbon, two of the other guests—an engineer, a Mr. White, and the manager of the coal mine—recognized Lyell, having met him previously in England. White took the Lyells and their guides to Pottsville to visit an anthracite coal mine. To enter the mine, they walked down a steep incline to a horizontal gallery, or drift, where they boarded a small wagon in which they rode along the gallery, stopping frequently to examine the coal. The strata were almost vertical. On the southern wall, representing the underclay, they saw *Stigmariae,* while on the opposite wall were impressions of leaves and stems of fossil plants. The coal itself also contained impressions of ferns.

At Tamaqua they visited the Lehigh Summit Mine on the top of Bear Mountain. It was an open-pit mine; by "scalping" the mountain of its surface rock and soil, the miners had exposed a nearly vertical seam of anthracite coal forty-five feet thick. Lyell walked along the open fissure created by removal of the coal and observed on one side the coal roof, the rock stratum originally above the coal, and on the other side the underclay, which had lain beneath it. As at Blossburg, the coal roof consisted of shales with impressions of fernlike leaves, while in the underclay were many *Stigmariae,* often several yards long and with leaves or rootlets attached to them.

The consistent presence of a bed of clay containing *Stigmariae* be-

lege of Princeton. A gentleman of leisure, McIlvaine spent his summers on his country estate at Burlington, New Jersey, and his winters in Philadelphia. See Ord, "McIlvaine."

61. Lyell, *Travels,* 1:82.

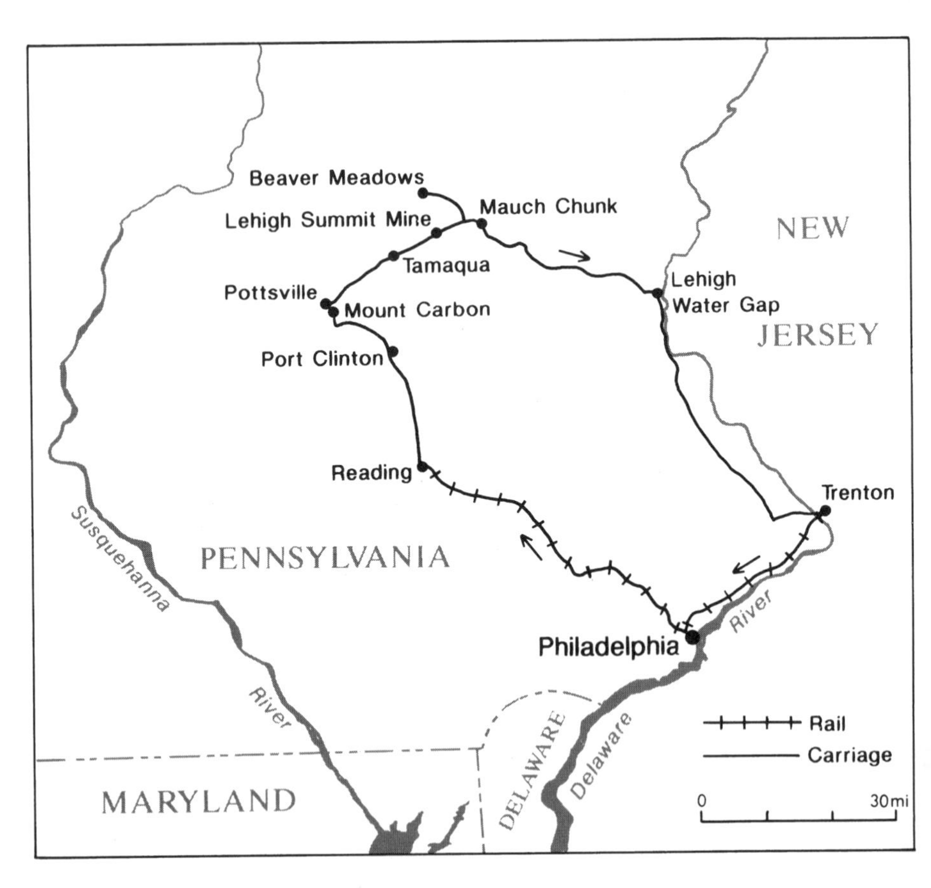

Lyell's Tour through the Coal-Bearing Region of Eastern Pennsylvania, September 1841. Map by Philip Heywood.

neath each bed of coal had been noted in the Welsh coal seams by William Logan only the year before and had thrown new light on the origin of coal.[62] William Buckland and others had previously suggested that the plants found in coal had been "torn away, by the storms and inundations of a hot and humid climate, and transported into some adjacent Lake or Estuary or Sea."[63] There, after floating for a time, Buckland postulated, they sank to the bottom, where they were gradually converted into coal. However, since underneath the coal beds lay beds of clay with *Stigmariae* in them, Logan argued that the *Stigmariae* represented the roots of the plants that had produced the coal. The coal plants, therefore, had grown where the coal had accumulated and now occurred.

When Lyell found in Pennsylvania at both Blossburg and Pottsville exactly the same underclay containing the same *Stigmariae* beneath each coal bed and realized that these two sites, 120 miles apart, were both more than 3,000 miles from the coal beds of Wales, he was astonished at the similarity between the coal beds on the two sides of the Atlantic.[64]

On his return to Boston, Lyell sent Dr. William Henry Fitton an account of the Pennsylvania Coal to be read at the Geological Society.[65] In addition to noting the exact analogy between the coal beds of Pennsylvania and those in South Wales, Lyell described the gradation from the nearly horizontal beds of bituminous coal at Blossburg on the northern edge of the Pennsylvania coalfield to the beds of anthracite coal among the steeply tilted, sometimes inverted, strata of the southern Appalachian ridges at Pottsville and Bear Mountain. The conversion of bituminous coal into anthracite appeared proportional to the degree of disturbance of the strata. In the disturbed regions, the sedimentary strata were also much intruded by veins of syenite and other igneous rocks. Lyell concluded that the conversion of bituminous coal into anthracite was a result of disturbance of the strata accompanied by igneous activity. On this point, Lyell differed from Henry Rogers, who did not recognize any true igneous rocks among the disturbed strata of the anthracite district, evidently subscribing to the older Wernerian view that crystalline rocks such as syenite had been deposited from water.[66]

62. Logan, "Beds of Clay."

63. Buckland, *Geology and Mineralogy*, 1:482.

64. Wilson, "Lyell on North America and Europe."

65. Lyell, "Carboniferous and Older Rocks of Pennsylvania."

66. H. D. Rogers, "Origin of the Appalachian Coal Strata."

At the Lehigh Summit Mine, mules pulled the cars loaded with coal up railways and out of the mine. The loaded cars were then allowed to roll by gravity nine miles down into the Lehigh Valley, where the coal was loaded onto canal barges. The empty cars were drawn back up to the top of the mountain the next morning by mules. "These animals," wrote Mary, "are so accustomed always to go up hill that they refuse to go down a descent. They are sent down in cars in the evening & we saw them start, about sixty mules, four in each car, standing abreast & feeding out of a manger all the way."[67] Like the mules, the Lyells and their guides rolled down the nine miles of railroad in an open car; they then walked down a steep mountain path to an inn in the village of Mauch Chunk.

After visiting another anthracite coal mine at Beaver Meadow, where a layer of blue clay containing *Stigmariae* formed part of the coal roof as well as its floor, the Lyells and Henry Rogers traveled down the Lehigh and Delaware Valleys in brilliant October sunshine. The fall colors on the mountain sides struck the Lyells as almost incredible. "No one would believe the brightness, who had not seen it," wrote Mary.[68] Even as Lyell delighted in the autumn splendor of the mountains, he reflected on the astonishing discovery that the coal beds in southern Pennsylvania were the same in every way as those at Blossburg and across the Atlantic. His mind burning with new ideas about the ancient swamps in which the coal had formed, they left Philadelphia for Boston.

Boston and the Lowell Lectures

Lyell and Mary both thought Boston the most congenial city they visited in America. They stayed there longer than at any other place and made genuinely close friends in the families of George Ticknor, a former professor at Harvard College now at work on a history of Spanish literature, and William Hickling Prescott, already famous for his *History of the Reign of Ferdinand and Isabella* and, though nearly blind, then at work on his *Conquest of Mexico*. They were welcomed also into a much larger circle. Boston, with its long-established families, its wealth, its education, and its ordered life, was perhaps the closest counterpart to London that America could offer. Nevertheless, its Puritan inheritance and the origin of many of its distinguished people in the New England countryside gave the life of Boston a quiet simplicity that London lacked. Its richest men lived

67. Mary Lyell to Fanny [Frances] Lyell, 8 October 1841, Kinnordy MSS.
68. Ibid.

with little ostentation. Moreover, they distributed their wealth in philanthropic schemes usually marked by shrewd practicality and economy. Such a philanthropic scheme was the cause of Lyell's coming to America. The Lowell Institute had been established under the terms of the will of John Lowell Jr., who, after the death of his wife and two children, had bequeathed half his estate to establish a trust to provide free public lectures. He died in Bombay in 1836 at the early age of thirty-seven. His cousin, John Amory Lowell, appointed as trustee under the will, then became the sole director of the Lowell Institute, which he administered for more than forty years.

John Amory Lowell was the epitome of the enlightened Yankee businessman. Hardheaded, strict, and authoritative, yet withal imaginative and liberal, he directed the affairs of the Lowell Institute in the same efficient and unobtrusive way that he ran an industrial empire. All the lectures were given at the Odeon Theatre, rented for the occasion. John Amory Lowell ensured that the income of the trust was kept primarily to pay the lecturers. In this way, he could attract the most distinguished scholars to give lectures and to pay them so liberally that the trust served also to assist their researches. Lyell, Louis Agassiz, and many other scientists benefited from the Lowell Institute. It is fair to say that without the institute, Lyell could not have visited America. Certainly he could not have come so soon or so often, remained so long, or traveled so widely.

The Lyells were scarcely settled at the Tremont Hotel on the morning of Thursday, 14 October, when Dr. Benjamin Cotting, assistant curator of the Lowell Institute, called on them to discuss the arrangements for Lyell's lectures. Mary was writing letters to catch a departing steamer, and Lyell was writing his *Stigmaria* paper to send to Dr. Fitton. George Ticknor called on them, and Charles Sumner, whom the Lyells already had met at London, dropped in to chat. Saturday evening they spent in Ticknor's library, furnished with casts of ancient Greek and Roman statuary standing on top of the bookcases.

The Lyells immediately felt at home with the Ticknors. They had many mutual friends and acquaintances in London and a common experience of travel, life, and culture in Europe. During his European sojourns, Ticknor had met and talked with many famous literary and scientific figures. Lyell found that "whether you spoke of Byron or Scott, Guizot, Mde. de Stael, Schlegel, Humboldt, or almost anyone in our times in any country, including Spain & Italy, you find him full of anecdotes from personal intimacy."[69] On Ticknor's bookshelves Lyell discovered a rich collection of both the works of Dante and the literature about him, a subject in which Lyell's father was deeply interested. In a French work on Dante that had just arrived from Paris, Ticknor showed them an engraving of the elder Lyell's drawing of the bust of Dante.[70] The culture and comfort of the Ticknors' house impressed Lyell, especially the hot water heating system that kept even the halls and passageways pleasantly warm on a chilly fall night. It was evidence of the progress and wealth of America. "We have never once been asked for alms in America," he wrote, "& we are told not to give, if asked, as no one need be a beggar. It is the paradise of the poor whatever some may think it for the rich, & when you see the splendour of the merchants here & their houses, pictures, furniture, gardens & carriages, one begins to think that <u>even</u> they may make it out tolerably."[71]

On Tuesday, 19 October, Lyell gave his first lecture in the Odeon Theatre before an audience of a thousand people on the geology of the Auvergne district of France. Lyell was not a good lecturer, nor a showman. He spoke slowly, often hesitating as he turned from one illustration to another to demonstrate a point. Yet, despite his faults as a public speaker, his seriousness, the clarity of his reasoning, the beauty of his illustrations, and the force of his examples carried his audience with him. He described the successive periods of volcanic eruptions in the Auvergne and presented the evidence for long periods of rest between them. He reconstructed the probable appearance of the country at successive stages, including the kinds of animals present at each stage. He described the series of freshwater strata with their characteristic fossils. The whole lecture was designed to show how a geologist might reason from geological evidence to reconstruct the detailed history of a region.[72] To illustrate the lecture, he used an impressive large-scale painting of Mont Dor and Tartaret, supplemented by diagrams mounted on a green baize curtain that could be raised or lowered on pulleys.[73] Despite a bad cold, Lyell made himself heard throughout the theater, and the lecture seemed to go reasonably well. Mary sat with John Amory Low-

69. Lyell to his mother, 5 December 1841, Kinnordy MSS.

70. Montor, *Dante Alighieri,* 618.

71. Lyell to Frances Lyell, 15 October 1841, Kinnordy MSS.

72. This interpretation of the content of Lyell's first lecture is based on Mary Lyell's letters and on Henry Raymond's transcript of Lyell's lectures as they were delivered later at New York. See Lyell, *Lectures on Geology;* cf. Dott, "Lyell in America."

73. Mary Lyell to Susan Horner, 26 October 1841, Kinnordy MSS.

ell and his family in a box close to the stage. Benjamin Silliman, who came from New Haven for the lecture, was impressed by the illustrations, which he thought "very magnificent in dimensions & in execution." He made an astute assessment of Lyell's public speaking abilities: "He is certainly not a fluent & easy lecturer, but his dignity, simplicity, truth, great personal experience, curious & logical exactness have won for him the confidence & esteem of a thinking people—disciplined—even in the middle rank, to strict intellectual attention."[74]

The next day Lyell repeated his lecture to a second audience. James Hall and Ebenezer Emmons arrived from Albany to attend and called on the Lyells at the Tremont Hotel before the lecture. Hall was to spend the next six weeks at Boston, comparing Lyell's collection of European fossils with his own from New York State.[75]

The Tremont was comfortable, but the Lyells found it too expensive. At the end of the week they moved to the more modest Pearl Street House, where they remained for the rest of their stay in Boston. The next few weeks passed swiftly. As a consequence of repeating each lecture, Lyell was delivering four lectures a week, on Tuesdays, Wednesdays, Fridays, and Saturdays. Betweentimes he was busy with preparations for them. George Ticknor saved them as much as possible from social engagements by inviting them sparingly himself and by discouraging other Bostonians from deluging them with invitations. Mary fulfilled many of their social obligations. Sometimes Anna Ticknor accompanied her as she paid calls. John Amory Lowell invited them to dinner, where the many courses and variety of wines created an impression of great luxury. There was only one such occasion.

After a cold and squally period, the weather turned warm and bright and the Lyells experienced a New England Indian summer. On Sunday, 30 October, they went in a carriage to Cambridge, where, after breakfasting with Jared Sparks, they attended service at the Harvard College Chapel. Henry Ware, then within a year of his death, preached on the text "thou shalt love thy neighbour as thyself." Mary noted in her diary:

> The college was filled with students, fine young men & professors & their families. We afterwards made a number of visits to the Nortons, Websters, whom we interrupted in their dinner &

74. Silliman to Mantell, 14 December 1841, Silliman MSS.
75. Clarke, *James Hall*, 121.

> shared it with them . . . & Professor Lovering who has a number of good books round him & showed us through the observatory where they are making magnetic observations. Came back to tea and went to bed early.[76]

One Sunday the Lyells attended service at Trinity Church on Summer Street. Wherever they were, they usually attended church. In contrast to the Unitarian sermon in the Harvard Chapel, the Episcopalian service had no sermon at all. Mary admired the beauty of the service, and they remained afterward for the Communion Sacrament. She noted in her diary, "My feelings were hurt more than they have been since we landed by seeing after all had communicated, seven coloured women, well dressed as any one, go up by themselves. What a time to make colour a distinction!"[77]

The Boston people continued to be attentive. At Ticknor's house, they met Daniel Webster,[78] who afterwards came to hear Lyell lecture. William Appleton,[79] the publisher who had printed the second edition of Lyell's *Elements of Geology* and the sixth edition of his *Principles* in American editions, called regularly. Old Dr. James Jackson, professor of medicine at Harvard Medical School, came to see them. In 1800 Jackson had helped introduce the practice of vaccination to America, and in 1821 he had helped found Massachusetts General Hospital. He was an uncle to John Lowell Jr., whose estate had established the institute. Mary found him "a very pleasant old man." Dr. Augustus Gould lent them specimens from the collections of shells at the Boston Natural History Society.

Gradually, Lyell's lectures improved. He set before his audience all of the most striking facts that he had observed and interpreted over the years in the Auvergne, Italy, Sicily, and Sweden. He described the remarkable changes of the level of the land revealed by the Temple of Serapis at Pozzuoli near Naples on the Bay of Baiae;

76. Mary Lyell, Diary, 1841–42, 7, Kinnordy MSS. Andrews Norton (1786–1853), from 1819 to 1830 Dexter Professor of Sacred Literature at Harvard Divinity School, was the author of *The Evidence of the Genuineness of the Gospels.* John White Webster (1793–1850) was a professor of chemistry at Harvard College. In 1850 Webster was hanged for the murder of Dr. John Parkman, to whom he owed money. Joseph Lovering (1813–92) was the Hollis Professor of Mathematics and Natural Philosophy in Harvard College.

77. Mary Lyell, Diary, 1841–42, 8, Kinnordy MSS.

78. Daniel Webster (1782–1852), lawyer and statesman, was in 1841 secretary of state for the United States in the administration of President John Tyler.

79. William Appleton (1814–99) had in 1838 formed a partnership with his father, Daniel Appleton, as D. Appleton and Co. The firm published the works of many scientists during the nineteenth century.

the account of its discovery and excavation was particularly fascinating to an audience already deeply under the spell of classical civilization. On 11 November, Mary wrote to Marianne Lyell:

> The lectures go off so well. It quite puts one in spirits. Even I did not suppose he could deliver them half so well & rivet the attention so completely. He has an excellent audience, the best that ever attended these lectures Mr. Lowell told me, that is to say, many literary & other rich people, who have not been in the habit of going there, have been attending regularly. Some of the scientific go both to the evening & afternoon course. He has just finished the 7th [lecture] on the Cheirotherium footsteps which excited great interest & we are now preparing the coal.[80]

On 15 November, John Amory Lowell took the Lyells by railway to the town of Lowell to see the great cotton manufactory there. In 1813 his uncle, Francis Cabot Lowell, had introduced the power loom to America with machines of his own devising and, with his associates, had created some of the earliest integrated textile mills. The Lowell mills had grown into an industry that for European visitors was one of the wonders of the Western world. At Lowell they dined with William Boott and after dinner walked through the Boott mill, the Merrimack mill, and the machine shop. "They are kept beautifully clean," Mary observed, "all the windows perfectly bright & not a broken pane of glass to be seen—all comfortably warmed but not in the least close & the people, both men & women had a thoroughly healthy good-looking appearance."[81] There were about seven thousand young women working in the mills. Mary was impressed by the appearance of the women, whom she considered "nice as lady's maids," and by the order and comfort of the mills, which she thought "something quite surprising." Most of the young women were the daughters of New England farmers, intelligent and respectable. They worked the usual twelve-hour day and were able to save money. Generally they worked for only a few years before they married. Lyell was impressed by the fact that the cleanliness and order of the mills were part of their efficiency. They were run to make high profits and did so.

Lyell thought that although a similar system of management ought to be possible in England, it could not work so well there, because of the severe lack of education among the working classes. In

80. Mary Lyell to Marianne Lyell, 11 November 1841, Kinnordy MSS.

81. Mary Lyell to Leonard Horner, 16 November 1841, Kinnordy MSS.

New England, all children were required by law to attend school at least three months of the year. A tradition of education and a respect for learning pervaded the population. Lyell noted that each year the city of Boston spent for public education the equivalent of thirty thousand pounds sterling, roughly equal to the amount the British parliament appropriated for education for the whole of England.[82]

Lyell was constantly busy with his lectures and could often be found at the Odeon, arranging giant illustrations. On 17 November he spent the whole morning arranging the illustrations for his lecture on the recession of Niagara Falls, while Mary and Mrs. Ticknor went for a walk in bright sunshine around Boston Common. In the afternoon Mary went to tea at the home of Thomas Perkins, a rich merchant who, Mary observed, "spends his money like a prince, endowing hospitals, or anything that will be useful to his fellow men."[83] That evening at Ticknor's house, where they usually went once or twice a week to dine and spend the evening, they attended a party much like those in England. A number of young people danced to the music of a piano, while in a room to one side oyster stew, champagne, and ice cream were served. It was at this party that the Lyells met Prescott, who had just returned to Boston from the country. Mary found him "an interesting, delightful person & very lively."[84] Prescott's mother, a vigorous lady of seventy-six, had gone the Saturday before to hear Lyell's lecture.

On 19 November, Lyell lectured on the geology of Niagara Falls, presenting the evidence of its steady recession. The very large panoramic view of the falls that he had prepared was greatly admired. Joseph W. Ingraham, who had collected historical descriptions of the falls, drew Lyell's attention to the description and drawing of Niagara Falls provided some time before by Father Louis Hennepin. First published in 1678, Hennepin's account showed the existence at Niagara of a third cascade, formed by a projecting rock. Lyell found that by 1751, when Peter Kalm visited Niagara, this rock had disappeared, but Kalm was told of its former existence. This historical documentation for the wearing away of the falls over the preceding two centuries strengthened the evidence for the long, continued recession of Niagara.

82. Lyell, *Travels,* 1:119.

83. Mary Lyell to Marianne Lyell, 22 November 1841, Kinnordy MSS. Thomas Handasyd Perkins (1764–1854) was a merchant, philanthropist, and officer in the Massachusetts militia.

84. Mary Lyell, Diary, 1841–42, 8, Kinnordy MSS.

On 23 November, Lyell lectured on glaciers. The next day he and Mary attended a public ball for the Prince de Joinville in Faneuil Hall, and on Thanksgiving Day, 25 November, they walked with the Ticknors to the Prescotts' for Thanksgiving dinner. After dinner they had "a famous game of blind man's bluff" with the children, after which the elders went into Prescott's library to see his books and method of writing. Mrs. Prescott was a pleasant woman, but in delicate health. In contrast, Prescott was "just like a boy, so full of spirits & gaiety, talking all kinds of nonsense with plenty of sense interspersed."[85]

The day after Thanksgiving, Lyell delivered his final evening lecture, again on glaciers. He suggested that the erratic boulders scattered over the northern area of North America as well as Europe had been transported by ice, either icebergs or glaciers. At the conclusion of the talk, Dr. Jackson and the lawyer John Pickering proposed a vote of thanks for the whole series of Lyell's lectures. This manifestation of enthusiasm was "speedily & wisely stopped by Mr. Lowell lest the Institute should become a place for voting resolutions."[86] Everyone expressed regret that the lectures were over.

The *Boston Evening Transcript* reported that Lyell's Lowell lectures were everywhere deemed to be "of unusual merit."

> His subject was to him *all in all*, and he told a plain tale with all the earnestness of a fond student of nature.
>
> Deeply impressed, however, with the fact of the entire infancy of the science of which he at times so enthusiastically treated; and, careful of being led away by the crude theories of others, he seemed to labor only to set before his hearers the *results*, so far as they might be fully matured, of the careful observations to which the greater portion of his life thus far has been devoted.

The reporter admired the "geological beauty" of the Auvergne region in France and, after describing several other of Lyell's lectures, returned to Mont Dor. "Mr. Lyell has proved that Mt. Dor had ceased to throw out lava, and had even wholly cooled down before Etna began to burn; and yet the records of our race cannot compass even this *latter* period! To what immense *ages* of preparation then does not geology point back to make the earth a fit residence for man." The reporter commented also on Lyell's platform style:

85. Mary Lyell to Frances Horner, 23–26 November 1841, Kinnordy MSS.

86. Mary Lyell to Frances Horner, 23–26 November 1841, Kinnordy MSS. John Pickering (1777–1846) was a Boston lawyer and philologist.

> Though the sometimes slowness and hesitancy of his manner, compelled to extemporaneous speech as he was, by the constant necessity of appealing to and explaining the diagrams by his side, may have caused many frequently to anticipate his conclusions, and wonder at the useless pertinacity with which he piled proof upon proof, yet he deserves our highest thanks for the zeal with which he labored to explain what to *him* rather must have been the *tiresome minutia*, but which he well knew was so necessary for the comprehension of all.[87]

The Lyells' last days in Boston were busy with packing and taking leave of many friends. Snow fell, followed by bright sunshine and a hard frost; the Lyells heard for the first time the sound of sleigh bells in the streets. They attended a breakfast at Francis Calley Gray's house,[88] a dinner at the Ticknors, and a ball given by Mr. and Mrs. John Amory Lowell for their daughter's coming out. So many pleasant occasions and the kindness of so many people made them sorry to leave. Mary thought that Boston would always be one of her favorite places. On 3 December when they left for the South, the Ticknors came to the station to see them off and brought books to amuse them on their journey.

87. *Boston Evening Transcript*, 1 and 6 December 1841.
88. Francis Calley Gray (1790–1856) was a legislator and Harvard benefactor.

CHAPTER 2

Pinelands and Plantations: The Southeast

HIS LECTURES finished, Lyell wished to travel to the South to examine the strata of the coastal plain. Henry Darwin Rogers and William Barton Rogers had classified strata in Virginia as Eocene and Miocene. In South Carolina, Timothy Abbot Conrad thought he had detected a formation intermediate between the Eocene, the oldest Tertiary formation, and the Cretaceous, the youngest of the Secondary formations. If Conrad were right, such a formation would be especially interesting, because in Europe there was nothing intermediate between the Chalk and the Eocene. Most of all, Lyell wanted to learn how American Tertiary formations corresponded to those of Europe.

When the Lyells left Boston by train on 3 December, their luggage consisted of a large black trunk, a black portmanteau, a square box covered with cloth, a black bonnet box, a dressing box, two carpet bags, a hat box, basket, deal box, cloaks, and umbrella.[1] With this array of impedimenta they were to travel through the unending pine forests and along the rivers and estuaries of the southern United States. After stopping overnight at Springfield, Massachusetts, they traveled by stagecoach to Hartford, Connecticut, and then by train to New Haven. On this visit to New Haven, they were guests of Benjamin Silliman Jr. He worked as his father's assistant, both in teaching and in editing the *American Journal of Science,* and lived next door to the elder Silliman.

There was no snow in New Haven, so Lyell took the opportunity to make a geological excursion with the younger Silliman; his brother-in-law, Oliver Hubbard, professor of chemistry at Dartmouth College; and the geologist James Whelpley to see fossil fish in the red Connecticut sandstone beside a small waterfall at Middlefield. The fossil fish had been identified as belonging to the genus *Palaeoniscus,* and for this reason the Connecticut sandstone was believed to belong to the Lower New Red Sandstone, or Permian, group of formations. However, William Redfield thought that the

1. Lyell, Notebook 94, 91, Kinnordy MSS.

Connecticut fossil fish might belong to a new genus, distinct from *Palaeoniscus*. In a similar formation in Virginia, William Rogers had discovered fossil shells that would place the sandstone formations of Connecticut and Virginia in the Upper New Red Sandstone, or Triassic, group of formations. The Connecticut sandstone also contained numerous very large footprints of what appeared to be birds; Edward Hitchcock, president of Amherst College, had studied them extensively.[2] Although identified later as the footprints of dinosaurs, at this time Hitchcock, Lyell, and others believed them to be bird tracks. They were clearly the tracks of a two-legged animal and, apart from their size, resembled modern bird tracks.

After three days at New Haven, the Lyells continued by steamboat to New York City. At the Carlton Hotel, they saw Jared Sparks of Boston, a welcome encounter that made them feel less alone in the city. Daniel Wadsworth, the great landowner of Geneseo, and his daughter were also staying at the hotel. When Lyell and Mary left, William Redfield accompanied them as far as Newark. He showed them a letter from Sir Philip Egerton expressing his pleasure with some fossil fish that Redfield had sent him. In August, Lyell had suggested to Redfield that he send specimens to Egerton so that Egerton could compare them with others in his collection.[3]

At Philadelphia, Lyell arranged with Henry Rogers and William McIlvaine to give a course of lectures there the following spring. He also lent his copies of Roderick Murchison's *Silurian System*, John Phillips's *Palaeozoic Fossils*, and Alcide D'Orbigny's *Paleontologie Française* to Timothy Conrad to use while he was in the South.[4] One evening they went to McIlvaine's house for tea.[5]

The railway cars in which they left Philadelphia on 13 December were drawn by horses. After they crossed the Schuylkill River into the countryside, a steam locomotive was attached in place of the horses. They passed through nearly level country, crossing various estuaries, including the Susquehanna River. Mary admired the beauty of the winter scenery. "There are so many pines & other evergreens that it enlivens the landscape," she noted.

2. Edward Hitchcock (1793–1864); see Gloria Robinson, "Edward Hitchcock," in Wilson, ed., *Silliman and His Circle*, 49–83.

3. Mary Lyell to Leonard Horner, 12 December 1841, Kinnordy MSS.

4. Lyell, Notebook 94, 130, Kinnordy MSS.

5. Mary Lyell to Leonard Horner, 12 December 1841, Kinnordy MSS.

Baltimore and Washington

At President Street Station in Baltimore, the locomotive was disconnected from the railway cars, which were again drawn through the city by horses. That evening they dined on canvasback duck, which Mary thought a sign that they were entering the South. Maryland was the first slave state they entered. Above a slave agent's office they saw a sign: "Men and Women bought and sold here." They began to see signs of poverty and people in ragged clothing. The evidence of poverty in Baltimore reminded them of how much of America—New England, New York, and Pennsylvania—they had traveled through without any indication of poverty. In the North, even free blacks dressed better and looked more prosperous than many of the working people in large towns in England.[6]

While at Baltimore, Lyell examined a collection of Tertiary fossils gathered by Julius Ducatel from the Maryland beds. Between 1835 and 1840 Ducatel had carried out a geological survey of Maryland.

From Baltimore the Lyells made the short journey by rail to Washington. At the newly founded Smithsonian Institution, Lyell saw the many chests containing the collections brought back from the Pacific by the American exploring expedition commanded by Captain Charles Wilkes. While dining with Daniel Webster, then secretary of state in the Tyler administration, Lyell remarked that in Washington everyone seemed to live a mile apart from each other. Webster replied that "Correia, the Portuguese ambassador, called Washington the city of magnificent distances." They visited the Capitol, a building "fine & princely both outside & in,"[7] and attended sessions of the Congress and the Senate. The members of Congress seemed to Mary quite gentlemanlike in manner and appearance, except for their habit of chewing tobacco and spitting out the juice. She was repelled by the spittoons, which became more common as they traveled farther south.

Webster's son took them to the White House to meet President Tyler. Mary thought the White House handsome and well furnished, with fine, large rooms. President Tyler's son and daughter received them, the president himself being engaged in a cabinet meeting. However, as they were going downstairs at the end of their

6. Mary Lyell to Caroline Lyell, 19 December 1841, Kinnordy MSS.
7. Lyell to Caroline Lyell, 19 December 1841, Kinnordy MSS.

visit they crossed paths with Tyler, were introduced, and shook hands.

Afterwards they paid a visit to the British embassy. Although it was half past two in the afternoon, the ambassador's servant said that the ambassador was not yet up, and he did not know when he might be.[8] Evidently, American-British relations were quite calm.

American Tertiary Geology

In 1841 the question at issue in American Tertiary geology was whether American Tertiary formations of the Atlantic coastal plain—all then known—were true equivalents of those of Europe and whether Lyell's classification of European Tertiary formations was valid for America. In 1823, John Finch recognized the existence of Tertiary formations in America. He suggested that various formations of the Atlantic coastal plain corresponded to formations of the Paris Basin. Although he had visited only a few of the geological localities in the Atlantic coastal plain, he thought that between the Allegheny Mountains and the Atlantic Ocean there might be as many as eight or ten distinct formations.[9] Before 1823 geologists usually accepted William Maclure's interpretation. Early in the century he had designated the beds of the Atlantic coastal plain as Alluvial. The strata were not usually consolidated into rock, and Maclure knew little of their fossils. At St. Mary's City, Maryland, Finch made a large collection of fossil shells, which he gave to the Philadelphia naturalist Thomas Say to identify. Say described shells belonging to twenty-eight genera and nearly one hundred fifty species, remarking that the fossil shells were nearly as perfect "as the recent ones of the coast."[10] Say noted also that the formation from which they came probably extended much farther south than Maryland, because Stephen Elliott of Charleston had sent him several fossil shells collected near the Santee River in South Carolina and one of them was identical to a species in Finch's collection.[11]

8. Mary Lyell, Journal, 1841–42, 116, Kinnordy MSS.

9. Finch, "Tertiary Formations in America."

10. Say, "Fossil Shells of Maryland." Thomas Say (1787–1834), entomologist and conchologist, was in 1824 a professor of natural history in the University of Pennsylvania and the curator of the American Philosophical Society. In 1825 he joined Robert Owen's ideal community at New Harmony, Indiana, where he married and lived for the remainder of his life.

11. Stephen Elliott (1771–1830) was the president of the Bank of the State of South Carolina at Charleston. He was the author of *Sketch of the Botany of South Carolina and Georgia,* based on firsthand studies made on his plantation at Beaufort, South Carolina, between 1800 and 1808. He was one of the founders of the Literary and Philosophical Society of South Carolina.

Lardner Vanuxem had described some years earlier the Secondary and Tertiary formations of the Atlantic coastal plain, identifying as Secondary a broad band of strata extending across New Jersey, Delaware, and northeastern Maryland. In the South, Vanuxem maintained, the Secondary strata were overlain by the Tertiary, except in a few places where they came to the surface. Vanuxem's evidence for the Secondary age of the strata was the occurrence in them of Belemnites, and sometimes Exogyrae. He believed they were the same age as the Chalk in Europe. Vanuxem described the Tertiary formations as extending from Martha's Vineyard south to Georgia. Among the fossil shells described by Say from St. Mary's County, Maryland, Vanuxem noted that all of the fossil genera were still represented among living shells. The Tertiary fossils were numerous and all of littoral or shallow-water species, whereas the Secondary were restricted to a few deep-sea species.

Samuel George Morton presented to the Academy of Natural Sciences of Philadelphia both Vanuxem's observations and a paper of his own in which he argued that the Secondary formation of New Jersey corresponded to the Ferruginous Sand formation underlying the Chalk of England. He noted that in France, Alexandre Brongniart considered the Chalk and the Ferruginous Sand to belong to one series of strata (known later as the Cretaceous).[12] In New Jersey, the Ferruginous Sand underlay a Tertiary formation that Morton referred to as the Plastic Clay. Some geologists confused the Plastic Clay of New Jersey with the underlying Secondary beds, whereas others considered it equivalent to the London Clay. Morton agreed with Jeremiah Van Rensselaer's identification of the fossil shell beds of Maryland as equivalent to the Upper Marine formation of the London Basin. The Reverend William Conybeare and William Phillips described the London Basin formations in 1822 in their *Outlines of the Geology of England and Wales.*[13] Unlike the Upper Marine formation, some seventeen species of fossils from the Maryland beds were still living in the coastal waters.

In Maryland near the mouth of the Potomac River, Timothy Conrad found an abundance of fossil shells, almost all belonging to species still living on the coast of the United States. He thought that

12. Morton, "Secondary, Tertiary and Alluvial Formations of the Atlantic Coast"; Morton, "Fossil Shells Which Characterize the Atlantic Secondary Formation."

13. Van Rensselaer, *Lectures on Geology,* 261; Conybeare and Phillips, *Outlines.* Conybeare and Phillips included in the Upper Marine formation the Crag, Bagshot Sand, and Isle of Wight strata and listed the fossils characteristic of each.

the Maryland bed represented the most recent geological formation yet discovered and might be even younger than the Upper Marine formation of Europe. Fifteen miles farther north, in the banks of the St. Mary's River, Conrad found a very different set of fossils. One abundant shell was *Voluta lamberti.* It appeared also in the Crag of Suffolk, part of the English Upper Marine formation, confirming Morton's earlier assertion that the St. Mary's beds belonged to the Upper Marine of England. Other beds near Fort Washington on the Potomac River, just below Washington, Conrad believed to be older than the Upper Marine and more like the London Clay. He found in them two fossils, *Turritella mortoni* and *Cucullaea gigantea,* that occurred in no other beds, and *Venericardia planicosta,* characteristic of the London Clay. Conrad was convinced of the distinctness of the Maryland beds from those of New Jersey, since not a single fossil species was common to both formations. Furthermore, all the genera of the Maryland fossils were still living, many species being found in the waters along the coast.[14]

In the 1820s American geologists were thus beginning to correlate the Tertiary formations of the coastal plain of the United States with those of the Paris Basin and England. Although these tentative correlations were based essentially on fossils, nevertheless geologists usually described the physical characters in some detail, assuming that strata that looked alike would represent the same formation even in widely separated districts. But in 1828 Morton noted that geological analogies between America and Europe founded on fossil organic remains were nearly as obvious in the Secondary and Tertiary formations as in the Primary formation.[15]

The publication of the third volume of Lyell's *Principles of Geology* radically altered the whole effort to correlate American Tertiary formations with those of Europe. Lyell showed that the Tertiary formations of the Paris and London Basins were only one part of a much longer succession of Tertiary formations. He introduced the terms *Eocene, Miocene,* and *Pliocene* to designate distinct periods during the Tertiary. Then he described each period, specifying type formations and characteristic fossils. The whole series of formations of the Paris Basin belonged, he argued, to the earliest Tertiary period, the Eocene. The Miocene period was represented in France by strata at Bordeaux, Dax, and in Touraine, as well as in many other Eu-

14. Conrad, "On the Geology . . . of Maryland."

15. Morton, "The Ferruginous Sand, Plastic Clay and Upper Marine Formations of the United States."

ropean localities. The Pliocene included formations in Italy and Sicily and the English Crag. As an appendix to his third volume, Lyell published tables of Tertiary strata of fifteen different European localities and listed some three thousand species of fossil shells contained in them. The tabular arrangement of so many Tertiary fossil species made it possible to correlate Tertiary strata in other parts of the world with the series of Tertiary formations established by Lyell in Europe.

In the spring of 1833, Isaac Lea returned to Philadelphia from London, where he had obtained a copy of the third volume of Lyell's *Principles* published there in April. At Philadelphia, Lea began to study a collection of fossils from Claiborne, Alabama, sent to him by the Alabama planter Charles Tait. The high sand and limestone bluff on the Alabama River at Claiborne was part of the shell-limestone of southern Alabama, an extensive formation.[16] Among the Claiborne fossils, Lea identified 250 species of shells, 219 of them new. The genera to which they belonged showed that the Claiborne formation corresponded to the London Clay and the *Calcaire grossier* of Paris. Therefore, they were Eocene, as Lyell defined it. None of the new species from Claiborne seemed to belong to living species. The Claiborne formation appeared to be the same age as strata near Fort Washington, Maryland. Lea also mentioned that the strata at Vance's Ferry on the Santee River in South Carolina contained a characteristic Eocene fossil, *Venericardia planicosta.* Eocene formations seemed to extend over a wide area of the United States.

In 1834, based on his own extensive collection from Claiborne, Timothy Conrad confirmed that the Claiborne strata were Eocene.[17] Apart from the bluff at Claiborne, Conrad noted that most of the surrounding country in Alabama was Secondary. In the Claiborne strata Conrad also noted certain species of fossils considered to be Secondary fossils. The supposed Secondary fossils disturbed him. Lyell had written that all of the species, and many of the genera and larger groups, of Tertiary fossils differed from those of Secondary rocks. Although Conrad was confident that the Claiborne formation was Eocene, he thought the American Eocene must be older than that of Europe. He believed that the formation contained some fossils identical with those of Cretaceous beds in Alabama. Similarly, Conrad thought that if the deposits on the Po-

16. Lea, *Contributions,* 21–28.

17. Conrad, "Tertiary and More Recent Formations of the Southern States."

tomac River near Fort Washington were also Eocene, they must be younger than those of Claiborne. Their fossils were generally different from those of Claiborne, and, unlike Claiborne, the Maryland deposits contained no Secondary fossils. Conrad also thought that in the southern states Pliocene beds were of limited extent with only two possible localities, one at Wilmington, North Carolina, and another in South Carolina at Vance's Ferry on the Santee River. He believed that the strata in South Carolina between Charleston and Eutaw Springs were entirely Secondary. In Maryland he considered the more recent deposits on the Potomac River to be Miocene.

In a second paper, Conrad described a number of new fossil shells from the southern states.[18] At Claiborne he had found only a few shells of two species of *Cerithium*, whereas European Tertiary strata contained many species of this genus in abundance. Living species of *Cerithium* occurred in brackish water near the mouths of rivers. The scarcity of *Cerithium* species at Claiborne suggested to Conrad that the Claiborne strata were an open-sea deposit.

In a further study of the occurrence of Pliocene strata, Conrad found two groups of beds above the Eocene located between Chesapeake Bay and the Potomac River in Maryland. The older deposit was characterized by the fossil *Perna maxillata* and contained fewer living species. Conrad decided that American Tertiary formations did not correspond exactly to those of Europe. Nevertheless, he proposed to use Lyell's terms "inasmuch as they are equally descriptive of, and applicable to, the American formation."[19]

In the United States, there was a marked distinction between Eocene and Pliocene deposits. "No one species of shell is common to both," wrote Conrad, "whilst the former contains a few secondary fossils."[20] In contrast, Gerard Paul Deshayes's tables of European Tertiary fossil shells showed many fossil species common to two or more Tertiary formations. In a second paper in 1835, Conrad discussed the distribution of fossil shells that would exist if the present Atlantic coastline of the United States were elevated as much as the Pliocene beds had been elevated during the geological past. He showed that the fossils would be distributed through the strata in much the same way as in the Pliocene strata.

Conrad expressed even more clearly his uncertainty whether American and European Tertiary formations were of the same ages.

18. Conrad, "New Tertiary Fossils from the Southern States."

19. Conrad, "Tertiary Strata of the Atlantic Coast," 106.

20. Ibid.

Conrad's Classification of American Tertiary Formations

Newer Pliocene	Nearly all recent species of the coast
Medial Pliocene	About 30 percent of recent species St. Mary's River, Maryland; Yorktown, Virginia; James River, near Smithfall [Smithfield], and Suffolk, Virginia
Older Pliocene	*Perna Maxillata*—Very few recent species Potomac River, Maryland
Miocene	Probably wanting
Eocene	No recent shells of our coast and a few secondary species Vance's Ferry, South Carolina; Shell Bluff on the Savannah River, Georgia; Claiborne, Alabama

By adopting Lyell's terms, Conrad said that he did not mean to imply that American Tertiary formations were of the same ages as those of Europe, merely that they lay in the same relative positions. The earth movements that elevated them had probably been restricted to the American continent.[21]

Conrad's uncertainties, and his changes in the names assigned to various Tertiary formations, provoked a protest from William and Henry Rogers. When describing in 1835 the Tertiary formations of Virginia, they noted that Conrad had assigned various strata in Virginia, along with the beds at St. Mary's, Maryland, to his *Medial Pliocene,* describing the *Medial Pliocene* as containing 30 percent of recent species. The Rogers brothers pointed out that by Lyell's classification of Tertiary formations, strata containing 30 percent of recent species should be considered Miocene. In fact, they found about 18 percent of recent species among nearly a hundred species from the Tertiary beds of Virginia, a proportion almost exactly that of European Miocene strata, showing, they wrote, "a remarkable, though no doubt accidental coincidence."[22]

The Rogers brothers also showed that Virginia possessed an extensive Eocene formation, occurring in a broad band from the Potomac to the Roanoke Rivers. The best locality for collecting the fossils of the Virginia Eocene was at City Point at the junction of the Appomattox and James Rivers. The City Point beds contained *Cardita (Venericardia) planicosta* and other characteristic fossils,

21. Ibid., 282.
22. W. B. Rogers and H. D. Rogers, "Tertiary Formations of Virginia," 334.

known from both the Paris Basin and Claiborne, but they also contained a number of new fossil shell species. At Coggin's Point, farther down on the south bank of the James River, the Rogers brothers found Miocene strata, containing *Pecten* and *Pectunculus* shells and a multitude of sharks' teeth, overlying the Eocene in the same cliff.

Conrad was curiously reluctant to use Lyell's term *Miocene* to describe American formations. In his publications after 1835 he no longer used the term *Medial Pliocene* for the strata that the Rogers brothers had shown were Miocene; instead, in 1840 he used the term *Lower Tertiary*.[23] In 1841 he used this same term to describe Eocene formations.[24]

In 1839 James T. Hodge, a young geologist from Massachusetts, published an account of the geology of the southern Atlantic states. He used Conrad's terms *Upper, Medial,* and *Lower Tertiary,* rather than Lyell's *Pliocene, Miocene,* and *Eocene.* Hodge explained that he preferred terms that indicated merely the relative positions of the beds. In America, he argued, both living shells and fossil shells were as yet so little known that it was premature to use Lyell's terms.[25] In England Lyell's use of the proportion of living species to extinct species among fossils to classify formations had suffered attack. Although by 1840 Lyell had proved that it was an accurate guide to the relative ages of strata in the complex question of the age of the Crag formations, Conrad and Hodge may have been more aware of the criticisms of Lyell's classification than of its vindication. Hodge's paper was also a superficial survey of the whole coastal plain from Virginia to Georgia, in contrast to the detailed and accurate study of Virginia made by the Rogers brothers.

In view of the confusion resulting from the divergent opinions of Conrad, the Rogers brothers, and other American geologists, Lyell was eager to examine for himself the Tertiary formations of the southern coastal plain and to visit as many as possible of the sites where American geologists had collected fossils.

Virginia and North Carolina

As their steamboat dropped down the Potomac River on the evening of 17 December, it was snowing. Next morning the Lyells landed at Acquia Creek in bright sunshine. They traveled the nine

23. Conrad, "New Fossil Shells from N. Carolina."

24. Conrad, "Fossil Shells in the Medial Tertiary Deposits of Calvert Cliffs, Maryland," 33.

25. James T. Hodge (1816–71); Hodge, "Secondary and Tertiary Formations of the Southern Atlantic States," 101–2.

miles to Fredericksburg, Virginia, through a snow-covered countryside "in a wretched stage with many jolts & bumps."[26] After crossing the Rappahannock River, they continued by railway to Richmond, a handsome town on hills overlooking the rapids of the James River.

At Richmond, Lyell had his first opportunity to study the Virginia Tertiary formations. He was interested especially in the ravines around Richmond, which were cut through Tertiary strata. From beds of red clay in the banks of Shockoe Creek, he collected casts of shells characteristic of the European Miocene.[27] By microscopic examination, William Rogers had found that a bed of yellow clay beneath the red was composed almost entirely of minute siliceous infusoriae also of Miocene species. Lyell mistakenly thought that the beds of clay were the same age as the underlying marl, which in fact was Eocene.

On 19 December the Lyells went by rail from Richmond to Petersburg and on to City Point, a small landing place on the south bank of the James River. They stayed at Moody's Hotel, situated on a narrow point between the Appomattox River and James River, "a good style of country inn looking full upon the broad estuary." As the sun was setting, they took a walk along the sandy beach below the bluffs. All traces of snow had disappeared. Along the densely wooded banks of the James, not a house was in sight. The air was "perfumed with a delightful fragrance decidedly like the hyacinth," but they could not learn from what tree or shrub it came. As they walked admiring the magnificent sunset and the wild beauty of the broad river, Lyell and Mary enjoyed imagining that they were with Captain John Smith on his first landing in Virginia. During their walk, they went into a humble cottage where three black men were "eating a comfortable supper of meat & squash." On their return to the inn, they had tea with the landlord. The courtesy of the slave servants, who seemed of much higher quality than those in their hotel at Washington, impressed them. At meals they were served generous quantities of meat and hot bread. "Here in Virginia," Mary wrote, "they bake three times a day & during the meals a second hot loaf is brought in as the first has cooled too quick."[28]

26. Mary Lyell, Journal, 1841–42, 118. Charles Dickens also gave an eloquent account of this stagecoach ride in his *American Notes,* 2:8–15.

27. Lyell to Caroline Lyell, 19 December 1841, Kinnordy MSS. On 20 December, Lyell went to the Capitol at Richmond to examine William Barton Rogers's geological map of Virginia made during the geological survey from 1834 to 1837.

28. Lyell and Mary Lyell to Caroline Lyell, 19–20 December 1841, Kinnordy MSS.

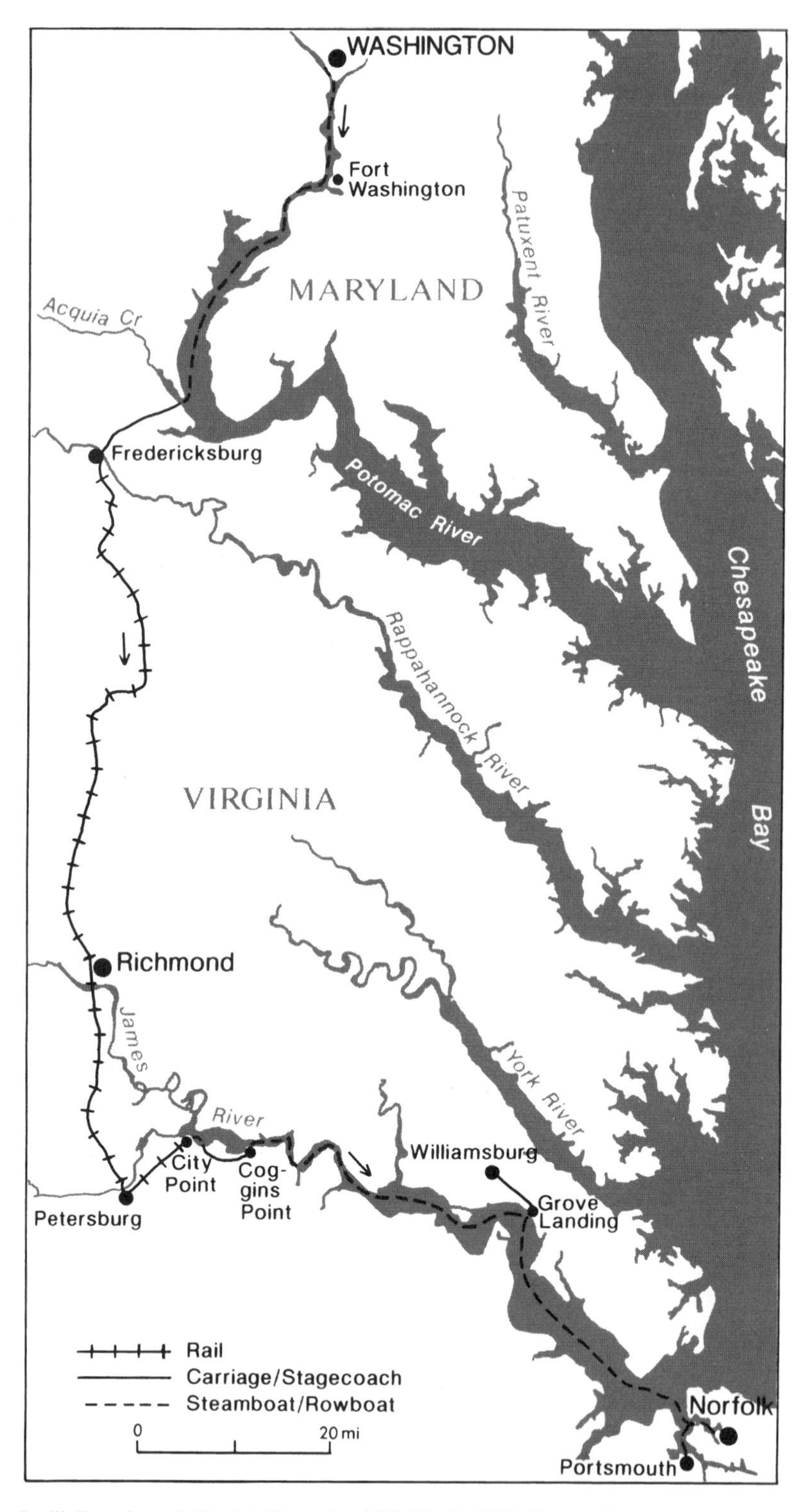

Lyell's Tour through Virginia, December 1841. Map by Philip Heywood.

In the bluffs at City Point, Lyell found Eocene beds of dark green clay overlain by Miocene beds of red, ochreous, yellow, and white sands, as had been described by the Rogers brothers. In the Miocene beds, the species of shells were very similar to those of the Suffolk Crag. The Virginia Miocene overlay the Eocene clay just as the Crag overlay the London Clay along the coast of Suffolk in England. On the James River, Lyell was finding all that he could wish for in the form of series of sediments.

From their room at the inn next morning, Lyell and Mary saw the sunrise. Taking their baggage, they set off downriver in a rowboat, following the south bank of the James. At Evergreen plantation, about two miles below City Point, they stopped to collect fossils. In the Miocene beds, Lyell was struck by the abundance of one fossil, an *Astarte,* much like one of the most common fossils of the Suffolk Crag in England, *Astarte bipartita.* Other fossil shells in the riverbank at Evergreen similarly reminded him both of the Crag and of the faluns of the Loire in France. He also found a fossil coral and some fossil sharks' teeth.[29]

Through the morning they collected fossils along the shore. When it began to rain, they went up to the mansion of Evergreen, occupied by Edmund Cocke. Although Lyell had no letter of introduction, Cocke invited them in to dinner most hospitably. After the morning on the wintry bank of the James River, they enjoyed the hot mince pie and hot corn bread they were served. After dinner, Cocke lent them his carriage and coachman to drive some seven miles through the fragrant woods of long-needled pine and holly to Shellbanks, the plantation of Edmund Ruffin Jr., standing on the high bluff of Coggin's Point overlooking the James estuary.

Ruffin's father had introduced Virginians to the use of marl to improve impoverished soils and had thereby helped to revive agriculture in the state.[30] The elder Ruffin lived at Petersburg, Virginia, where he edited the *Farmer's Register,* while his son operated Shellbanks. Despite having no notice of their coming, the younger Ruffin welcomed them into a hospitable plantation house with "large wood fires & smoking hot wheaten bread at tea." It was, nevertheless, "a cold house with great draughts." Ruffin was supervising the salting down of pork to feed his slaves through the winter, as, he said, "all Virginia farmers must." Before breakfast next morning he took

29. Lyell, "Miocene Tertiary Strata of Maryland, Virginia, and of North and South Carolina," 416.

30. Edmund Ruffin (1794–1865); see Ruffin, *Essay on Calcareous Manures.*

Lyell about a mile and a half from the plantation house to a place where shell-marl was dug to put on the land, "just as in Suffolk and on the Loire."[31] In the marl pit, Lyell collected more than thirty species of fossil shells, beautifully preserved. Most of the genera, and many of the species, were the same as those he had collected in 1840 in France from the faluns of the Loire.

As the sun was rising next morning, Mr. and Mrs. Ruffin accompanied the Lyells to the water's edge, where they said good-bye. Four of the Ruffins' servants rowed the Lyells about two miles out into the broad James estuary in the bitterly cold air to wait for the steamboat. As she shivered in the cold, Mary noted that the slaves' clothes were "ragged and patched." To her sister Susan she wrote:

> Charles and I could not help laughing & being glad papa had not Ulfrida's glass to see us sitting in a large boat, with 4 niggers, tied to a post in the middle of the James river wrapped in cloaks & bear skins & a blanket to keep out the bitter North East wind. We were delighted to get on board the steamer to a comfortable fire.[32]

The steamboat took them down the James River past the remains of old Jamestown, the first permanent English settlement in America, marked only by a crumbling church tower. They disembarked at the Grove landing, where Lyell found multitudes of Miocene fossil shells washed out of the sandy bluff.[33] From the Grove, they took a stagecoach seven miles to Williamsburg, the former colonial capital. It was an old-looking town with cottages scattered around a green and the old Bruton Parish Church. At one end of Duke of Gloucester Street was the College of William and Mary, the second oldest college in America, founded in 1683. They stayed at Mr. Southall's Inn.[34] In the half hour remaining before dusk, Lyell went to see a marl bed at the College Mill that the Rogers brothers had mentioned as a Miocene deposit. In the evening he called on John Millington, professor of chemistry at the College of William and Mary.[35]

31. Mary Lyell, Diary, 1841–42, 218, Kinnordy MSS.

32. Mary Lyell to Susan Horner, 23 December 1841, Kinnordy MSS.

33. Lyell, *Travels,* 1:36.

34. This inn was formerly Weatherburn's Tavern and is so designated in the Colonial Williamsburg restoration.

35. John Millington (1779–1868) was appointed professor of chemistry, natural philosophy, and engineering at the College of William and Mary in 1835. Millington was born at Hammersmith, England, and until 1830 had lived at London, where he must have known Lyell at least slightly, since they both belonged to the small community of scientists and were members of the Linnean Society.

Accompanied by Mr. Roswell, a New England schoolmaster and itinerant lecturer on phrenology, who was also staying at Southall's Inn, Lyell went the next day to Burwell's Mill, another locality for Miocene shells the Rogers brothers had mentioned. Despite bad weather, they obtained more than seventy species of shells and seven species of corals. Several *Pectens* and other large shells had full-grown barnacles and corals attached to them, indicating that the marl deposit had accumulated slowly and gradually. About one fifth were species still living in the neighboring sea.[36] The bad weather kept Mary in their room at the inn, writing letters. That evening Mr. Southall's Negro Jesse came to help them nail down a wooden box full of fossils to be shipped to Philadelphia. The next day, Christmas Eve, was bright and sunny, and after breakfast Mary visited Bruton Parish Church.

> The bricks were all brought from England as is the case with several other churches in this section of the country, but the inside has been repaired & modernized. Saw the tombs of some of the Governors & a President of the College who married one of the Evelyns of Surrey. Went to the old College of William & Mary which looks venerable like some of the side buildings at Hampton Court, but is sadly falling into decay in consequence of political prejudice, the professors not being democratic enough. Two old evergreen oaks in front bearing a small black acorn the leaf being narrow & long. A statue of Lord Botetourt, a former governor, injured by time & neglect. Walked into the library. Books look old, yet there were some quite modern.[37]

That afternoon, accompanied by Roswell, the Lyells took the stagecoach back to the Grove landing, where all three collected fossil shells along the riverbank until the steamboat arrived. In the confusion of boarding, their basket of fossils was almost left behind, but Roswell dexterously threw it to them. The engineer on the steamboat was a Thomas Lyell from Dundee; though he had never seen Kinnordy, he knew it was the Lyell estate. On the steamboat, a black man named Lord Wellington took charge of their baggage and, wrote Mary, "as they all look a good deal alike to us, when another man came up Charles said 'Are you Lord Wellington?' 'No, sir, Julius Caesar.'"[38]

36. Lyell, "Miocene Tertiary Strata of Maryland, Virginia, and of North and South Carolina," 417.

37. Mary Lyell, Diary, 1841–42, 220–21, Kinnordy MSS. Norborne Berkeley, Baron de Botetourt (1718–70), was appointed governor of Virginia in 1768 and died at Williamsburg of a fever.

38. Mary Lyell to Caroline Lyell, 31 December 1841, Kinnordy MSS.

With the short daylight of winter, it was sunset by the time the steamboat arrived at Norfolk. Lyell had to drive about the town for some time to find an office where he could leave his boxes of fossils to be shipped back to Philadelphia. As a result, they missed the ferry to Portsmouth and had to cross the water from Norfolk to Portsmouth after dark in a rowboat with some other gentlemen.

Upon landing they walked to the Crawford House, said to be the best hotel, but on first appearance it seemed very cheerless. It was Christmas Eve, and most of the servants were away; from the bar came sounds of drunken revelry. Finally a polite mulatto woman provided them with a bedroom, where she built a good fire in the fireplace and brought them hot tea. In the streets all evening, boys were throwing firecrackers and firing pistols, as was the custom in the South at Christmas. On Christmas morning, when the same woman came to light the fire in their bedroom, she told them that she was hired out to the hotel by her owner for fifty dollars a year. "She brought us each before breakfast a tumbler of egg nogg, rum, sugar & a beat egg, very good," Mary noted in her diary.[39] After breakfast they hurried to catch the train for North Carolina, which was supposed to leave at nine o'clock. When they were seated in the railway car, however, they learned that the train was to wait for the Baltimore boat, which would not arrive for two hours.

Two hours later the boat had not yet come, but the train went anyway. The railway crossed a very flat country, passing through part of the Great Dismal Swamp. On either side of the line stretched the unending pine forest, with "the most beautiful underwood—a kind of bay tree in great profusion with a very sweet smell, a perfect tangle of canes, [and] a beautiful moss pendulous from the branches of the trees."[40] The Spanish moss (actually a lichen) hanging from the pine trees reminded Lyell of old man's beard, a lichen on the trees of the New Forest around his boyhood home in Hampshire.

At midday they stopped to eat at a log house by the side of the railway line. The landlord, looking at the grey, lowering sky, predicted ominously that they would not reach their destination, Weldon, North Carolina, that night. During the afternoon rain began to fall, changing to sleet and then snow. As the rails became covered with ice and snow, the engine wheels gradually lost traction. "Well we went on slower & slower," Mary wrote, "sometimes going some way

39. Mary Lyell, Diary, 1841–42, 224, Kinnordy MSS.

40. Mary Lyell to Caroline Lyell, 31 December 1841, Kinnordy MSS.

backwards to get a momentum & making a little progress that way till just at nightfall the Engineer came to a full stop."[41] Only the Lyells and two young men were left on the train. They had come to a halt at a water station. From the windows of the car they could see near the track a small, two-story farmhouse. The four passengers plus the engineer and fireman went there in search of shelter. As the freezing rain and snow fell outside, the refugees from the storm ate "a good supper—quantities of hot bread, sausages & tea & coffee" before a blazing fire. The farm family gave up their own bedroom to the Lyells and found beds for the others. Their kind hosts also built a fire for the Lyells in their bedroom, but, since the walls of the house were lathed but not plastered, they could look out through the cracks to see the snow on the ground outside. Their bed was clean and comfortable with plenty of bedclothes. As Mary lay in bed her nose, she said, "was like an icicle & I had to hold it to keep it warm."[42] The two young men, along with the engineer and fireman, slept in an adjoining room.

Although the weather the next morning was bright and frosty, the sun soon melted the snow on the rails. By ten o'clock the train was able to start. On the way they passed through more of the Great Dismal Swamp, a region suggesting to Lyell the type of forest in which the plants that formed the ancient coal deposits had grown. About noon they reached Weldon, a small cluster of houses beside the Roanoke River in North Carolina. After the Lyells were settled at Mr. Spurrel's inn, they took a walk along the banks of the Roanoke, admiring the bay trees and pines, with green mistletoe and leafless grape vines twining about them. To study it firsthand, Lyell went into a swamp, where, Mary wrote, "he scratched his face, wet his feet & tore his coat."[43]

At Weldon they rose at two in the morning to take the train to Wilmington. Again they passed through a level country covered with tall, long-needled pines, many of them stripped of their bark for about six feet up from the ground for the collection of turpentine.[44] At Wilmington the Lyells boarded a steamboat for Charleston. As they dropped down the Cape Fear River with beautiful low cliffs along either shore, Lyell planned on their return to examine the cliffs for fossils.

41. Ibid.

42. Ibid.

43. Mary Lyell, Diary, 1841–42, 226, Kinnordy MSS.

44. Mary Lyell to Caroline Lyell, 31 December 1841, Kinnordy MSS.

South Carolina and Georgia

At Charleston next morning Lyell and Mary went to the comfortable, old Charleston Hotel.[45] They had left winter behind. Roses and other flowers were still blooming in the gardens, and many trees, such as the magnolia, live oak, and long-needled pine, were green. Beneath the pines, palmettos gave a subtropical appearance. In 1841 Charleston was a busy port from which much rice and cotton were shipped. It possessed a small but active community of naturalists. George Ticknor had given Lyell letters of introduction to several Charleston naturalists and scholars, and at Washington, Lyell had received more letters of introduction to southerners.

Despite its being a rainy day, the Lyells went to see Dr. Edmund Ravenel's collection of fossils. Dr. Ravenel was away, spending the Christmas season at his plantation on the Cooper River, but his young nephew, Dr. St. Julien Ravenel, showed them the collection.[46] Lyell was especially interested in Ravenel's fossils, because Timothy Conrad considered them to be from beds intermediate in age between the Cretaceous and Eocene. Lyell wanted to examine the localities from which the fossils came, but since Ravenel could not easily be reached, he decided instead to visit Shell Bluff and other fossil localities along the Savannah River. He wrote to Dr. Ravenel that he had collected shells from the marls along the James River in Virginia but so far had not obtained any Lower Tertiary fossils. He hoped to collect some at Vance's Ferry on the Santee River or along the Savannah River. He did not have time to go to Alabama, where the richest source of Eocene fossils was in the bluff at Claiborne. Although anxious to collect fossils, Lyell emphasized that his chief purpose was to see for himself the geological position of American Tertiary rocks.[47]

After the Lyells returned to the hotel, they received visits from Dr. Samuel Dickson, a physician and former student of Benjamin Silliman at Yale, and Thomas King, a lawyer and congressman who owned Retreat plantation on St. Simon's Island. In the afternoon the Kings sent their carriage to bring the Lyells to tea at their house, where Mary admired the beautiful library with statues.

On 31 December, the Lyells boarded the train to Augusta, Georgia, where Lyell went to call on Richard Henry Wilde, to whom he

45. Mary Lyell, Diary, 1841–42, 227, Kinnordy MSS.

46. Edmund Ravenel, M.D. (1797–1871), physician and planter, had preceded Charles Upham Shepard as professor of chemistry at South Carolina Medical College.

47. Lyell to Ravenel, 30 December 1841, Ravenel Collections.

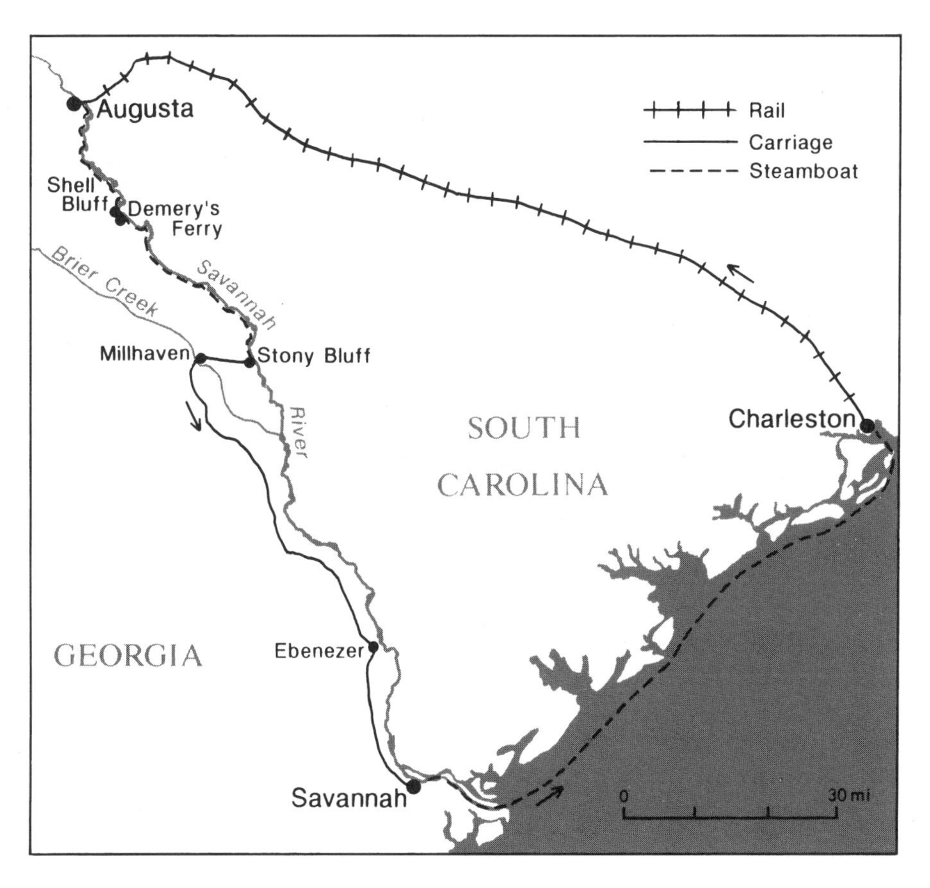

Lyell's Tour through South Carolina and Georgia, December 1841–January 1842. Map by Philip Heywood.

had a letter of introduction. Wilde, a lawyer and former congressman, had recently returned from five years of travel and study in Europe, mainly in Italy. Deeply interested in Dante, he had discovered, in the Bargello at Florence, Giotto's previously unknown portrait of Dante as a young man. The next day, New Year's Day of 1842, Wilde took the Lyells for a drive to see the rapids of the Savannah, marking the line separating the Appalachian upland from the coastal plain. Lyell planned to examine the banks of the Savannah River as it cut through the Tertiary strata of the coastal plain from the edge of the granite region to the sea. They also visited some quarries and "a pretty village of villas," where planters went during the summer months to escape the heat and fevers of the Low Country. For the first time, they saw cotton growing in the fields.

On Sunday morning, Wilde saw the Lyells off on the steamboat *Santee* to go down the Savannah River. They were the only passengers aboard. As the steamboat dropped down the river, they passed an occasional bluff, large swamps of cypress trees with their branches draped with Spanish moss, and canebrakes with, Mary noted, "some of the canes 10 feet high, very handsome." They landed at Demery's Ferry, just below Shell Bluff. The high water on the river had flooded the ground back of the landing, but a strong Negro carried Mary in his arms over the water, then took Lyell on his back, and finally carried over the luggage. Mr. Hill, the storekeeper at Demery's Ferry, came down to meet them and they left their luggage at his store.

From the landing they walked a mile upriver to Shell Bluff. They hoped to stay overnight at a small house at the top of the bluff, but the people were unwilling to put them up. Lyell therefore sent a note to Mr. Nance, the overseer at Mr. Harris's plantation, three miles inland, asking him to send someone to the landing for their luggage and some means of transportation for themselves. While they awaited an answer, they collected fossils along the bluff. It soon began to rain. Mary took shelter in the house at the top of the bluff, but Lyell continued to hammer away in the pouring rain. As dusk began to fall they still had not heard from Nance. They set out to walk to the plantation but had gone only about half a mile when they met a Negro leading two horses for them, one with a sidesaddle. They mounted and in pouring rain rode what seemed a long way through beautiful pine woods, crossing two streams. Twice tree branches knocked Lyell's hat into the mud. Just as it became wholly dark, they saw the lights of Nance's house. It was "a regular Georgia house, an immense wood fire on the hearth & two doors to the room opening back & front to the open air day & night so that there is a constant through draught." The Nances welcomed them warmly. They had five young children, a girl and four boys. After a good supper—"coffee & pork & sausages & corn bread & excellent milk"—the Lyells retired to a comfortable bed in a room where two of the Nances' little boys also slept. "Our means of ablution were not very ample," Mary noted, "but it is astonishing how necessity is the mother of invention."[48]

The next morning was bright and beautiful, putting an end to the wet, cloudy weather they had endured since arriving at Charleston. Lyell rode to Shell Bluff to spend the whole day collecting fossils. He found more than forty species of shells, mostly casts, but all tending to show that the beds were Eocene. Mary remained behind at the plantation and took a walk with the Nances' little girl, Elizabeth, to look for fossils. In the cool winter weather, Nance was killing hogs about a mile away from the house. The carcasses were sent up to the house in cartloads to be smoked. Returning on horseback from Shell Bluff in the evening, Lyell saw the slave women rendering lard in huge iron kettles.

> The red glare of the fire was reflected from their faces and . . . they reminded me of the scene of the witches in Macbeth. Beside them, moving slowly backwards and forwards in a rocking chair, sat the wife of the overseer muffled up in a cloak, and suffering from a severe cold, but obliged to watch the old slaves, who are as thoughtless as children, and might spoil the lard if she turned away her head for a few minutes.[49]

Observing the Nances with their slaves during the day, Mary gained new knowledge of the realities of slavery and slave owning. The Lyells had noted the ragged and patched clothes of the Negroes in Baltimore, their introduction to a slave-owning society. At Washington, they had found the slave servants in the hotel slovenly and inefficient. By contrast, in Virginia the hotel servants were of much higher caliber, polite and helpful. At Williamsburg the servants at Southall's Inn were slaves hired out to Southall—the head man, Jesse, at the rate of fifty dollars per year—and seemed proud that they were worth a high rent. Jesse had been about to marry a free woman, which would mean that his children would be free. He had been married before, but when he went to New Orleans with

48. Mary Lyell, Diary 1841–42, 229, Kinnordy MSS.

49. Lyell, *Travels,* 1:125.

his master he was separated from his wife. On his return to Williamsburg four years later, his wife had married someone else, even though she and Jesse had a daughter. The casual dissolution of marriage was to Mary a shocking aspect of the degradation of slavery. At Charleston, Mrs. King had told Mary how difficult household slaves were to manage. If a slave took a purgative, he or she went to bed, and after childbirth a slave woman did no work for four weeks. Thomas King did not allow his children to slap black children and as a result, complained Mrs. King, the little blacks were very impudent.[50] Now, with the Nances, Mary saw how slavery enchained both master and slave. She wrote to her sister Katherine:

> It is the last life I should like to spend & I feel sure that if I had Negroes under me I must become stern & harsh. They are so troublesome & difficult to manage, a life in which every motive to exertion is taken away. No rewards, nothing but fear of punishment to make them work & what an education for white children, allowed to tyrannize over the little Negroes whom they look upon but as animals on the farm. Mind I have <u>seen</u> no cruelty & the manner to the slave is generally gentle, but the whole system is much worse than I imagined in its effects on all.[51]

On 4 January, the Lyells rose when the Nances did, at half past five and long before daybreak. Immediately after breakfast they rode on horseback to Demery's Ferry to wait for a steamboat. Mr. Nance rode part of the way with them and sent their luggage in a light cart. At the ferry, Mrs. Hill, the storekeeper's wife, told them they might have to wait as long as seven days for a boat, which, Mary said, "would have been a treat." She and Lyell occupied the morning by walking along the bank part of the way to Shell Bluff looking for fossils. At midday they returned to the landing, where they dined with Mrs. Hill on wild turkey. Mary thought it excellent, in flavor somewhat like pheasant. Through the afternoon they continued to wait.

> At last at three o'clock a steamer came in sight towing two large cotton boats fixed on each side. We made a sign & she put in much to our joy. There were no passengers but there was the captain of another boat on board, an Englishman from Dorchester. It was a dirty boat, but they made the little cabin at the end the ladies cab-

50. Mary Lyell, Diary, 1841–42, 220–21, 228, Kinnordy MSS.
51. Mary Lyell to Katherine M. Horner, 1–11 January 1842, Kinnordy MSS.

> in, parted off for Charles & me & we drank tea & breakfasted with the two captains. I really think if Mr. Dickens wants to see life he ought to join us for we certainly have a variety. It was a beautiful evening & we admired the trees on both banks. The cypress trees which are very common, have a peculiar angular growth.[52]

As the steamboat descended the Savannah River, Lyell tried to read the geology of the bluffs they passed. In the Long Reach, about eight miles below Demery's Ferry, he saw in the London Bluff the same stratified beds of white shell-limestone as at Shell Bluff. At the top of the bluff, the limestone was covered with red clay and loam. The captain anchored for the night and the Lyells slept on the boat in the ladies' cabin, on a very hard bed. Next morning, they arrived at Stony Bluff, thirty miles downriver from Demery's Ferry. Here the beds of shell-limestone had disappeared to be replaced by a fine sandstone called burrstone. The beds of red clay and loam that had appeared at the top of the London Bluff were now visible at the base of Stony Bluff. Though the beds appeared horizontal, they actually dipped slightly toward the sea. As the steamboat descended the Savannah River, the beds in the bluffs became higher and younger. The river was so high that the captain doubted whether he could land them at Stony Bluff, but he managed to do so. At nine o'clock in the morning, Lyell and Mary found themselves at the foot of Stony Bluff, alone with their baggage "& not a soul in sight."[53]

Carrying as much of their luggage as they could, they walked up the road to a small house at the top of the bluff. As they approached, they saw a woman sitting in front of the house. She did not rise to greet them and they thought at first she was unfriendly.

> But, poor creature, she had only just recovered from the fever & looked at me with perfect astonishment when I told her I had never had either the fever or the chill. "Oh," she said, "there is not one in this section of the country that could say that. I should like to live in your country." It is fearfully general & gives them a great appearance of languor but to make up for this consumption is nearly unknown.[54]

On closer acquaintance Mary found the lady, Mrs. Roe, quite pleasant. Her husband was absent just then. Lyell wanted to obtain a car-

52. Ibid.
53. Ibid.
54. Ibid.

riage to drive about ten miles to Millhaven, the plantation of Colonel Seaborn Jones, to whom Richard Wilde had given him a letter of introduction. He walked to another house, but its owner also was away. Returning to the Roe house rather tired, Lyell spent some time examining the burrstone of Stony Bluff with his geological hammer. In it he found many corals and tiny shells as well as other fossils that on later microscopic examination proved to be sponges. Stony Bluff, he decided, was Eocene, like Shell Bluff.

Roe soon returned home and agreed to rent his barouche and horse to Lyell to drive to Colonel Jones's. Roe could provide no one to drive the carriage, so Lyell was to drive it himself, with Mary watching out for stumps. A young man who was traveling part of the same road on horseback agreed to show them the way. He was going to teach school that winter, but he looked to Mary more like a schoolboy himself. "It was a most lovely day," she wrote, "as warm I think as any day in summer in Scotland, the sky the clearest blue with the bright green pines against it." The first part of the road was good, but after their guide left them, it became deeply rutted, winding among stumps. At one place their path was blocked by a fallen tree that they had to maneuver around; at another, they had to ford a creek. Finally they came to Millhaven plantation, a pleasant-looking white house with a veranda, standing on a knoll among trees, surrounded by slave cottages. Lyell went to the door with his letter of introduction. As soon as Colonel Jones saw Lyell's name, he said, "I shall not read this letter, you need no introduction. I cannot say how happy I am to have you here. I was in the North when you landed & was very near going to Boston to look at you. I have your book on my shelves."[55] With such a welcome, they immediately felt at home.

Jones and his household were at their midday dinner, and he invited the Lyells to join them. Mrs. Jones and part of the family were away at Augusta, but his son, a boy of fifteen, was with him, and a housekeeper presided at meals. The plantation overseer and the storekeeper also dined with them. In addition to his plantation, Jones operated a mill and kept a store. After dinner, Jones took them across Briar Creek in a flatboat to look for fossils along the banks. The creek cut through the same kind of burrstone or sandstone they had seen at Stony Bluff. He also gave them many fossils from Jacksonborough, lower down Briar Creek. Mary thought Mill-

55. Ibid.

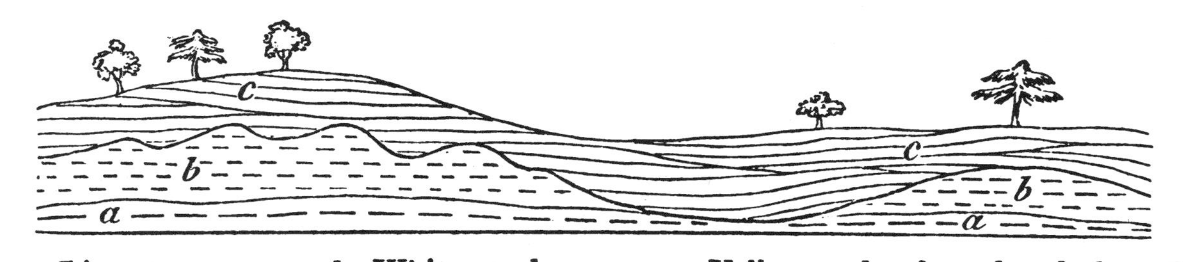

Section of Eocene Strata on the Right Bank of Beaverdam Creek, Scriven County, Georgia. Lyell, "Observations on the White Limestone" (1845).

haven beautifully situated on the creek, with its trees and palmettos growing about the house. Next morning she wrote, "We did enjoy our bed so much last night. Certainly roughing it a little does make one enjoy one's comforts very much." She also remarked, "After waiting on myself so long it seemed quite strange to have a Negro woman last night to fold up my clothes."[56]

That day, 6 January, Jones and Lyell rode on horseback to various places to collect fossils. Jones was hospitable in the customary manner of southern planters. He was also an enthusiastic amateur geologist who took pleasure in helping Lyell. They had many long rides together through the sunlit pine woods, where there was little undergrowth to prevent a horse from galloping freely.[57] Near Jacksonborough, at the junction of Briar and Beaverdam Creeks, they visited a limestone quarry. The limestone passed upward into marl and was overlain again by the burrstone of Stony Bluff and Millhaven. The sandy soil on which the pine forests of Georgia grew was derived, Lyell decided, from the disintegration of the burrstone, and throughout the country the burrstone was underlain by the same limestone he had studied at Shell Bluff. Near Jacksonborough, Jones showed him several lime sinks—places where the underlying limestone had been dissolved away by underground water and the overlying sand had fallen into the cavity thus produced. The fossils that Lyell obtained from Colonel Jones and those that he collected himself all tended to show that the burrstone, like the limestone of Shell Bluff, was Eocene.

While Lyell and Colonel Jones were exploring the country, Mary remained at Millhaven. She labeled the fossils they had collected and then went to the plantation store for a wooden box in which to pack them. She drew a sketch, wrote some letters, and finally went across Briar Creek by herself on the flatboat to collect fossils. She saw many hogs running in the woods, and many slaves passed by where she was collecting. The men said nothing to her, but the women generally stopped to ask her, "What you do, Missus?"[58]

The old Negro woman who folded her clothes at night had been a nurse to all the children in the Jones family. She talked to Mary a great deal.

> [She] had been in the family before Col. Jones himself was born & said she belonged to them & they belonged to her. Her chief

56. Ibid.

57. Lyell, *Travels,* 1:127.

58. Mary Lyell, Diary, 1841–42, 234, Kinnordy MSS.

occupation now is attending to all the births, black & white, in the neighborhood & she seemed very proud of her occupation. I saw three little black infants under a month old & there were three or four more just expected. It was very pleasant to see Col. Jones's manner to his Negroes, so very mild.[59]

On 7 January, the Lyells left Millhaven. Colonel Jones borrowed a carriage for them from his brother-in-law, his own carriage being with his wife at Augusta, and accompanied them on horseback for some distance. As they were driving along, they came suddenly to a dead halt. The Negro coachman said that he had dropped one of his white gloves on the road and would have to drive back to try to find it. The Lyells were anxious to get on, but the coachman would not proceed until finally Colonel Jones took off his own gloves and gave them to him. They passed through Jacksonborough and at three o'clock reached Denton's Tavern. Since the railroad had been built from Charleston to Augusta, the road along the Savannah River was not much traveled. The inns were not usually prepared to receive guests, but the Lyells were able to stay at Denton's Tavern and sent the carriage back to Millhaven. For the rest of the afternoon, Lyell went on horseback, accompanied by a small boy, to collect fossils at the bluff along the river. That night it was very warm and bats flew about inside the inn, for every window had broken panes of glass.

The next day, Saturday, Lyell hired Denton's barouche and a horse and mule for three days to drive to Savannah. Denton's younger brother was to drive them and bring back the carriage. The weather continued warm and beautiful. "In travelling through the woods," Mary wrote, "the birds were singing & grasshoppers & frogs cheeping & we saw three or four butterflies & moths."[60] From time to time, they left the road to go down to the Savannah River to study the geology of the bluffs. At Tiger Leap, on a stretch of the river known as Hudson's Reach, Lyell found a particularly fine section of the Eocene beds, including beds of fuller's earth formed from multitudes of diatoms. That evening they stayed with Mrs. Denton's mother, Mrs. Blake. Ducks, turkeys, chickens, pigs, and dogs wandered about the yard. There was no glass in the windows. Despite the fact that Mrs. Blake was just recovering from an illness, they received a very good supper and a clean bed.

59. Mary Lyell to Katherine M. Horner, 1–11 January 1842, Kinnordy MSS.
60. Ibid.

On Sunday the Lyells, accompanied by young Denton, drove on to Mr. Grubenstein's, "a very neat tavern in [a] grove of live oaks."[61] A slave guided them from the tavern down to Red Bluff, one of the high bluffs of loam, clay, and sand on the Savannah River. When they returned to the road, they met a procession of people on horseback returning from church in the hamlet of Ebenezer. Proceeding on toward Ebenezer, the Lyells and Denton crossed a causeway through a swamp where the water reached above the middle of the wheels. On either side tall cypresses grew in the water. At the end of the causeway, Lyell paid a half-dollar toll. Late in the afternoon, as the shadows were lengthening, they arrived at Ebenezer. Founded in 1736 as a settlement of Protestant refugees from Germany, the hamlet had kept its old customs and Lutheran faith. In this peaceful place, they found a small church and pastor's house on a green surrounded by cedar trees and a half-dozen goats grazing. The church, built in 1747, was the oldest in Georgia.

Beside the green, they found a former inn owned by a Mr. Blance. At first, Blance was unwilling to give them lodging; his wife was away at church and he was uncertain when she would get home. However, their mule was tired and after some persuasion, Blance invited them to stop and sit. For more than an hour they sat with him on the porch in front of the house, enjoying the beauty of the evening, while frogs and grasshoppers hopped about in the yard. There was no sign of supper, and, uncertain of their welcome, the Lyells and Denton dared not ask for it. To entertain them in their starving state, Blance filled the empty stillness of the late afternoon with hunting stories. A few weeks earlier he had shot a large bear that was carrying off one of his hogs, and a little later he had killed an alligator, also in the act of taking a hog. The alligator had been difficult to kill, but Blance had shot him with a rifle, and when measured, he was fourteen feet long.

At long last, Mrs. Blance and the children returned in a carriage from a distant church. Immediately all was bustle, and presently Mrs. Blance invited them into the house to an abundant supper of ham, bacon, and other forms of pork, chicken, waffles, cakes, and other dishes, urging them to eat "most earnestly." After supper they again sat outside in the warm darkness, looking at the multitude of stars overhead. Struck by the beauty and strangeness of the country, Mary observed:

61. Ibid.

> This is the country for an astronomer, so many more stars can be seen here than with our cloudy sky. The vegetation too is very different, a sort of fan palm growing about three or four feet high, one of the commonest weeds in the swamps, long canes, the cypress tree & immense pines. Great part of the road lay through a forest the trees growing in water, with a most luxuriant undergrowth. One of the most beautiful trees is the live oak, but it all looks green as if it were summer. I cannot realize, as they say here, that we are in the middle of winter. The Fauna is as distinct as the Flora, opossums, raccoons, bluebirds, redbirds, a kind of mole they call a Salamander, land turtles.[62]

About two o'clock the next day, the Lyells and Denton entered Savannah, driving along sandy streets bordered by trees and houses surrounded by gardens. After settling at the Pulaski Hotel, they went to call on Dr. Thomas Habersham, who gave Lyell a turtle skull from Shell Bluff. Habersham also told Lyell of a place at Beauly on the Vernon River, about twelve miles southeast of Savannah, where he had found mastodon bones. The site could be explored only at low tide. Next morning Lyell rode out before dawn, accompanied by a Negro guide provided by Habersham to reach Beauly at seven o'clock for low tide. His guide had a passport stating the exact streets he was to pass along to meet Lyell at the Pulaski Hotel. If he departed from his route, he might have been arrested and placed in the guard house, precautions adopted after a slave insurrection incited by abolitionists from the North.[63] Reflecting on the possibility of slave insurrection, Lyell thought of Colonel Jones and his family at Millhaven and of the Nances near Shell Bluff, living among their slaves on isolated plantations. The implications for them of a slave uprising were appalling.

Lyell and his guide rode out to Heyner's bridge on White Bluff Creek and then took a boat down a tidal creek to the clay bank where Habersham had found the mastodon bones. With the help of the slave, Lyell found a molar tooth of a mastodon. The bed of clay containing the mastodon bones and teeth overlay a bed of sand containing fossil seashells of species still living along the coast. As a finer sediment, the clay must have been deposited in deeper water than the sand. Lyell concluded that although the whole southern Atlantic coast of North America had been raised from beneath the

62. Ibid.
63. Lyell, *Travels,* 1:134.

sea in recent geological times, and at a time when the species of seashells were the same as the present, this general elevation had occasionally been interrupted by smaller movements of subsidence. Lyell noted that the flat coastal marshes on the landward side ended in a steep bank or bluff that marked the line of an old coastal cliff. Other lines of former coastal cliffs existed further inland, indicating long interludes of rest between successive periods of elevation.

In addition to mastodon bones, the bones of other mammals, including the *Megatherium, Mylodon,* elephant, and horse, appeared in the beds along the coast of Georgia and the Carolinas. Most astonishing was the horse. Before the arrival of the Spaniards, no species of horse had been known to live in either North or South America. At Philadelphia, Timothy Conrad had given Lyell a fossil horse's tooth from Newberne, North Carolina, that proved remarkably similar to one found by Charles Darwin on the Rio de la Plata in South America. Like Darwin, Lyell marveled that so many species of mammals had become extinct in North and South America, while the species of seashells living along the coast had remained the same. The extinction of the mammals could not, he thought, result from human action. The Danish naturalist Peter Lund had found in Brazil that during the same period, even mammals as small as the rat also became extinct.

The day after Lyell's expedition for mastodon bones, he and Mary called on Bishop Stephen Elliott, son of the late Dr. Stephen Elliott, naturalist of Charleston.[64] The Elliotts were pleasant people with a fine library and a fondness for things English. They invited Mary to stay with them while Lyell went to visit James Hamilton Couper's plantation on the Altamaha River, seventy miles to the south. Lack of time forced Lyell to give up the expedition. Early the next morning, 13 January, Lyell and Mary left Savannah on the steamboat *General Clinch* to return to Charleston.

The Cooper River, Santee Canal, and Eutaw Springs

Upon arriving at Charleston, Lyell went to call on Dr. Edmund Ravenel, whom he found a small, slim, active man. Ravenel, after receiving his medical degree from the University of Pennsylvania in

64. Stephen Elliott (1771–1830), author of *Sketch of the Botany of South Carolina and Georgia;* see Hennig, *Great South Carolinians,* 1:215–27.

1819, returned to Charleston to practice medicine. In 1823 he began to spend his summers on Sullivan's Island at the entrance to Charleston harbor. During the summer many planters took their families to the island to escape the heat and fevers of the interior. Using his vacations to collect and study, Dr. Ravenel published a catalogue of South Carolina seashells in 1834. During the winters, in addition to his medical practice he had taught chemistry in the Medical College of South Carolina until 1835, when ill health forced him to resign. He then purchased The Grove, a plantation of more than three thousand acres on the Cooper River, where he raised cotton. He had also purchased the South Carolina rights to use Sawyer's brick-making machine and was manufacturing bricks on his plantation.

On 14 January, the Lyells went to Dr. Ravenel's house to study his fossil shell collection while his small steamboat was being got ready. At four o'clock they set off from Charleston in the steamboat, "a sort of toy affair, which just held us."[65] While Edmund Ravenel's nephew, Dr. St. Julien Ravenel, steered, two slaves fired the boiler. In the winter dusk they steamed up the Cooper River, reaching The Grove after dark. Mrs. Ravenel and a large family of children and servants welcomed them with a cheerful fire burning in the fireplace, a comfortable bedroom, and a good supper.[66]

Next morning Edmund Ravenel took Lyell to examine the pits where they dug clay for his brick-making operation. The beds of clay and sand contained many trunks of cypress and other trees in an upright position, indicating that they were embedded where they had grown, yet they were now as much as sixteen feet below high-tide level. The Low Country of South Carolina thus appeared to have undergone recent subsidence.

At noon the Lyells and Edmund Ravenel re-embarked in the little steamboat and continued up the Cooper River, with beautifully wooded banks on either side. Unfortunately, the burning of pitch pine in the firebox choked up the funnel of the steamer with soot and tar. The little steamboat made slower and slower headway against the current, and for a while the Lyells got off to walk along the bank. After dark, they reached Dean Hall plantation, where, as they approached, they heard slaves singing in their cabins. Dr. Ravenel requested hospitality for himself and the Lyells from the

65. Mary Lyell to Leonora Horner, 8 February 1842, Kinnordy MSS.
66. Mary Lyell, Diary, 1841–42, 239, Kinnordy MSS.

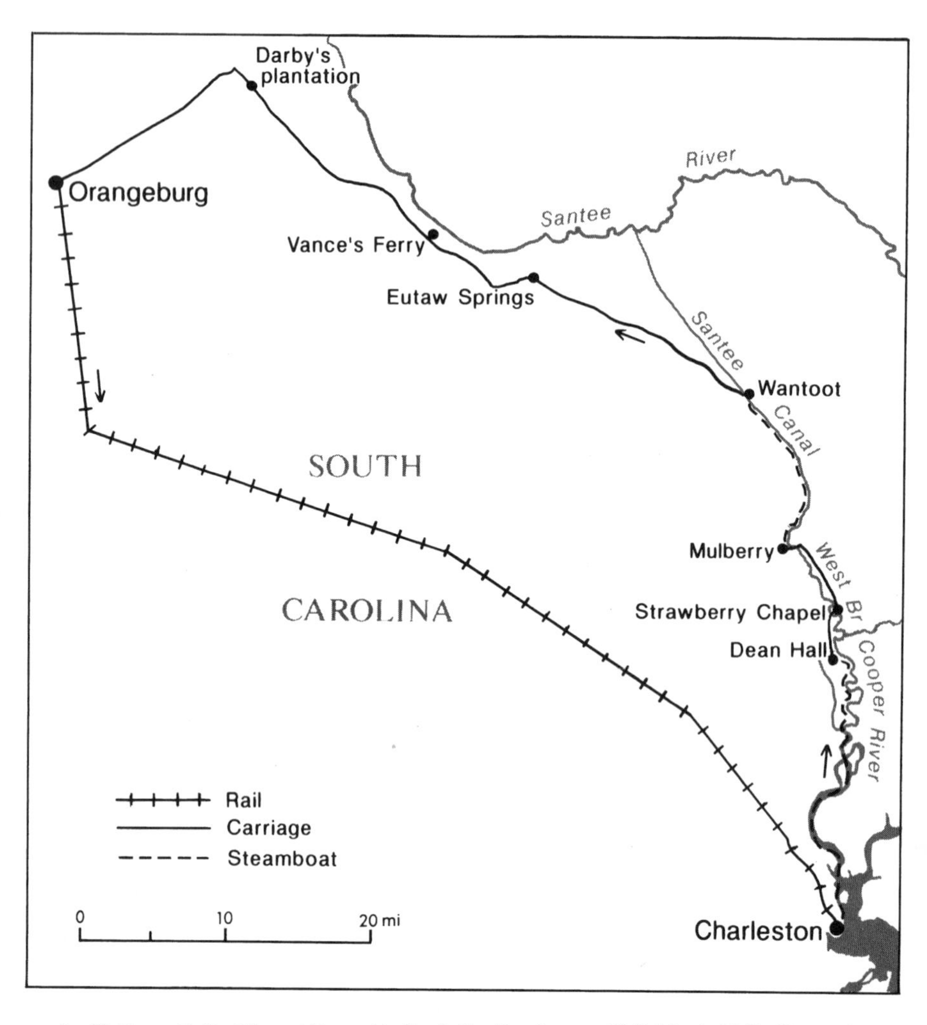

Lyell's Tour with Dr. Edmund Ravenel in South Carolina, January 1842. Map by Philip Heywood.

owner of Dean Hall, Colonel William Carson, who welcomed them warmly. Planters like Carson were so accustomed to sudden demands on their hospitality that they kept spare rooms prepared, each with a wood fire laid ready to light. Dean Hall was a large, three-story brick house with a slate roof, entirely surrounded by a broad piazza supported by brick arches that in summer protected the house from the heat of the sun and provided a vantage point from which Carson could look out across his broad rice fields.[67]

Dean Hall was settled first in 1725 by a Scotsman, Alexander Nesbit, who had named it for his native village of Dean in Scotland. William Carson had been a student at Harvard College in 1816 when his father, a rich Charleston merchant, died. Returning to Charleston to look after his mother's property, young Carson became fascinated with rice planting, and in 1821 he and his mother bought Dean Hall from Nesbit's descendants. With great intelligence and energy, Carson had directed his slaves in clearing the cypress swamps, digging canals, and building water gates, dikes, and roads—all in a solid and permanent fashion—so that the fields could be flooded or drained as needed for rice cultivation. In 1842 Dean Hall, with its great brick plantation house, its rows of neat, white cottages for the Negro slaves, its rice mill, and threshing machine, had the appearance of a prosperous village.[68]

A few weeks before the Lyells and Dr. Ravenel arrived, Carson, at the age of forty-one and after many years of bachelorhood, had surprised his friends by marrying the beautiful Caroline Petigru, daughter of James Louis Petigru, a leading attorney in Charleston. The colonel and his new wife now entertained their guests at supper, a hearty meal that included rice and hominy, various kinds of bread and cakes hot from the oven, and sausages, ham, and eggs.[69]

Reflections on Southern Society

As they traveled through the South, the Lyells were deeply interested in every aspect of slavery and race relations. Before they came to America, they shared the attitudes of educated, liberal-minded British people toward slavery. In 1807 Mary's uncle, Francis Horner, as a member of Parliament, supported the abolition of the trans-Atlantic slave trade. In 1833 the British Parliament, urged by the

67. Stoney, *Plantations of the Carolina Low Country*, 79–222.

68. Irving, *A Day on Cooper River*, 25.

69. Mary Lyell to Leonora Horner, 8 February 1842, Kinnordy MSS.

forces of liberal thought and evangelical Christian piety, abolished slavery in the British colonies, indemnifying the slave owners for the loss of their property. Christian thought and common humanity joined to insist that slavery was wrong. Furthermore, descriptions of slavery commonly depicted the slaves as overworked, subjected to many cruelties, and intensely unhappy.

When the Lyells saw slavery firsthand on plantations in Virginia, Georgia, and South Carolina, they gradually came to think that their former impressions were inadequate to comprehend the true situation of slaves in the South. They saw immediately the inefficiency of slavery and its impoverishing influence on the South, in contrast to the rapid growth and great prosperity of the North. Nevertheless, they saw little evidence that slaves were treated cruelly or overworked. Instead, at Dr. Ravenel's house at Charleston, Lyell had been present at a slave wedding in the kitchen, an evidently happy occasion. As Dr. Ravenel's small steamer docked at Dean Hall, they heard through the darkness the sound of joyful singing coming from the slave cottages. The black house servants were well dressed and well trained, like servants at well-ordered English country houses. To the Lyells, the slaves' physical circumstances appeared to be as comfortable as those of English house servants.

Lyell learned that the slave population was growing rapidly, a fact that by itself argued that the slaves were not unduly oppressed. From an economic standpoint, he noted that slavery gave black people an almost complete monopoly of the labor market in the South. At the same time, he observed that the increase in the number of slaves on a plantation created a dilemma for the planter. To house and feed a growing number of slaves, a planter must either expand his plantation, acquire additional plantations, or migrate with his slaves to new lands farther west. The alternative was to sell some of his slaves to other planters. Either way, the growing number of slaves required the expansion of southern agriculture to provide them with a livelihood. Yet, as Lyell noted, the economy of the South, dependent on slave labor, could not develop the manufacturing industry, the towns and cities, and the roads, schools, and bridges that were appearing so rapidly in the North.

The economic and social liberation of the South seemed to require the liberation of the slaves, but when Lyell reflected on how they might be set free, he saw that the question had no simple answer. If he himself had inherited a plantation with slaves and had set them free, he would still by law be compelled to support them. If he decided to educate them before setting them free, so that they might be prepared to earn a living as free people, he would be faced with the obstacle that, in response to the abolition movement, the southern states had passed laws making it a felony to teach slaves to read and write. Although Lyell saw no clear way to solve the problem of slavery, he thought that the southerners, who tended to overestimate the number and influence of abolitionists, ought to proceed to educate their slaves as rapidly as possible.[70]

At Dean Hall the Lyells saw slavery in the most favorable circumstances, on an exceedingly well-managed plantation. Yet, even when viewed in such light, slavery revealed its darker shadows. Though the house servants at Dean Hall were well dressed and well trained in their duties, very different from the slaves the Lyells had seen at small inns in the pine woods of Georgia, Mary considered the position of the slave as at best a miserable one. The bonds of the slave also tied the master. Each day at Dean Hall the slave children, some forty in number, came up before the door of the house to have their dinner, so that Carson might be assured that they were fed properly. "It is really like taking charge of so many animals on a farm," wrote Mary, "but I fear their moral & religious state is sadly neglected. Indeed they are designedly kept in ignorance." Caroline Carson told Mary that her mother was the lady at Charleston whose slave woman had attempted to poison her—an incident described by the English writer Harriet Martineau.[71] The woman later had been sold and taken to Augusta, Georgia. "The only consolatory view," Mary decided, "is that the slaves are better off than they were in Africa & I have quite made up my mind that we in England have no business to interfere. We sh[oul]d look to our own poor at home whose sufferings I believe are infinitely greater than the Negroes!"[72]

Lyell, too, was impressed by the relative physical comfort of the slaves at Dean Hall. Dr. John Irving noted Lyell's visit to the plantation, referring to him as the "distinguished nobleman."

> After scrutinizing, as was his wont, with an inquisitive eye, all things appertaining to the habits, food, clothing and treatment of the slaves, [the visitor] voluntarily tendered this honest conviction of his heart. "It is impossible for me an Englishman to say

70. Lyell, *Travels,* 1:144–55.

71. Martineau, *Society in America,* 2:318–19.

72. Mary Lyell to Leonora Horner, 8 February 1842, Kinnordy MSS.

I am a convert to your institutions, but I candidly confess from all I have seen, *my prejudices have been entirely eradicated.*"[73]

Fossil Hunting on the Santee Canal

On Sunday morning, Carson sent the Lyells in his carriage with a fine pair of horses up the Cooper River to Strawberry Ferry. They crossed the river on the ferry to Strawberry Chapel, built in 1725, and described by Mary as a "pretty old church with splendid live oaks." A country surgeon and friend of Edmund Ravenel's lent them his carriage to travel along the bank of the Cooper to a point opposite Mulberry plantation. There they crossed the river in a dugout canoe, "one man paddling very ingeniously," wrote Mary, "first on one side & then on the other." In the riverbank they examined beds abundant in fossils, including *Cyrenae carolinensis* and oysters, mixed with fragments of Indian pottery. The owner of Mulberry, an Irish gentleman named Milliken who had come to South Carolina from Belfast thirty-five years before, came down to the river and invited them up to the house, Mulberry Castle. The plantation house, built in 1714, was situated on a high bluff. It was an unusual square brick house with a broad gambrel roof and small square towers with large windows at each of its four corners. "The house [is] very old," observed Mary, "tradition says there were loopholes to defend themselves against the Indians, but this is doubtful—large rooms, the place going to decay, but might be made very pretty."[74]

Edmund Ravenel's steamboat, having had its funnel cleaned, joined them at Mulberry. They went aboard and continued along the Cooper River into the Santee Canal. Completed in 1800, the Santee Canal was twenty-two miles long and thirty-five feet wide at the surface of the water. It connected the Cooper River with the upper part of the Santee River and provided a direct line for the transport of cotton and other crops from the interior country to Charleston. From the Santee River the canal rose by locks thirty-four feet to the summit level and then descended by seven locks sixty-nine feet to the Cooper River. The canal was intended for barges, drawn by mules that walked along a towpath on the bank. The barges were of graduated sizes. After hauling their cargoes of cotton to Charleston, they could be fitted one inside another for

73. Irving, *A Day on Cooper River,* 26.

74. Mary Lyell, Diary, 1841–42, 240–41, Kinnordy MSS.

the return journey, in order to reduce the tolls at the locks. The slaves, who were fond of cockfighting, kept gamecocks that perched on the bows of the barges, looking to Mary like ensigns.

Steamboats were normally forbidden to use the canal, but Edmund Ravenel had obtained special permission to take Lyell along it. The banks of the Santee Canal were densely wooded with a great variety of trees, including the gum tree, its branches bare at that season, but its trunk green with mistletoe. At several places where they landed, Mary picked a blue violet and collected several land snails. Raccoons and opossums were abundant in the woods. The Santee Canal passed through the Biggon Swamp, in which the remains of mastodons had been found when the canal was cut. Lyell thought that animals might still become mired in the swamp and their bones preserved.

In the winter dusk, Edmund Ravenel took the steamboat through several locks. After it was quite dark, they reached Wantoot plantation, which had belonged to Edmund Ravenel's father and where his mother and two sisters now gave them a hearty welcome. The Ravenels were Huguenots who had come to South Carolina from France after the revocation of the Edict of Nantes in 1712. Wantoot reminded Mary of houses in the south of France. In the garden the next morning, she saw a mockingbird "the size of a large thrush," she noted, "gray & white, but elegantly shaped & a very tame bird fond of the neighborhood of houses." She also admired a shrub that they called the wild orange, which was not an orange but a *Prunus*. The Ravenels assured her that the shrub was very hardy, and Mary collected some of its berries to take with her.[75]

Dr. Ravenel had arranged to have his carriage follow them by land from The Grove to Wantoot, and on 17 January they continued their journey by land in the carriage. On the road they passed a clearing in the woods where camp meetings were held. Along each side of the clearing was a row of log houses and at one end was a large building. In the middle was a pile of wood for a great bonfire to light up the whole clearing at night during the meetings. They stopped and walked into the various houses.[76]

At Eutaw Springs, the site of a savage battle fought during the American Revolution, a large spring, or subterranean stream, ap-

75. Mary Lyell to Leonard Horner, 8 February 1842, Kinnordy MSS. In 1865 Wantoot house was burned by Federal troops, and in 1940 the plantation was flooded by a hydroelectric power development on the Santee and Cooper Rivers.

76. Mary Lyell, Diary, 1841–42, 243, Kinnordy MSS.

peared at the bottom of a limestone sink. The limestone, containing a great many corals, was an excellent place for collecting fossils. They stayed overnight at Eutaw plantation, belonging to William Sinkler. In its grounds and approach, with tame deer grazing on the lawn in front of the house, Eutaw reminded the Lyells of an English country house. Sinkler was hearty and pleasant, "just like a country gentleman with race horses & a very pleasing daughter who manages his house."[77]

On 18 January, the Lyells and Dr. Ravenel went to Vance's Ferry on the Santee River. Lardner Vanuxem and later Timothy Conrad had collected fossils at this site. Lyell and Mary did likewise. Nightfall found them still in the woods and at several houses where they stopped to ask for lodging, they were refused. Finally at one small tavern in the woods, the woman agreed to take them in, because the party included a lady. "This was a place we paid at," said Mary, "but it was curious to have to coax her to receive us."[78] That night they heard horns in the woods and were told that it was neighboring farmers hunting foxes with torches.

The following day was "a beautiful day, almost too hot for exertion." They visited "a great lime sink which communicated with a deep cave & a romantic glen about 2 miles long to the Santee river,"[79] where they found many fossils. Dr. Ravenel was not acquainted with anyone in the area, and as evening came on they needed to find somewhere to stay the night. In the gathering dusk, they finally saw a plantation house. When they drove up to the door and sent their compliments to the master with a request for a night's lodging, the master turned out to be Dr. Artemus Darby, a former student in Dr. Ravenel's chemistry class at Charleston. He and his wife welcomed them hospitably. The next morning, the Lyells parted from Dr. Ravenel and drove into Orangeburg, where they took the railway back to Charleston.

Lyell's examination of fossils from the beds along the Cooper River and Santee Canal, at Eutaw, and at other places along the way convinced him that all were of Eocene age and not intermediate between the Eocene and Cretaceous, as Dr. Ravenel and Timothy Conrad previously had thought. The beds were essentially an eastward continuation of the formation at Shell Bluff on the Savannah River and at Jacksonborough, Georgia. Both the fossils and the lime-

77. Mary Lyell to Leonard Horner, 8 February 1842, Kinnordy MSS.

78. Ibid.

79. Ibid.

stone rock in which they were found reminded Lyell strongly of the limestone he had seen along the Timber Creek in New Jersey. Samuel George Morton had determined that the Timber Creek beds were of Cretaceous age. Lyell could understand how easily the Tertiary formation of South Carolina and Georgia had been confused with the Cretaceous. The resemblance between the South Carolina beds and those of New Jersey caused Morton to list some of the fossils from the South Carolina limestone as Cretaceous. These fossils were in turn used to identify the rocks themselves as Cretaceous—or intermediate between the Eocene and the Cretaceous. Lyell found that all the fossils of the South Carolina and Georgia limestone were distinct from those of the New Jersey Cretaceous formation and belonged to genera and species characteristic of the Tertiary. The southern formation was, therefore, unquestionably Eocene. The confusion over its age was another illustration of the deceptiveness of the appearance and lithological characters of rocks and the need for detailed examination of the fossils from different strata.[80]

The next two days the Lyells spent at Charleston, arranging and packing their collections of fossils. They again saw their friends the Shepards and the Kings. One evening they went to the house of Dr. John Holbrook, professor of anatomy at the South Carolina Medical College and a devoted zoologist, who was just preparing to publish his work on the reptiles of America.[81] "You cannot think how kind the people were," wrote Mary. "Mrs. King wanted me to drive about all day in her carriage & Mrs. Holbrook offered me a maid of her own to help me all the time we were at Charleston."[82]

Return to Philadelphia

On Saturday, 22 January, the Lyells started north again. Traveling by steamboat from Charleston, they arrived Sunday morning at Wilmington, North Carolina. After attending service at the Episcopal church, they looked for fossils along the beach of the Cape Fear

80. Lyell, "White Limestone and Other Eocene or Older Formations of Virginia, South Carolina and Georgia"; cf. Lyell, *Travels,* 1:140.

81. John Edwards Holbrook, M.D. (1794–1871) was born at Beaufort, South Carolina, and educated at Brown University in Providence, Rhode Island. In 1818 he received his medical degree at the University of Pennsylvania, and in 1822, after a period of travel in Europe, he began to practice medicine at Charleston. In 1824 he became professor of anatomy in the Medical College of South Carolina. Holbrook's work on reptiles established him as a leading American zoologist. Holbrook, *North American Herpetology.*

82. Mary Lyell to Leonora Horner, 8 February 1842, Kinnordy MSS.

River, accompanied by a black man and two white boys. They found an abundance of both Miocene and Eocene forms. They also explored the riverbank, where several young men joined them in the search for fossils. With so many helpers, they collected a large number, including many fish teeth and casts of shells.[83]

Monday morning they took a train twenty miles to the Rocky Point depot, where they visited the farm of a Dr. McRea on the northeast branch of the Cape Fear River. In the afternoon Lyell rode out to collect fossils. On the banks of the river he found Cretaceous green marls similar to those in New Jersey.

After spending the night at Dr. McRea's, they returned to the Rocky Point depot and continued by train to South Washington. There they went out to the farm of a Mr. Carroll, Lyell walking, Mary on horseback. Again Lyell went off to look for fossils along the riverbank, and in the afternoon they returned to the depot. When the train came from Wilmington, the conductor was the same young man with whom they had spent the night in the snow on their way south. At eleven o'clock that night they reached Weldon, where they crossed the Roanoke River in a steamboat. In the moonlight they walked from the boat a quarter of a mile up an inclined ramp to board the train. "[The] cars had sofas & beds in them," Mary noted, "so we might have slept very well [but] for the badly laid road, the roughest we have been upon."[84] Next morning at six o'clock they reached Petersburg, Virginia, where Edmund Ruffin Sr., the father of their host at Coggin's Point earlier, had reserved a room for them at French's Hotel. After sleeping until midday, they went to the senior Ruffin's house in the afternoon.

Ten years earlier, Ruffin had become well known through the publication of his book on the use of marl to restore the fertility of soils in Virginia. The application of marl doubled the yield of crops on Ruffin's plantations, and his advocacy of marl did much to restore Virginia's agricultural prosperity. In 1833 he had launched an agricultural journal, the *Farmer's Register,* to promote the use of marl and other agricultural improvements, publishing it at first from his plantation Shellbanks. In 1835 he moved to Petersburg, where he continued to publish the *Farmer's Register.* Ruffin had made a fossil collection from his finds in the Virginia beds and gave Lyell many specimens.

Friday morning the Lyells also spent studying Ruffin's fossil

83. Mary Lyell, Diary, 1841–42, 247, Kinnordy MSS.

84. Mary Lyell to Leonora Horner, 12 February 1842, Kinnordy MSS.

collection. In the afternoon Lyell went with Michael Tuomey, a schoolteacher and amateur geologist, to collect fossils, of which they gathered enough to fill a box. Lyell also examined the fossils in Tuomey's collection. Tuomey, who had missed Lyell when he passed through Petersburg earlier on his way south, was delighted to have someone with whom to discuss fossils.[85]

On 29 January the Lyells took the train again to Fredericksburg, and the stagecoach to Acquia Creek. "The road," Mary observed, "in much better condition than when we travelled it before, but sufficiently bad." From the steamboat on the Potomac River, they had a good view of Mount Vernon. At Washington, Lyell saw Joseph Nicollet, who had recently returned from the Far West. Nicollet had collected many fossils showing that Cretaceous strata occupied an immense area in the upper Mississippi and Missouri Valleys. With no more than a brief pause for a night's rest, the Lyells went on to Philadelphia, arriving the evening of 31 January. After securing a bedroom and parlor at Jones's Hotel, "Charles went to Mr. McIlvaine's," wrote Mary, "& brought home all our dear English letters & a box containing the Old World from Frances & Susan."[86]

Lyell's winter journey through the coastal plain from Virginia to Georgia had shown him that American Tertiary formations were closely comparable to those of Europe. Their fossils were strikingly similar, often being if not identical, merely variants of European species or similar species of the same genus. In the Miocene of Virginia, the proportion of living species was similar to that of such European Miocene formations as the Suffolk Crag and the faluns of the Loire Valley. Finally Lyell had demonstrated the continuity of Eocene formations through Georgia and South Carolina. In doing so, he dispelled the chimera of a Cretaceo-Eocene formation, thereby rescuing good geologists, such as Timothy Conrad and Edmund Ravenel, from errors into which they had strayed. Lyell had confirmed the validity of his classification of Tertiary formations for America and had placed the geological study of the southern coastal plain on a firm foundation.

85. Ibid.

86. Mary Lyell, Diary, 1841–42, 249, Kinnordy MSS.

CHAPTER 3

Lectures and Summer Travels

AT PHILADELPHIA during the next six weeks, Lyell delivered a course of twelve lectures at the Philadelphia Museum, essentially repeating the lectures he had given at Boston. William McIlvaine helped with the arrangements for the lectures, advertising them so successfully as to attract an audience of four hundred persons, each of whom paid four dollars to hear the series. Even after all expenses were paid, Lyell received a handsome sum.

In his first lecture, delivered on Wednesday, 2 February, at seven o'clock in the evening, Lyell explained how the relative ages of geological formations were determined, using again a large drawing of the Mont Dor district in the Auvergne. He discussed the formation of valleys and the successive changes in the species of plants and animals through geological time. The *Philadelphia Public Ledger* reported Lyell's lecture at length. The reporter was scientific enough or good-humored enough to describe Lyell's appearance from a phrenological point of view:

> Mr. L. is in the prime of life, perhaps five feet ten inches in height, and well formed. The features of his face with the exception of his chin are rather small. The forehead is large, particularly in the perceptive department, and presents just the proportions of Locality, Individuality, Eventuality, Comparisons and Ideality which a skilled phrenologist would be likely to assign to it, judging from Mr. L's scientific character. The whole head is rather large, with a fully proportionate development of the region of Firmness, Self-esteem and Approbativeness. His temperament is sanguine, with a good share of nervous and bilious. His articulation is very distinct and his countenance is expressive, not only of perseverance and earnestness, but of a considerable degree of enthusiasm. He is formed in short, not only to enjoy the romance of science himself, but to desire and to enable others to enjoy it with him.[1]

1. *Philadelphia Public Ledger,* 4 February 1842.

In his second lecture, Lyell discussed volcanic phenomena, using examples taken principally from the Auvergne. On 9 February, he discussed the significance of rock strata for preserving the evidence for subsidence and later elevation of the sea bottom, dramatically illustrated in the Himalaya Mountains at heights as great as sixteen thousand feet. He described his classification of Tertiary strata into Eocene, Miocene, and Pliocene groups and concluded with an introductory description of the volcano of Etna in Sicily. Lyell's fourth lecture dealt with the formation of Etna and the evidence for its enormous age, but in his fifth lecture, on 16 February, he changed scenes. The large drawings of Mont Dor and Etna were replaced by a panoramic view of the Bay of Naples. Lyell described the submergence and later elevation of the Temple of Serapis on the shore of the Bay of Baiae and the evidence the temple gave for numerous movements of the land. He then referred to a drawing of the geological column. The Tertiary strata he had been discussing represented only a small portion of the whole succession of geological formations. From this drawing his audience appreciated how immense was geological time. The reporter for the *Philadelphia Public Ledger* wrote:

> These periods are immeasurable; how incomprehensible then must be the periods occupied in the formation of the whole! He compared those spaces of time, to the inconceivable distances which separate the different stars of creation, and concluded with some eloquent remarks on the mutual adaptation of the earth's surface at different periods, like so many different worlds, and the different beings which were to inhabit it.[2]

Lyell went on in the series to discuss coral islands and coral reefs. He showed a large map of the distribution of coral reefs prepared by Charles Darwin to illustrate his forthcoming book on the subject.[3] He also showed a beautiful drawing of Crescent Island, an atoll in the Dangerous Archipelago, explaining the origin of coral reefs in terms of Darwin's new theory. Lyell explained how Darwin's theory indicated that large areas of the floor of the Pacific Ocean had subsided through thousands of feet, while other smaller areas had been elevated. As evidence of subsidence during the geologi-

2. Ibid., 18 February 1842.
3. Darwin, *Coral Reefs*.

cal past, he cited the presence of ripple marks originally formed on the surface of a beach but now in strata thousands of feet below the surface. Fossil footprints in strata provided evidence of the same kind. Lyell described how on the coast near Savannah he had seen fresh opossum tracks filled with sand in less than three hours.

The formation of coal was Lyell's next topic. He said that European geologists were eagerly awaiting the publication of the geological surveys of Pennsylvania, with its great coal deposits, and of New York. Within the interior of the earth, he postulated the existence of seas of molten rock, although to a depth of at least four hundred miles, the interior of the earth must be solid. Movements within the interior seas of molten rock caused volcanic eruptions and earthquakes and the elevation or subsidence of related areas of the earth's surface. Yet, through all the changes that had occurred through geological time, conditions on the earth's surface had remained remarkably steady. "The eyes of trilobites and the bones of other animals found in very ancient rocks prove," said Lyell, "that when they lived, light, air, water and food existed then with the same peculiarities which distinguish them now."[4]

To illustrate his lecture on the recession of Niagara Falls, Lyell presented a colored bird's-eye view of the Niagara River and escarpment from the north, showing Lake Erie in the distance. This drawing, which the *Public Ledger* described as "very large and beautiful," was the work of the Glasgow-born Philadelphia artist and scene painter Russell Smith.[5] On 9 March, Lyell discussed the formation of the Niagara escarpment. He interpreted the escarpment to be an ancient sea cliff and therefore to constitute evidence of the elevation of eastern North America. The subject of sea cliffs led him to discuss the wastage of coastlines by the sea and the wearing away of the whole land surface by running water. Lyell concluded by describing the transportation of boulders by glaciers and the formation of terminal moraines.

His final lecture, on 12 March, was delivered "to the usual large and profoundly attentive audience." Lyell continued to discuss glacial phenomena, the transport of boulders by ice islands, and the

4. *Philadelphia Public Ledger*, 4 March 1842.

5. Russell Smith (1812–96) came to America from Scotland with his parents in 1819 at the age of seven and grew up at Pittsburgh, Pennsylvania, where he became a scene painter. He later painted scenes for theaters at Philadelphia, Baltimore, and Washington, illustrations for scientific lectures, and drawings for geological surveys, including the geological survey of Pennsylvania. I am indebted to the late John W. Wells, professor of geology at Cornell University, for drawing my attention to Smith as Lyell's illustrator.

geological distribution of transported boulders. He noted that in his tour through the South he had seen no such boulders until, on his return north, he crossed the Susquehanna River. Erratic boulders were clearly a northern phenomenon. At the end of the lecture, Lyell thanked his audience for attending so faithfully. The *Public Ledger* reported that "the audience for a while seemed reluctant to leave their seats"[6] and seemed genuinely regretful that so interesting a series of lectures was over. Mary wrote that the last lecture "was much admired & thought one of the best of the course."[7] Lyell's lectures at Philadelphia were clearly a great success.

Since Lyell had his lectures already well prepared, he spent much time comparing the fossils he had collected in the South with those of Timothy Conrad, Samuel George Morton, and Isaac Lea.[8] He was impressed by the richness of Lea's collection in new North American species, seeing a relationship between the large number of species of *Unio* and *Melania,* two genera of freshwater shells, and the large number of rivers in North America. The relationship between number of species and the distinctive geography of North America was striking. "It reminds the geologist," said Lyell, "of the different states of the animal creation which have accompanied the successive changes of the earth's surface in former ages."[9]

With Conrad's assistance, Lyell was able to identify twelve species of shells from the limestone of Shell Bluff on the Savannah River. All were Eocene species and all characteristic of the Eocene beds of Claiborne, Alabama. The fossil casts that Lyell had collected from the limestone at Jacksonborough, together with those given him by Colonel Jones of Millhaven, all proved to be species new to Conrad, but the genera were characteristically Eocene. Lyell also found that the proportion of living species in the American Miocene formation was the same as that in European Miocene formations. Furthermore, just as living species among European Miocene fossils belonged to species living in the warmer coastal waters of Europe, the living species among American Miocene fossils belonged to species living in American coastal waters.[10] Such results, Lyell suggested, showed the accuracy and reliability of species determinations for

6. *Philadelphia Public Ledger,* 15 March 1842.

7. Mary Lyell to Joanna Horner, 9–15 March 1842, Kinnordy MSS.

8. Isaac Lea (1792–1886) was a publisher and malacologist. Lea had made a very large collection of shells, especially freshwater forms. In his *Observations on the Genus Unio,* he described and illustrated many new species from the rivers and lakes of North America.

9. Lyell, *Travels,* 1:202.

10. Lyell, "Tertiary Formations and Their Connection with the Chalk in Virginia," 739, 741.

fossil shells. They also demonstrated the usefulness of his percentage system, since it could be extended to cover wide areas of North America.

At Philadelphia, as at Boston, the Lyells observed and felt the various forces then vibrant in America. While they were at Philadelphia, the state of Pennsylvania repudiated its bonds, in effect declaring bankruptcy. Lyell consequently suffered a loss in changing the paper money he received for his lectures into gold, a loss compensated for by the friendliness of the Philadelphians. James Fenimore Cooper, the novelist, called on them, and William McIlvaine and Clement Biddle each invited the Lyells to his house. On 22 February, Washington's birthday, McIlvaine and Henry Darwin Rogers took the Lyells for a drive to see the buildings of Girard College, then under construction. They drove next to the prison, described by Charles Dickens, where prisoners were kept in solitary confinement, given useful work to perform, and serious books to read. Finally, they visited the pride of Philadelphia, the great water reservoir at Fairmount, to which water was pumped up from the Schuylkill River and thence sent through pipes to supply the houses of the city.

New York and New England

On Sunday, 13 March, the Lyells went to New York City, taking rooms at the Carlton House, preferring a hotel to a boarding house for its greater privacy. That week, Lyell gave the first of his course of lectures at the Tabernacle, his reputation as a lecturer preceding him from Philadelphia. To her sister Joanna Horner, Mary wrote:

> It is quite true, as one of the newspapers remarks on the close of the last lecture at Philadelphia, the audience seemed inclined to linger & quite unwilling to go. The first morning we breakfasted here we heard a gentleman say to a lady "Are you going to Lyell's lectures? Lyell, the great London lecturer, the President of the Royal Society."[11]

The *New York Tribune* published Henry Raymond's transcript of Lyell's lectures.[12] The lectures were also reported in other New York newspapers, but those accounts—although more detailed and factual—were less friendly than at Philadelphia.[13] On 17 March the

11. Mary Lyell to Joanna Horner, 9–17 March 1842, Kinnordy MSS.
12. Lyell, *Eight Lectures on Geology.*
13. See *New York Herald,* 17, 20, 21, 24, 31 March and 4, 7, 11 April 1842.

New York Evening Post reported that Lyell's lecture at the Tabernacle the previous evening had been well attended and was illustrated with splendid transparencies.[14]

From the windows of their room at the Carlton House, the Lyells looked out on Broadway. Mary admired the ladies' bright-colored dresses, which made the street look like a bed of tulips. They were struck by the overwhelmingly commercial character of the city. One man said to Lyell, complimenting him on his audience, "You have all the disposable intellect of the place at your lectures." Daniel Wadsworth and his daughter, who were also staying at the Carlton House, were friendly to the Lyells. The Wilkeses, American relations of Francis Jeffrey, former editor of the *Edinburgh Review,* invited them to dine, as did other New York families. Mary appreciated the kindness they were shown. "If my letters appear too much coleur de rose," she wrote to her sister-in-law, "you must remember that wherever we have staid we have been received with open arms by one or two families & become almost domesticated among them, so that I doubt whether I shall ever be able to look upon America with impartial eyes."[15] On 21 March, Lyell went with William Redfield to Long Island, where they examined excavations made in the Brooklyn Navy Yard. He also went one day with Daniel Wadsworth to dine with John Jacob Astor, seventy-eight years old and the richest man in America. Still very alert, Astor had just given $350,000 to endow a public library in New York City.

When his lectures were finished, Lyell decided to look again at the older rocks of New York State. On 12 April he and Mary left New York City, traveling by steamboat up the Hudson River as far as Hudson City, where Lyell examined the Silurian slates and limestones. The next day they traveled from Hudson City on a newly built railroad eastward across the Taconic Hills to Springfield, Massachusetts. Ebenezer Emmons had asserted that the strata of the Taconic Hills constituted an independent series or system, older than the Siluri-

14. "Mr. Lyell's Lectures on Geology," *New York Evening Post,* 19 March 1842, p. 2, col. 3. On 18 March, the *New York Herald* observed that a great complaint had been made because members of the public were expected to pay three dollars for the whole course of eight lectures before they would be permitted to hear any of them. The *Herald* thought this was wrong and that persons ought to be able to buy tickets to individual lectures for perhaps fifty cents per lecture. On 19 March, the *Evening Post* reported that Lyell had preferred that only tickets for the whole course of lectures be issued, because he thought that to derive benefit from the lectures, his listeners needed to follow them from beginning to end. At Boston and Philadelphia, no tickets for individual lectures had been issued. *New York Herald,* 18 March 1842, p. 2, col. 2.

15. Mary Lyell to Sophia Lyell, 19 March 1842, Kinnordy MSS.

an strata of the New York system. Edward Hitchcock and Henry Rogers held the opposite view—the Taconic strata were simply metamorphosed representatives of the Silurian system. After examining various outcrops, Lyell inclined to the latter view.

At Springfield, Lyell hired a carriage to drive up the Connecticut Valley, stopping overnight at the village of South Hadley. In the morning they crossed Mount Holyoke to Amherst to visit Professor Hitchcock at Amherst College. A former student of Benjamin Silliman's at Yale College, Edward Hitchcock had carried out the geological survey of Massachusetts, the final report of which had just been published.[16] He had also participated in the early stages of the New York geological survey. Hitchcock took Lyell for a drive to see various ridges and hillocks of "drift," which were in fact glacial deposits. Both men were willing to attribute them to ice, but in the form of icebergs rather than glaciers. Hitchcock also showed Lyell the places in the Red Sandstone on the Connecticut River that contained innumerable birdlike footprints, later identified as dinosaur tracks. These remarkable footprints had been discovered first by James Deane, but Hitchcock had published the principal descriptions of them. Lyell noted that the distance between the fossil footprints was proportional to their size.[17]

From Amherst the Lyells returned to Springfield. The next day they proceeded by railway to Worcester, where Lyell examined a deposit of impure graphite interstratified with mica schist and clay slate. These rocks were, like the Taconic strata, metamorphic, but Lyell thought (mistakenly) of Carboniferous age. Lyell collected samples of the graphite to take back to England for analysis. On 16 April, the Lyells took the afternoon train into Boston and were met at the station by George Ticknor, who took them to his home on Park Street. Their time with the Ticknors was filled with Boston friends paying calls. They also saw John Amory Lowell, who had rescued the American edition of Lyell's *Principles* from its bankrupt publisher while the Lyells were in the South.

On 19 April, Lyell went alone to examine the Tertiary strata on Martha's Vineyard, an island off the south shore of Cape Cod. That evening he examined the Gayhead at the western end of Martha's Vineyard. The cliffs, more than two hundred feet high, contained steeply inclined, brightly colored strata consisting of bright red clay

16. Edward Hitchcock (1793–1864); Hitchcock, *Geology of Massachusetts.*
17. Lyell, *Travels,* 1:253.

mingled with white, yellow, and green sands and some beds of black lignite. They reminded him of the cliffs of Alum Bay on the Isle of Wight. Aided by some Indians, he collected many fossils that showed that the beds must be Miocene, rather than Eocene, as Hitchcock had thought. Some of the sharks' teeth belonged to the same species that Lyell had collected in the marls of the faluns of Touraine and the Suffolk Crag. He also found vertebrae of dolphins and whales. Near the lighthouse, in a great fold of the strata, one conglomerate bed was particularly rich in bones. From a fisherman Lyell bought a fossil skull that had fallen out of the conglomerate onto the beach. It later proved to be a walrus skull.[18]

On his return to Boston, Lyell attended the annual meeting of the Association of American Geologists. The association had met for the first time two years before. It grew out of the meetings of the four geologists of the New York survey; each autumn they met to compare the results of their summer's work. American scientists were eager to communicate with one another.[19] The New York geologists frequently consulted members of other state surveys. In 1838, at the suggestion of Edward Hitchcock, they decided to open correspondence with geologists throughout the country. The response was so favorable that in April 1840, the geologists held their first national meeting at Philadelphia, meeting there again in 1841. The meeting at Boston in 1842 was particularly significant, because the American Society of Naturalists asked to meet with them. During the course of the Boston meeting, the two societies decided to fuse to form the Association of American Geologists and Naturalists. In 1847 it would become the American Association for the Advancement of Science. Lyell was much impressed by the number of geologists present at the Boston meeting, mentioning by name in his *Travels* some twenty of those present. Clearly, the state surveys had created a significant body of experienced geologists. Lyell did not deliver a paper himself, but he took an active part in discussions.

At the meeting, J. P. Couthouy, a naturalist on the Wilkes Exploring Expedition, described his observations of icebergs and their geological effects. Although icebergs often carried large boulders and masses of sand, gravel, and mud, Couthouy was certain they could not produce the parallel furrows and striations in rocks described by Hitchcock as an essential part of the drift phenome-

18. Ibid., 204–5, 256.

19. See Kohlstedt, *American Scientific Community*, 59–79.

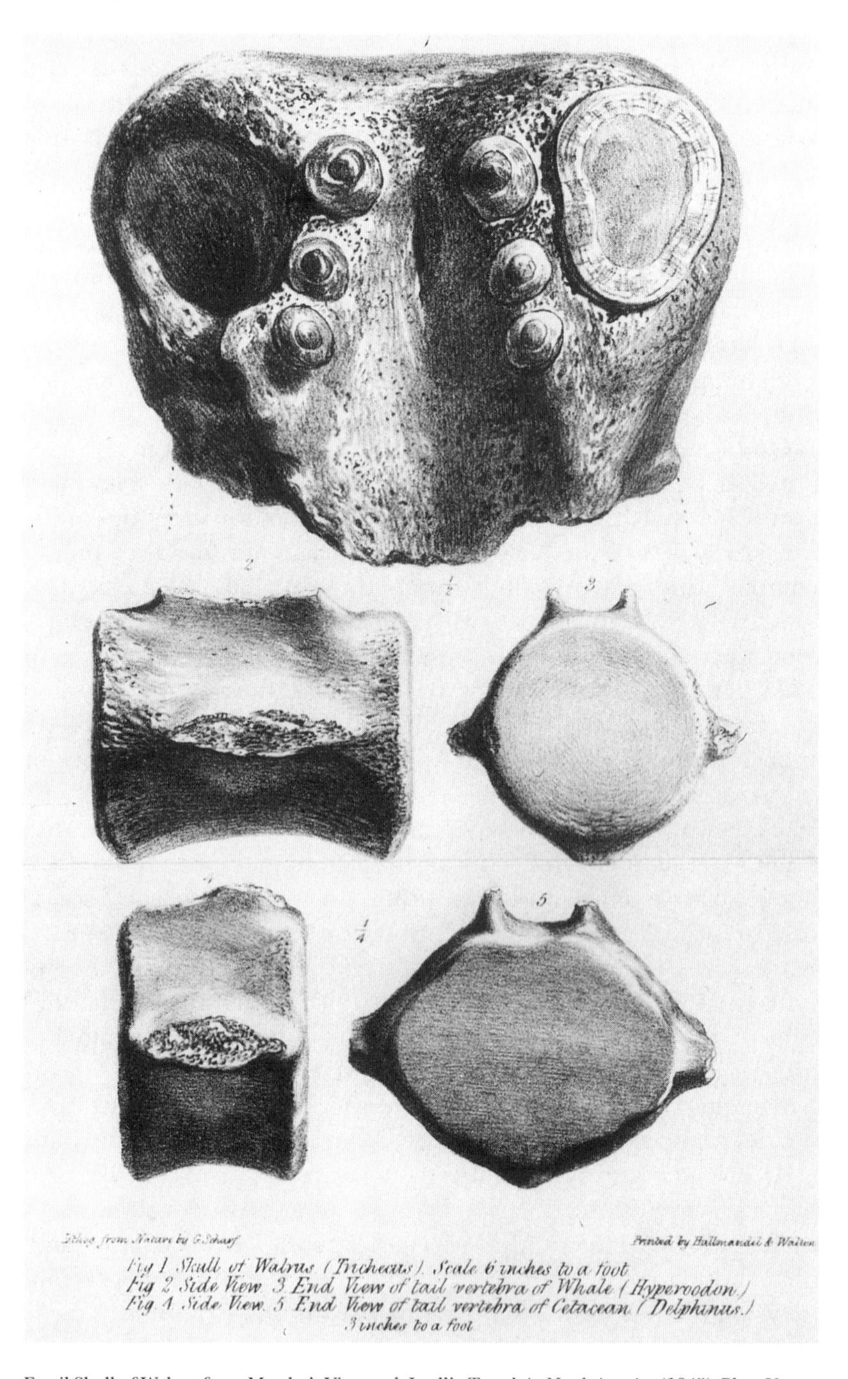

Fossil Skull of Walrus from Martha's Vineyard. Lyell's *Travels in North America* (1845), Plate V.

non. William Redfield described fossil rain-marks that he had observed with Lyell in the quarries of New Red Sandstone at Newark, New Jersey, and a new species of fossil footmark from the Red Sandstone of Connecticut.[20] The brothers Henry and William Rogers read a long paper containing a brilliant description of the Appalachian chain as a series of long and extraordinarily straight anticlinal and synclinal folds that appeared to be the result of lateral pressure. They had been elevated without any apparent accompaniment of volcanic activity. The distribution of sediments among the Appalachian formations suggested the former existence of a continental land mass to the southeast, where now the Atlantic rolled.

Lyell enjoyed meeting the American geologists assembled at Boston, but the meeting was not free of tension. In one session, Lyell got into a slight argument with Henry Rogers by mentioning that the present dip of the Red Sandstone of Connecticut was "due in part at least, to an uplifting of the strata."[21] Rogers attributed the dip of such beds in Pennsylvania to their sediments' having been deposited on a sloping surface. Lyell suggested that the direction of the dip, which was transverse to the ancient estuary in which the sediments had been laid down, and its steepness, sometimes as great as twenty degrees, argued for uplift. The incident illustrates Henry Rogers's catastrophist tendency as a theorist.

Of greater significance for Lyell and his association with some of the American geologists was a newspaper article that appeared about a month before the Boston meeting. While he was at New York, quite by chance Lyell saw at Daniel Wadsworth's the 31 March issue of the *Albany Evening Journal.* To his astonishment, it contained an anonymous article signed "Hamlet," reprinted from the *Boston Daily Advertiser* of 26 March, with an editorial endorsement by the Albany paper. The article charged Lyell with misappropriating the work of American geologists in his lectures and in his plans for publication on American geology.[22] Lyell wrote immediately to James Hall at Albany about the article and asked if he knew who was responsible for the editorial remarks in the Albany paper.[23] Hall responded that he could not say anything about the source of the article. He commented that it was rumored that Lyell had already

20. Association of American Geologists and Naturalists, *Proceedings and Transactions,* 65.

21. Ibid., 63.

22. "Hamlet" [James Hall], *Boston Daily Advertiser,* 26 March 1842.

23. Lyell to Hall, 1 April 1842, Hall MSS.

arranged to publish on American geology and was planning to extend his visit for a year to prepare himself to do so. If the rumor were false, Hall offered to arrange for a retraction by the Albany newspaper. Hall did not admit that he had had anything to do with the editorial accompanying the article.[24]

Lyell responded that Hall should have asked him directly about his publication plans if they were a concern. He declined Hall's offer to publish a retraction. Although the anonymous charges were not true, Lyell thought he had a perfect right to publish future editions of his *Elements of Geology* with references to American geology, just as he had included references to the geology of Europe, especially of countries he had visited. In 1842 he had no immediate plans for a new edition.[25] American publishers had brought out pirated editions of both Lyell's *Principles* and *Elements* without offering him any payment for them. Lyell was not going to allow an anonymous newspaper attack to restrict his plans for future publication, even though at the time he had not decided what those plans might be.

Lyell did not know then who his attacker was, nor what provoked the attack. He was assailed not for anything he had done, but for what his attacker feared he might do in the future. Lyell sought information eagerly and imparted his own knowledge generously. Various American geologists had shown him the districts where they worked. He was grateful for their hospitality and had learned much from them. In turn, they had learned much from him. From England Lyell had brought books and fossils from British Silurian strata, both of which he shared with American geologists, especially James Hall and Timothy Conrad. When Hall came to Boston to attend Lyell's lectures, he spent much time poring over Lyell's British fossils. At New York on 30 March in his lecture on coal, Lyell mentioned Henry Rogers, who had accompanied him in Pennsylvania, and Edmund Ruffin's observations of the Great Dismal Swamp in Virginia. He referred to Hall's map of American coal-bearing strata. Lyell said that he had been over much of the ground that the New York geologists had surveyed, and he testified "to the accuracy of their labors, to the great pains they have taken, and the science with which they have conducted the survey." The publication of the New York reports would herald "*an era in the advancement of science.*"[26] He had made these remarks before he had seen the article in the *Al-*

24. Hall to Lyell, 3 April 1842 [draft copy], Hall MSS.

25. Lyell to Hall, 4 April [misdated March] 1842, Hall MSS.

26. Lyell, *Eight Lectures on Geology,* 35–36 (emphasis in original).

bany Evening Journal and was glad later that he had done so at a time when they could not be considered a reply. In an editorial on 5 April, the *New York Tribune* announced that Lyell had no intention of bringing out new editions of his works with notes on American geology. The editorial cited Lyell's lectures, especially his lecture on coal, for evidence of Lyell's willingness to acknowledge the work of American geologists.[27]

During the Boston meeting, Lyell learned that Hall had written the Hamlet article. He gave no public indication of his feelings. After the warm friendship that he had developed with Hall during their weeks in the field the previous summer, Lyell must have been shocked and saddened by Hall's betrayal. Published under cover of anonymity, the Hamlet article was both cowardly and duplicitous. After Lyell learned of Hall's authorship, Hall attempted to wriggle out of responsibility for the article by hinting that others had encouraged him to write it. In his presidential address at the Boston meeting and in other ways, Benjamin Silliman sought to soothe hurt feelings and to maintain peace.

About the first of May, Hall sent Lyell a pathetic letter in which he regretted the breach in their friendship. While acknowledging responsibility for what he had done, Hall claimed that he had not been inspired by vanity or envy.[28] Lyell replied immediately: "In regard to the subject of your letter, I can merely reassure you that I feel and indeed never have felt any resentment & have given you credit from the beginning for showing a desire of not criminating any one else."[29]

When he was at Boston in July, just before departing for Nova Scotia, Lyell wrote again in response to a letter from Hall accompanying four boxes of New York State fossils that Lyell was to take to England.

> In reference to one subject in your letter I am bound to say that since my arrival here [Boston] I have learnt that the course which you took was encouraged by one so much your senior in age & scientific position that it ought to be duly allowed for in extenuation. I shall do my best, as indeed I have done since we parted, wholly to forget these matters.[30]

27. "Mr. Charles Lyell," *New York Tribune,* 5 April 1842, p. 1.
28. Hall to Lyell [1 May] 1842 [draft copy, partly illegible], Hall MSS.
29. Lyell to Hall, 3 May 1842, Hall MSS.
30. Lyell to Hall, 14 July 1842, Hall MSS.

The identity of the senior American scientist whom Lyell believed to have encouraged Hall remains uncertain, and perhaps better so.[31] Lyell held no grudge and would continue to correspond with Hall and join him in the field on later visits to America.

The Ohio Valley and Great Lakes

In May 1842, the Lyells set out again from their base at Boston, this time on a tour to the West. Lyell wished to see the great coal beds of the Ohio Valley and to explore the Great Lakes and the St. Lawrence River valley. He planned to cross the Appalachians in Virginia and Maryland, where he could see clear sections of their folded strata. As they crossed New Jersey by train, lilacs were just coming into bloom, apple orchards were white with blossoms, and dogwoods were in full bloom. From Baltimore they went by rail along the valley of the Patapsco River. With its multitudes of pink and red azaleas and white dogwood, they thought it one of the most beautiful valleys they had ever seen.[32] They traveled by rail to Frederick, Maryland, and thence by hired carriage to Harper's Ferry, Virginia, where they stayed overnight.

On the morning of 7 May they started north from Harper's Ferry in their carriage, traveling across successive ridges of the Blue Ridge through Antietam and Sharpsburg to Boonsboro in Maryland and via Clearspring to Hagerstown. As they went, Lyell drew sections of the strata exposed in the gaps through the ridges. From Hagerstown they followed the great National Road westward to Cumberland, traveling sometimes by stagecoach and sometimes by hired vehicle. The National Road was the main thoroughfare for the movement of people and goods between the eastern seaboard and the Ohio Valley. On the stagecoach they talked with two Kentucky farmers returning from Baltimore, where they had gone to sell their mules and cattle. Along the road, Lyell seized every opportunity to study the strata revealed by cuts made in road building.

They stopped at Frostburg, Maryland, for two days (10 and 11 May) while Lyell visited iron and coal mines in the neighborhood. The bituminous coal beds of the Cumberland coalfield at Frost-

31. The complexities of the affair are explored in Robert H. Silliman, "The Hamlet Affair," esp. 555. Silliman does not comment on the moral implications of Hall's actions nor his cowardice in resorting to anonymous publication. He seeks to suggest, as Hall did, that Lyell had a reputation for plagiarism. He did not, and the supposed examples Silliman cites do not bear examination. See Wilson, "Brixham Cave" and "The Gorilla and . . . Human Origins."

32. Lyell to Marianne Lyell, 6–7 May 1842, Kinnordy MSS.

Lyell's Travels through the Ohio Valley, the Great Lakes, and Canada, May–June 1842. Map by Philip Heywood.

burg, like those Lyell had seen in northeastern Pennsylvania the previous summer, were underlain by beds of clay containing *Stigmariae.* At one point, however, Lyell also found fossil leaves of what he thought were ferns in the same clay with the *Stigmariae.* Again he was struck by the identity between the coal plants of Pennsylvania and those in European coal. In the Carboniferous period, the floras of North America and Europe seemed to have been much more uniform than in modern times. From the inn at Frostburg they could look both southward into Virginia and northward into Pennsylvania.

Lyell and Mary admired the springtime beauty of the Appalachians—the soft green of the new foliage on the wooded mountains and the profusion of pink azaleas, sometimes covering whole hillsides. They delighted too in the splendid colors of the birds. "There are numbers of [a] black & white woodpecker with a head of the richest crimson velvet," Mary wrote, "the blue bird & the red bird [cardinal] & I think the Baltimore oriole which is black & orange." In the Ohio Valley the great size of the oaks, maples, and tulip trees in the forests and the luxuriant richness of the orchards and gardens in the settlements along the river, impressed them with the boundless fertility of the country. At Frostburg, Mary was also struck by the women's custom of riding horseback about the countryside, often by themselves, for it was their only means of getting about. On 12 May they continued westward along the National Road, the sole passengers in a mail coach, traveling all day up and down over a succession of steep hills covered with azaleas. Mary described the scene:

> When we reached the top of the highest, Laurel Hill, we stopped just as the sun was setting to take a view into the great valley of the West for the first time in our lives & to think that every streamlet flowed into the great Mississippi. We passed Braddock's grave, which is in a field by the roadside among these mountains, where he was brought after he was mortally wounded near Pittsburg then Fort Duquesne. How the troops ever made their way through these forests I cannot conceive. The great National road we traveled over is an immense work, cut winding along the sides of steep hills.[33]

From Laurel Hill they could see the pall of smoke over Uniontown, Pennsylvania, caused by the burning of bituminous coal. It re-

33. Mary Lyell to Caroline Lyell, 16 May 1842, Kinnordy MSS.

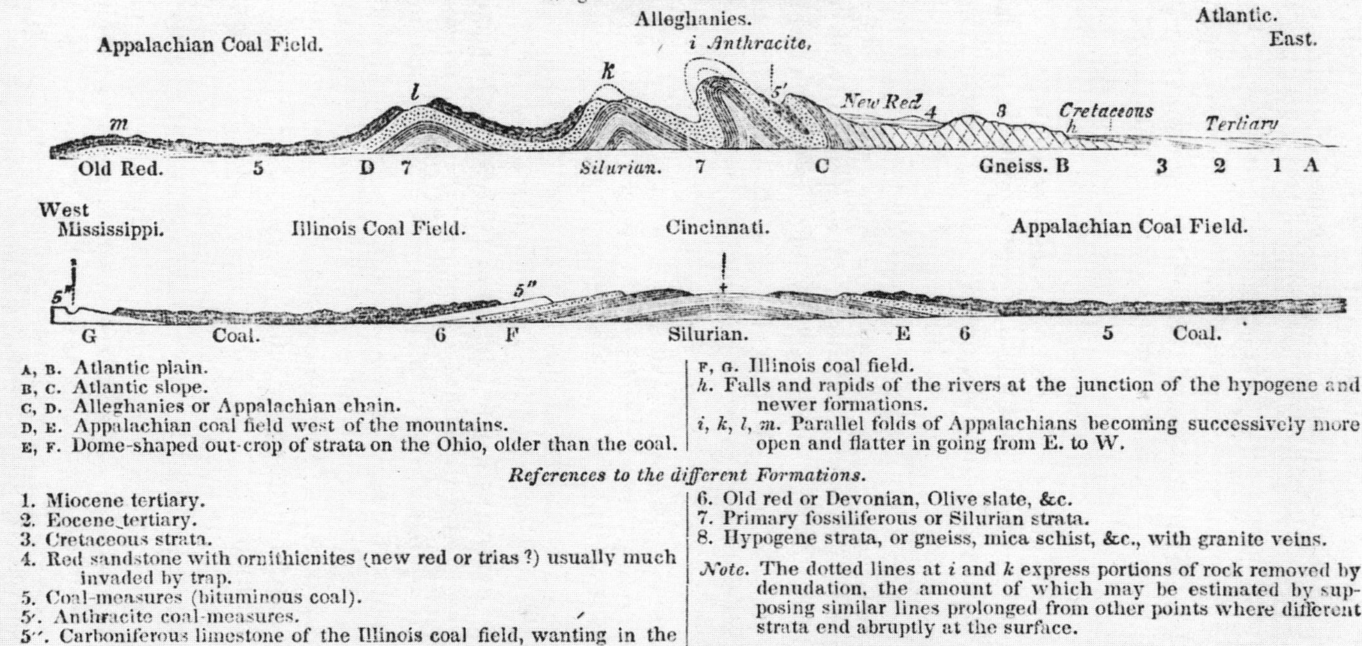

Ideal Geological Section of the Country between the Atlantic and the Mississippi. Lyell's *Travels in North America* (1845), Fig. 5.

minded them of English towns. At Uniontown, Lyell visited coal seams mined in open quarries. On 14 May they went on to Brownsville on the Monongahela River, where they saw the great Pittsburgh seam of coal, cropping out ten feet thick in the river cliffs near the water. At Brownsville they spent the afternoon walking along the banks of the Monongahela, where within five minutes they picked up six different species of the freshwater shell genus *Unio.*[34]

The next day they boarded a sternwheel steamboat to descend the Monongahela to Pittsburgh, "a most flourishing town," where twenty-two steamboats lay at the wharves. The steamboats leaving Pittsburgh to go down the Ohio were erratic and unpredictable in their departures, so on 17 May the Lyells set out by stagecoach for Wheeling, Virginia. After six hours and twenty-five miles, they reached Washington, Pennsylvania, where they found that the stage to go on to Wheeling would not arrive until the middle of the night. Indignant at this breach of promise, Lyell was mollified by the good-natured landlord. He called Lyell "Major" and said that the new stagecoach timetable was as inconvenient for him as for them.[35] The coaches in which they traveled were, wrote Mary, "very much like a French diligence to hold nine, three seats & a very long body which swings up & down like a ship & is open at the sides, with leather curtains if it rains."[36] They provided a rough, swaying ride.

At Wheeling where letters from England awaited them, they boarded a steamboat to go down the Ohio River to Marietta, Ohio, using the leisure of the boat trip to read their letters. At Marietta, Lyell had an introduction to Dr. Samuel Hildreth, a physician with a deep interest in both natural history and the history of the Ohio region. Hildreth had come to Marietta from Massachusetts in 1806 and had published many papers on the diseases of the Marietta district, on its geology, and on other regional topics. He took Lyell to see nearby rock strata belonging to the uppermost beds of the coal formation. Hildreth also gave Mary seeds of some of the native Ohio plants she admired.

The Lyells continued by steamboat down the Ohio River to Pomeroy, a village beneath wooded hills on the Ohio side of the river. Founded only five years before by a New Englander named Pomeroy, who with his sons and daughters and their families formed nearly the whole population, the village was established to

34. Lyell, Notebook 99, 132, Kinnordy MSS.

35. Lyell, *Travels,* 2:28.

36. Mary Lyell to Katherine Horner, 17 May 1842, Kinnordy MSS.

View of the Great Coal Seam on the Monongahela at Brownsville, Pennsylvania. Lyell's *Travels in North America* (1845), Plate VI.

mine a rich outcropping of coal in the riverbank. At Pomeroy the Lyells boarded another steamboat for Cincinnati. They had seen few blacks in Ohio, but now, on the steamboat, they witnessed the grim meaning of slavery. Mary described it in a letter.

> I saw such a painful scene the other day, four runaway slaves in the steamboat, brought back & landed & all chained together & two men with guns guarding them. They were on their way to Canada when they were caught. Poor creatures it made me feel quite sick when I saw them walk on shore into a slave state again. Yet, as Charles says, it is not much worse than deserters taken back to a regiment, but there was something in the crowd of people on the bank cheering at seeing them restored to captivity, which was inexpressibly painful & I c[oul]d not keep my indignation to myself.[37]

On Sunday morning, 22 May, they landed at Cincinnati in time to attend service at an Episcopal church. With its busy commerce, broad streets, and fine buildings, Cincinnati impressed them favorably. Lyell had introductions to two Cincinnati naturalists—a Mr. Buchanan and John G. Anthony, a businessman who was also an enthusiastic freshwater conchologist. On Monday morning, the Lyells set off with Buchanan and Anthony on a two-day expedition to Big Bone Lick in Kentucky, famous for the fossil bones of mastodons, elephants, the giant extinct sloth *Megalonyx,* and other animals. Crossing over the Ohio River into Kentucky, they rode through a forest of great trees, covering a nearly level tableland. Beneath the trees the undergrowth had been eaten down by cattle and replaced by grass, so they rode through an unending park or wood pasture of great oaks, maples, and tulip trees, the last in full bloom. Buchanan and Anthony pointed out to the Lyells the many brightly colored birds along their route. In the valley of the Big Bone Creek in former times, salt springs had attracted mastodons and other animals, some of which became mired in the soft ground. The American bison, or buffalo, used to come in numbers to the Big Bone Lick. Benjamin Finnell, owner of the land, showed the Lyells the traces of the old trails that were still visible, although no buffalo had come along them for sixty years. After spending Tuesday studying the lick, the party stayed overnight at the house of a Kentucky planter, where at breakfast the next morning they were of-

37. Ibid.

fered broiled squirrels. They returned that day by another trail through the forest to Cincinnati.

While Mary remained at the hotel in Cincinnati, Lyell went with Dr. John Locke, a physician and amateur geologist, on a geological expedition up the Ohio River to follow the succession of strata from the foot of the Carboniferous downward as they cropped out in the riverbanks. At Rockville, Ohio, they examined the Waverley Sandstone, a southwestward extension of beds that Lyell had seen the summer before in western New York. James Hall had recognized it as identical with the Portage and Chemung groups in New York. Below the Waverley Sandstone was the Cliff Limestone, which Hall considered identical to the Helderberg and Niagara Limestones of New York. Many of the formations when seen in New York State were composed of numerous thick strata with many subdivisions, but in Ohio they thinned out or disappeared. Lyell was impressed by the perfect preservation of the fossils in the rocks around Cincinnati, even though they were of Lower Silurian age and among the oldest sedimentary rocks. The character of the fossils suggested that they had been deposited in a deep ocean far from land. The oceanic character of Silurian deposits seemed reasonable to Lyell, because, he thought, it must require a long period of geological time to raise beds of sediment from the ocean bottom to form continents.[38]

After Lyell returned to Cincinnati, he and Mary set out by stagecoach for Springfield on the first leg of their journey northward to Cleveland. Hardly more than twenty-five miles out of Cincinnati, Lyell began to see patches of "northern drift" (i.e., deposits of glacial till and boulders) covering the countryside. At Springfield he saw many large boulders that must have come from north of Lake Erie. The reappearance of boulders and drift so close to Cincinnati struck Lyell forcibly. During his weeks of travel across the Appalachians and through the Ohio Valley, he had seen none.

After spending the night at Springfield, the Lyells drove in a hired carriage to Columbus, the capital of Ohio. From Columbus, they continued north by stagecoach through forested country just beginning to be settled. New clearings were being fenced and ploughed for the first time; in some, the settlers were burning trees chopped down during the preceding winter. In many clearings, settlers were building log houses. From the top of Stony Hill, at the edge of the Ohio tableland and sixteen miles out of Cleveland, they

38. Lyell, *Travels,* 2:47.

saw before them "a broad and level plain covered with wood; and beyond, on the horizon, Lake Erie, extending far and wide like the ocean."[39] By traveling only during the day, and alternating days on the stagecoach with days in a hired carriage, the Lyells had come across Ohio without mishap, despite the roughness of the roads, many of which were corduroy.[40]

At Cleveland, Lyell was fortunate to have Dr. Jared Kirtland serve as his guide to the local geological features.[41] Although his Yale degree was in medicine, Kirtland also had learned geology and mineralogy under Benjamin Silliman. In 1834 he had gained international scientific recognition for his discovery of the separate sexes of bivalve freshwater molluscs.[42] Kirtland had also taken part in the geological survey of Ohio. He took Lyell to see the Lake Ridges, two ridges of sand and gravel, each about fifteen feet high and running east and west parallel to the shore of Lake Erie. The first ridge was about half a mile back from the shore, the second another half mile inland. Still farther inland, two more ridges could be traced, but instead of running parallel to the present shore of Lake Erie, they followed the contours of the rising land of the interior. Lyell was certain that the ridges represented old lines of beach when Lake Erie stood at higher levels than at present. He thought they presented evidence for the elevation of the land. Cleveland itself appeared to be built on a raised delta deposited when the Cuyahoga River flowed at a higher level.

On 4 June the Lyells took a steamboat for Dunkirk, New York, where they were set down on the pier at two o'clock in the morning. With no one about, they walked to the inn, where the innkeeper gave them a room and brought them clean sheets for their bed. The next day, after a late breakfast, they walked in the bright June sunshine along the shore of Lake Erie and saw their first black squirrel. Later they went inland three miles to Fredonia, a pretty village with white houses around a green and six churches. A bubbling spring of natural gas on the outskirts was captured and piped into the village, where it was used for lighting houses. Lyell looked for the lake ridges but could not relate ridges at Dunkirk with those he had seen at Cleveland.

39. Ibid., 63.

40. Mary Lyell to Leonora Horner, 28 May–3 June 1842, Kinnordy MSS.

41. Jared Potter Kirtland, M.D. (1793–1877), was born at Wallingford, Connecticut, and in 1815 graduated with a medical degree from Yale. In 1823 he settled at Poland, Ohio, and in 1837 moved to a farm near Cleveland, Ohio.

42. Kirtland, "Sexual Characters of the . . . Naiades."

From Dunkirk they went on to Buffalo and to Niagara Falls. From the Clifton House on the Canadian side, the view of the falls—the snowy whiteness of the foam against the bright green water and the rainbow in the spray in the gorge below, all seen in the clear air and bright sunshine of early June—enchanted them.[43] During the next six days Lyell confirmed and extended his observations of the previous summer. At the Whirlpool four miles below the falls, Lyell found on the American side strata of sand and gravel containing freshwater shells, like those on Goat Island. Similar strata occurred on the Canadian bank. They demonstrated conclusively that the Niagara River had formerly flowed three hundred feet above its present level, before the river had excavated the gorge. Lyell also traced a former valley on the Canadian bank cut through the Silurian limestone from the Whirlpool to St. David's, three miles to the northwest.[44] One day he and Mary went with a guide behind the great arch of falling water that formed the Horseshoe Falls and afterwards received a printed diploma for having performed the feat.[45]

Canada and Northern New England

At Lewiston, near the mouth of the Niagara River, the Lyells took a steamboat to cross Lake Ontario. As the steamboat approached Toronto, they admired the beauty of its great harbor, protected from the open lake by wooded islands. The brick houses of Toronto were strikingly different from the white clapboard houses of New York towns. Although the colonial government had been moved to Kingston, Toronto looked prosperous, with many good shops.[46]

As soon as the Lyells arrived at their hotel, Thomas Roy, a Scottish civil engineer who had lived at Toronto since 1834, called on them.[47] In 1837 Roy had sent to the Geological Society of London an account of successive terraces running parallel with the north shore of Lake Ontario.[48] He postulated that a great inland sea of fresh water had once covered the site of all the existing Great Lakes and much of the surrounding land. The barriers retaining this sea broke down one after another so that the water subsided to successively lower levels until it reached the present level of the Great Lakes. Thus Roy was one of the first persons to study the vast con-

43. Mary Lyell and Lyell to Eleanor Lyell, 5–7 June 1842, Kinnordy MSS.
44. Lyell, "Ridges, Elevated Beaches."
45. Mary Lyell to Sophia Lyell, 20 June 1842, Kinnordy MSS.
46. Mary Lyell to Joanna Horner, 16 June 1842, Kinnordy MSS.
47. Legget, "Thomas Roy."
48. Roy, "Ancient State of the North American Continent."

sequences of the previous glaciation of North America and to envision the former existence of much larger lakes covering the site of the present Great Lakes.

Lyell and Roy rode northward on horseback from Toronto. About a mile from the lake, they came to the first ridge, 20 to 30 feet high, with its base 108 feet above Lake Ontario. The sandy crest of the ridge was marked eastward and westward by a narrow belt of pine trees growing along it. The second ridge, a mile and a half farther north, was fifty to seventy feet high, with a great many boulders strewn along its base.[49] Two and a half miles farther north they came to a third ridge, which was much less conspicuous than the first two. By taking levels, Roy had determined that the base lines of each ridge were extremely uniform as they ran eastward and westward. He showed Lyell eleven of these ridges, the highest of them 680 feet above Lake Ontario. Beyond the highest ridge, the land rose to 762 feet above Lake Ontario and then descended toward Lake Simcoe, forty-two miles north of Toronto. On the slope toward Lake Simcoe, Roy had found ridges at levels corresponding to those of the upper ridges on the southern side of the watershed. Lyell thought that the ridges could not be explained by the successive removal of barriers that had held back the waters of former lakes. He could not imagine where the land to form such barriers had been, nor, if it had existed, how it had been removed. Consequently, he thought the ridges must have been formed by the sea. They must represent ancient lines of coast and therefore constituted evidence for successive elevations of North America during the recent geological past.

To the north of Toronto, the country was being cleared and settled. It had changed so much since Roy had surveyed it two years before that he became lost in trying to find the inn where they intended to stay overnight. At a succession of log houses to which Lyell and Roy led their weary horses, they found they could not communicate with the people, because they understood neither English, French, nor German. Finally, late at night, they found the inn, and learned that the poor settlers from whom they tried to obtain directions were speaking Gaelic and were either Highland Scots, Welsh, or Irish. Although these new immigrants desperately needed work, and the Canadian farmers needed farmhands, the language barrier made it difficult to employ them. The obstacle of lan-

49. In the present city of Toronto, this ridge runs east and west a few blocks south of St. Clair Avenue.

guage showed Lyell how much the progress of a new colony such as Upper Canada might depend on elementary school education in the British Isles.[50] After their return to Toronto, Roy took the Lyells on a picnic to some quarries to collect fossils. The next day, the Lyells boarded the steamboat *Princess Royal* for Kingston at the eastern end of Lake Ontario.

From Kingston, Lyell rode twenty miles on horseback to Gananoque, a small Scottish settlement at the eastern end of Lake Ontario, to examine the geology of the country. On 20 June Mary boarded a steamboat at Kingston that stopped two hours later at Gananoque, where Lyell rejoined her. Then, partly by steamboat and partly by stagecoach, they traveled down the St. Lawrence River to Montreal. The great St. Lawrence, with its multitude of rocky islands covered with firs and white birches, reminded them of Norway. As they approached Montreal they seemed to be entering a province of France. "The peasantry in costume look as if they had just left Brittany," Mary observed. "An old beggar at the stage door might have been at Boulogne. We saw two crosses with the usual symbols by the roadside." Montreal she thought "a fine old looking town with many church towers built in the French style, with convents too. All the common people talk French & much better French than one usually hears in the provinces [of France] & they have the pleasant affable manners which are so often wanting in the Anglo Saxon race, though they are said to be very ignorant."[51]

At Montreal, Mary was confined to the hotel by a mild illness. Dr. Andrew Holmes, an Edinburgh medical graduate and professor of medicine at McGill University, took Lyell to see the geology of Mount Royal.[52] On the Côte de Neige about 540 feet above sea level and 306 feet above the level of Lake Ontario, Lyell found seashells in a gravel deposit of the "northern drift." The seashells were to Lyell a not unexpected find, for the terraces around Lake Ontario had suggested to him that North America had undergone recent elevation. Fossil seashells at such levels on the Côte de Neige suggested that in the recent geological past the St. Lawrence Valley had been elevated several hundred feet. In 1835 when Captain Henry

50. Lyell, *Travels,* 2:95.

51. Mary Lyell to Sophia Lyell, 20 June 1842, Kinnordy MSS.

52. Andrew Fernando Holmes (1797–1860) was born at Cadiz, Spain, of Scottish parents who had been passengers on a British vessel captured by the Spanish on its way to Canada. Holmes's parents reached Montreal in 1801, and he was apprenticed to study medicine there in 1811. In 1816 he began to study medicine at the University of Edinburgh, where in 1819 he was graduated M.D.

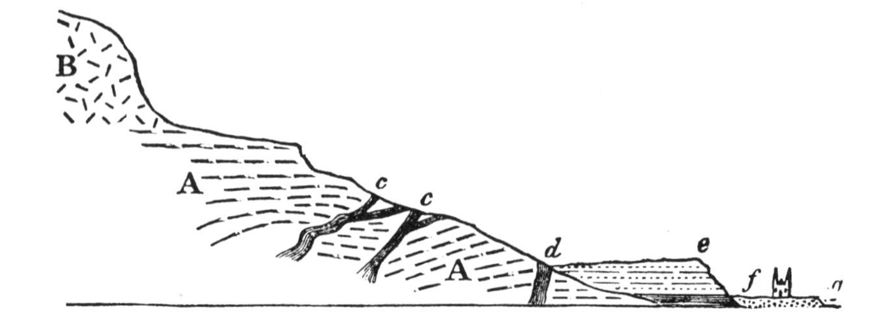

A. Silurian limestone.
B. Trap or greenstone.
c. Dykes of basaltic trap.
d. Dyke of felspathic trap, or claystone-porphyry.
d. *e*. Terrace of drift with shells.
f. Gravel, on which part of Montreal stands.
g. River St. Lawrence.

Section of Montreal Mountain with Shelly Drift. Lyell's *Travels in North America* (1845), Fig. 13.

Bayfield had sent Lyell fossil seashells collected at Beauport near Quebec City, Lyell was astonished to find all the species identical with ones he had collected the year before at Uddevalla in Sweden. The species of fossil shells from both places were still living in Arctic waters but were quite different from seashells living at present in the Gulf of St. Lawrence. In the drift on the Côte de Neige at Montreal, Lyell found the same assemblage of fossil seashells as those from Beauport and Uddevalla, all Arctic species. They showed that in the recent past the climate of the St. Lawrence Valley had been far colder.

Saturday evening, 25 June, the Lyells left Montreal by steamboat for Quebec City, where they arrived about seven o'clock the next morning. Mary described the city:

> The approach from the river is very striking. Indeed, I think it by far the most picturesque place we have seen in America. It stands on a rock with strong fortifications like Ehresbreitstein, but the St. Lawrence at that point is a much finer river than the Rhine. The quantity of shipping made it look very cheerful & above Quebec [were] a succession of coves or bays filled with rafts of timber ready to send off.[53]

At Quebec they attended service in the English cathedral, where for the first time they heard prayers said for the young Prince of Wales. After church, Lyell went to call on Colonel William Codrington of the Coldstream Guards stationed at Quebec.[54] Over the next few days, Codrington took the Lyells to see various places of geological and scenic interest around Quebec. They visited the falls of Montmorency, Beauport, and the spot on the Plains of Abraham where in 1759 General Wolfe fell. Mary thought the joint monument to Wolfe and Montcalm, erected by Lord Dalhousie, "in very good taste especially where the French population is so numerous." One evening, Colonel Codrington showed Mary the Citadel, the great fortress that had been strengthened during the rebellion of 1837 in Lower Canada. During a walk along the south shore of the St. Lawrence, the Lyells came upon a group of Indians lying asleep among their birch bark canoes, drawn up on the shore. Mary also observed the hostility in Quebec between the English and French:

53. Mary Lyell to Mrs. Leonard Horner, 4 July 1842, Kinnordy MSS.

54. Sir William John Codrington (1804–84) entered the British army in 1821 as an ensign in the Coldstream Guards. After successive promotions he was made lieutenant colonel in 1836 and later in his career a general.

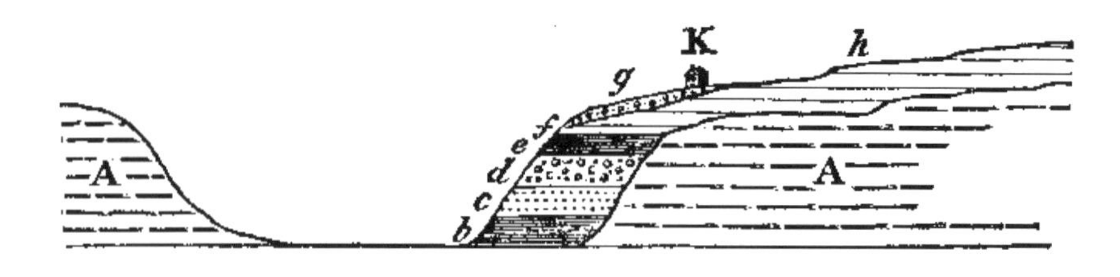

A. Horizontal Lower Silurian strata.
b. Laminated clay.
c. Yellow sand.
d. Drift with boulders.
e. *Mya, Terebratula,* &c.
f. Mass of *Saxicava rugosa.*
g. Gravel with boulders.
h. Clay and sand of higher grounds, with *Saxicava,* &c.
K. Mr. Ryland's house.

Position of Shelly Drift in the Ravine at Beauport. Lyell's *Travels in North America* (1845), Fig. 14.

> The loyalist or English party who are a very small minority find it difficult to get on at all, the French party which is so numerous being against improvement of every kind, railroads, steamboats, even Macadamized roads as we know to our cost. They are also opposed to all education, led by the priests, who are nevertheless a very respected class of men, but conscientiously think reading & writing a bad thing. The consequence is that there is no competition. The steamboats are overcrowded & always going wrong & the hotels in the large cities the worst we have been in, & all forming a lamentable contrast to the U.S. These habitants have however very pleasing polite manners reminding us constantly of Normandy & Brittany from whence they migrated. Indeed last Friday [30 June] when we stopped to dine at a small cabaret by the roadside & were asked whether we chose to eat "maigre ou gras" we were transported to our friends of 1840 [in Normandy and the Loire Valley of France].[55]

The Lyells left Quebec by stagecoach on the evening of 30 June and at midnight reached Three Rivers, where they stayed overnight. The next morning they continued along the bank of the St. Lawrence, driving themselves in a high two-wheeled calèche. After staying overnight at Maskinonge, they made a side trip upcountry to visit a waterfall in the midst of woods and were fiercely attacked by mosquitoes. From Maskinonge to Berthierville they drove through a heavy rainstorm that soaked them thoroughly. In their wet state, they were taken across the St. Lawrence River in a canoe to Sorel, where they stayed overnight at an inn. In the morning they took a steamboat to Montreal, where they rested over the weekend. On 5 July they crossed the St. Lawrence to La Ravie and the next day took a train to St. John on the Richelieu River. An American steamboat took them up the Richelieu and through Lake Champlain to Burlington, Vermont. In sharp contrast to the dirty, poorly built boats on the St. Lawrence River, overloaded with often drunken immigrants, the American boat was spotlessly clean. Part of the deck was covered by an awning, carpeted, and furnished with sofas like a drawing room. In this luxurious setting they enjoyed an elaborate tea set out with fruits and cakes.[56]

Lake Champlain, with its lofty mountains on either side, was extremely beautiful. Burlington, with its tree-lined streets and white

55. Mary Lyell to Mrs. Leonard Horner, 4 July 1842, Kinnordy MSS.
56. Ibid.

houses surrounded by gardens, was one of the prettiest towns they had seen. Lyell had a letter of introduction to George W. Benedict, professor of classics at the University of Vermont. He took them to examine deposits of loam thirty or forty feet above the level of Lake Champlain, containing shells of *Tellina groenlandica,* an Arctic seashell that occurred also at Montreal and Beauport, and at Uddevalla in Sweden.[57] Benedict also took them to see the falls of the Winooski River, where there was drift without shells. He gave Mary specimens of two rare freshwater shells from Lake Champlain.

On 8 July the Lyells crossed Lake Champlain to Port Kent, New York. Ebenezer Emmons had described a site near Port Kent where a small stream cut through a deposit of loam that contained Arctic species of seashells. From Lake Champlain to Keeseville, they climbed about five hundred feet over a series of terraces similar to those north of Toronto. Beneath the shell deposit, boulder clay formed a deep covering over the underlying rocks. At Keeseville, they descended into the chasm of the Ausable River, cut through strata of Potsdam Sandstone, the oldest geological formation then known in the United States.[58] Nevertheless, its characteristic fossil, *Lingula,* belonged to a living genus. "Throughout so vast a series of ages," Lyell commented, "has Nature worked upon the same model in the organic world!"[59] In the sandstone walls of the chasm, ripple marks were particularly well defined; that evening, as they returned across Lake Champlain, Lyell saw through the clear water near the shore similar ripple marks on the sandy bottom, a reminder of the similarity of conditions between the most ancient geological past and the present.

At three o'clock on Saturday morning, 9 July, the Lyells left Burlington. They traveled all day across Vermont on rough roads through wooded hills and green pastures with many swift-running streams with waterfalls. It was seven o'clock in the evening when they reached Hanover, New Hampshire. There they called on Professor Oliver Hubbard, Silliman's son-in-law, whom they had met earlier at New Haven, Connecticut. Hubbard invited them to accompany him and his wife to church on Sunday morning. In the evening the Lyells went on in a hired carriage to Enfield, New Hampshire, passing on the way a Shaker village. The Shakers did not allow strangers to visit them, but as the Lyells drove past one of

57. Lyell, "Ridges, Elevated Beaches," 20.

58. Mary Lyell to Frances Lyell, 4–15 July 1842, Kinnordy MSS.

59. Lyell, *Travels,* 2:132.

their houses, "the windows were all open & we heard singing," wrote Mary, "so we stopped & c[oul]d see them quite plainly men & women, jumping about in all sorts of strange ways & occasionally a preacher holding forth. I sh[oul]d have liked so to have gone in & seen them. Their houses are the most perfectly neat possible, but without a single ornament."[60]

On 11 July the Lyells continued by stagecoach from Enfield to Concord, New Hampshire, where they took the train to Boston. On their arrival they went to the Ticknors' house. The Ticknors were at the seashore, but their servants made the Lyells welcome. During the warm evenings of the next few days, their last in Boston, they delighted in the fireflies in Boston Common across Park Street. Each morning the iceman brought an abundance of ice to the house to provide cold drinks. At a wheelbarrow in the street Lyell bought for the equivalent of a shilling a West Indian pineapple that in London would have cost twelve shillings or more. The Ticknors came in from the seashore to say good-bye, as did many other of their Boston friends. On 16 July, Lyell and Mary boarded the steamer *Caledonia* to sail for Halifax, Nova Scotia.

Nova Scotia

Lyell wished to spend a month studying the geology of Nova Scotia—especially interesting to him because of its rich coal deposits and the many sections of strata exposed in the cliffs along its indented coastline. From the deck of the *Caledonia* on Monday morning, 17 July, the rocky, fir-covered hills around Halifax harbor again reminded them of a Norwegian fiord. The next day in a drizzling rain and fog they left Halifax by stagecoach for Windsor. After traveling about thirty miles and crossing a low range of hills, they emerged out of the fog and rain and fir-covered hills into a rich farmland with forests of maple and beech, bathed in warm sunshine. They stayed two days at Windsor, a pretty village surrounded by orchards on the Avon River, a tidal estuary.[61] At high tide, the Avon was broad and full of water; at low tide, it shrank to a rivulet between great mud banks. As on other rivers around the Bay of Fundy, the tide on the Avon came in as a high wave called a tidal bore.

At Windsor, Lyell met Judge Thomas Chandler Haliburton, author of *Sam Slick,* a humorous book about an itinerant clock seller

60. Mary Lyell to Frances Lyell, 4–15 July 1842, Kinnordy MSS.
61. Mary Lyell to Frances Lyell, 24 July 1842, Kinnordy MSS.

from Connecticut. The Nova Scotians struck the Lyells as very much like Americans: "quick, active, go ahead people with all the republican equality of manners so often attributed to democratic institutions," wrote Mary, "but which really belong[s] to the circumstances of a new country where people have to work their way."[62]

The Reverend John Pryor, a Baptist minister and professor at Acadia College in Wolfville, came to Windsor to meet them. On Friday, Pryor accompanied them by stagecoach to Wolfville and then on geological excursions. On the dry and hardened mud flats of the Bay of Fundy and in its tidal creeks, Lyell found tracks of marine worms and birds resembling various fossil footprints, including those of the Connecticut Valley sandstone. He separated and removed some slabs of the dried mud containing footprints, which he later presented to the British Museum.

Leaving Mary at Wolfville, Lyell went to Cape Blomidon, a magnificent promontory of columnar basalt resting on sandstone strata. In 1837 Lieutenant Richard Nelson of the Royal Engineers had described to Lyell these basalt cliffs, which he had visited in 1830.[63] On the beach below the Cape, Lyell was startled to find in the sandstone ledge recent furrows, exactly like the ancient grooves he had seen in rocks in both Canada and Europe and had attributed to glacial action. From his guide he learned that during the winter the Minas Basin frequently filled with ice. The packed ice, often fifteen feet thick, was forced by the rising tide over the sandstone ledges. Fragments of volcanic trap, fallen from the top of Cape Blomidon, often froze into the ice and were carried along with it. Lyell concluded that the fragments of volcanic rock, containing quartz crystals, were quite hard enough, when embedded in the ice, to make grooves in the soft sandstone. During the night at Cape Blomidon, where he had to sleep in a poorly kept house, Lyell was afflicted with the worst fleas he had experienced since traveling in Sicily in 1828.

The next morning he rejoined Mary at Wolfville, and from there they crossed the Minas Basin to Parrsborough in a schooner, a distance of about thirty miles. With a brisk wind the voyage took about five hours. "There we were," wrote Mary, "half the time at an angle of 45 degrees, I sitting on a box & holding on by the pump. I think

62. Ibid.

63. R[ichard John] Nelson to Charles Lyell, 14 September 1837, Lyell MSS. Lieutenant Nelson had visited Nova Scotia in 1830 while stationed with the Royal Engineers in Bermuda. He met Lyell in 1833 when, on his return from Bermuda, he was stationed for two years at Woolwich. Lyell cited Nelson's account of the coral reefs of Bermuda in the fifth edition of the *Principles of Geology*, 3:279.

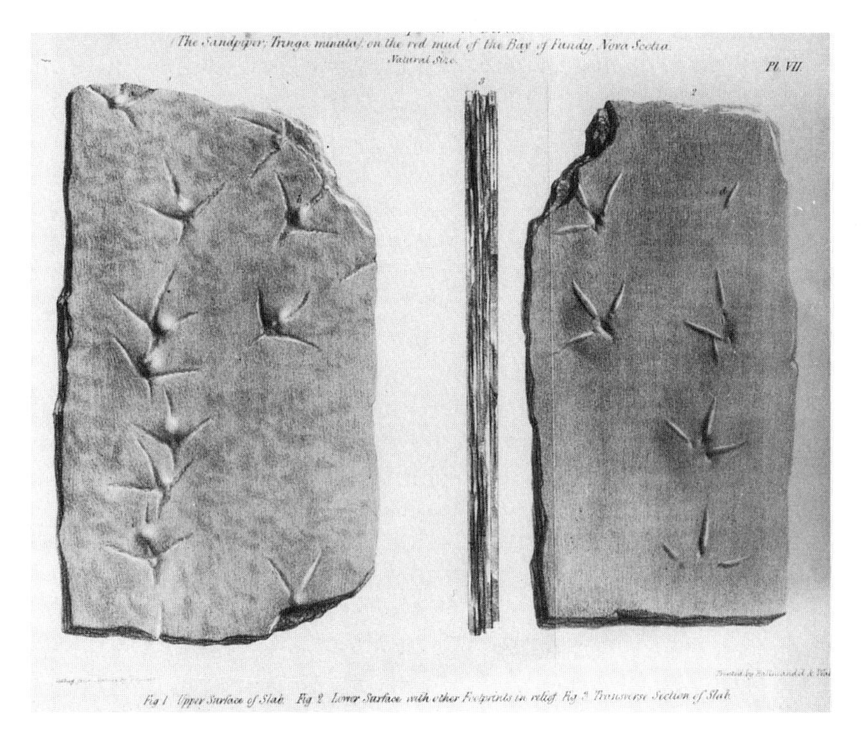

Bird Footprints on the Bay of Fundy. Lyell's *Travels in North America* (1845), Plate VII.

it is the first time in all my journeying that I have made so long a voyage with sails."[64]

At Parrsborough they met Dr. Abraham Gesner, provincial geologist of New Brunswick, who had come from Moncton to show Lyell the geology of the Joggins area on Chignecto Bay. Gesner had been born in Nova Scotia, the son of an American loyalist who had fought with the British during the Revolutionary War and, like many loyalists, settled in Nova Scotia after the Revolution. In 1825 Gesner went to London to study medicine and also attended lectures on mineralogy and geology. On his return to Nova Scotia, he began to practice medicine at Parrsborough. While walking or riding on horseback to visit patients, he made geological observations and collected specimens along the shores of the Bay of Fundy. In 1836 he published a small book on the geology and mineralogy of Nova Scotia.[65] It led to his being invited in 1838 to carry out a geological survey of the adjacent province of New Brunswick. At the time he met the Lyells, Gesner was in the fifth year of his survey and had published several parts of his New Brunswick report.

From Parrsborough, Gesner took the Lyells in a large double wagon to Minudie, on a tidal extension of Chignecto Bay. They stayed at Minudie for two days as guests of Amos Seaman, a landowner who had diked and drained the salt marshes to grow hay and raise cattle. He also quarried grindstones for export to the United States.[66]

To the southwest of Minudie, the South Joggins cliffs extended for about two and one-half miles along the southern shore of Chignecto Bay. About two hundred feet high, the cliffs contained a series of steeply inclined Carboniferous strata several thousand feet thick, including about nineteen seams of coal varying in thickness from two inches to four feet. At places in the South Joggins cliffs, the trunks of fossil trees appeared, always perpendicular to the plane of the strata in which they were embedded, showing that they were preserved where they had grown. The bases of the tree trunks always rested in a bed of coal or shale. Never in their upward extension did a tree trunk pass through a layer of coal, no matter how thin. All bore the same surface markings, resembling those of a roughly fluted column. In the two and one-half miles of the South Joggins cliffs, Lyell counted seventeen fossil tree trunks. All of them

64. Mary Lyell to Frances Lyell, 24 July 1842, Kinnordy MSS.

65. Gesner, *Geology and Mineralogy of Nova Scotia.*

66. Fergusson, *Amos "King" Seaman.*

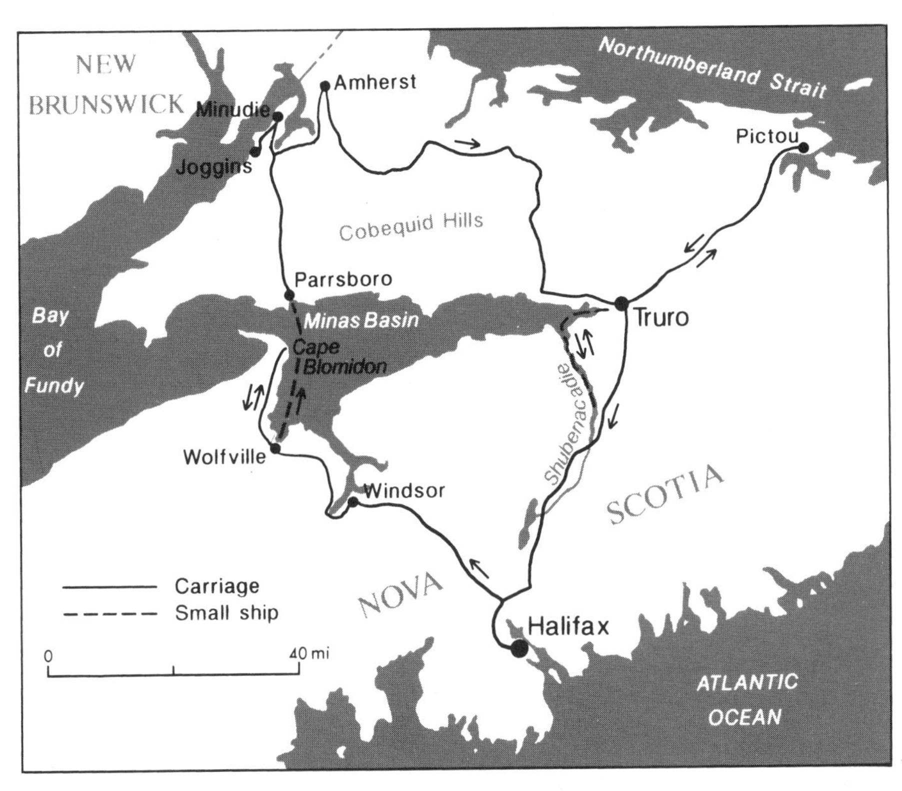

Lyell's Tour in Nova Scotia, July–August 1842. Map by Philip Heywood.

were different from those Gesner had seen in the cliffs when he had visited Joggins two years before. As portions of the cliff fell away by the continued wearing action of the sea, new trees were revealed. The largest tree that Lyell saw was about four feet in diameter and twenty-five feet long, but Gesner had once measured one forty feet long.[67] "This subterranean forest exceeds in extent & quantity of timber," wrote Lyell, "all that have been discovered in Europe put together."[68] The South Joggins fossil trees offered dramatic evidence that coal beds had formed from the accumulated vegetation of forests that had grown in the places where the coal now lay.

The position of the Joggins fossil trees at right angles to the strata demonstrated that the whole series of strata, now inclined at an angle of twenty-four degrees, was originally laid down in a horizontal position. The same must have been true of the deposition of the strata above and below the Joggins beds. Here was a formation laid down successively in a horizontal position to a total thickness of some fourteen thousand feet. The fossil forests of the Joggins coal had grown on an ancient land surface. The accumulation of so many strata required, therefore, a long period of subsidence, the subsidence and the deposition of sediment occurring simultaneously.

At Minudie the Lyells parted from Gesner and drove by themselves to the village of Amherst and thence through the thickly forested Cobequid Hills. Lyell noted that the hills consisted of granite and strata of clay-slate intersected by trap dikes, signs of former igneous action.[69] On 29 July, they reached Truro at the eastern end of the Bay of Fundy. The next day in pouring rain they rode in an open wagon that carried the mail through a seemingly endless fir forest to Pictou on the Northumberland Strait. They arrived thoroughly drenched. While they rested and dried out at the hotel at Pictou, Henry Poole, a young Englishman and superintendent of the Albion coal mines across the harbor, called on them. They returned with him to the mines, crossing the harbor in a steamboat and then riding on a railway used to haul coal. The year before in London, Samuel Cunard, who owned the *Acadia* on which the Lyells came to America, had arranged for them to stay with the Pooles at the Albion Mines. Cunard also had given them introductions to officials of the General Mining Association that worked the mines.

67. Lyell, Notebook 103, 142, Kinnordy MSS.

68. Lyell to Marianne Lyell, 30 July 1842, Kinnordy MSS.

69. Lyell, *Travels*, 2:191.

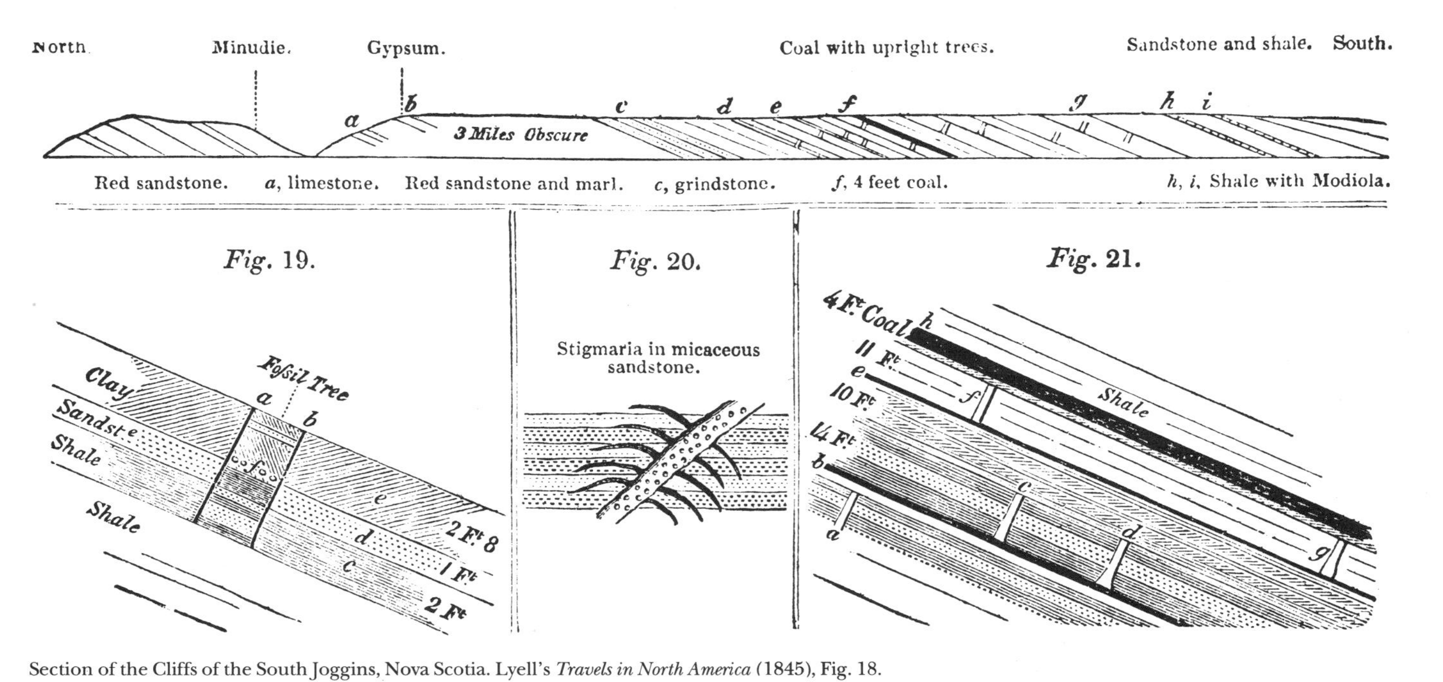

Section of the Cliffs of the South Joggins, Nova Scotia. Lyell's *Travels in North America* (1845), Fig. 18.

Poole introduced Lyell to a young Nova Scotian, John William Dawson, who had just returned from a year of study at the University of Edinburgh.[70] The son of a Scottish printer and publisher at Pictou, Dawson as a boy began to collect fossils from the coal-bearing rocks in the neighborhood and developed a strong interest in geology. He had visited the South Joggins, the Minas Basin, and other geological sites in Nova Scotia.[71] While Mary stayed with Mrs. Poole, Dawson showed Lyell his specimens and took him to Dickson's Mills, a short distance west of Pictou, where he had discovered sandstones and shales containing Calamites fossilized in an upright position. The reddish sandstone in the northern part of Pictou County, which had been called New Red Sandstone and thus had been thought to be much newer than the Coal, was, said Dawson, actually part of the Coal formation. On 1 August, Richard Brown, superintendent of the coal mines at Sydney on Cape Breton Island, came to Pictou to meet Lyell. The next day, Lyell, Dawson, Brown, and Poole examined the cliffs along the East River.

On 2 August, the Admiralty survey vessel H.M.S. *Gulnare,* commanded by Captain Henry Bayfield, sailed into Pictou harbor, where Lyell had arranged to meet it. For more than a quarter century, Bayfield had been engaged in the hydrographic survey of Canadian waters. He had found proofs of recent elevation of the land similar to the evidence that Lyell had found in southern Italy and in Sweden. In 1815 Bayfield was assigned to the Admiralty's hydrographic survey of Lake Ontario and in 1817, at the age of twenty-two, had been put in charge of the hydrographic survey of all the Great Lakes, a task completed in 1825. After two years at the Admiralty offices in London preparing charts of the Great Lakes, Bayfield had returned to Quebec to begin the survey of the river and the Gulf of St. Lawrence, on which he was still engaged in the summer of 1842.

Keenly interested in geology, Bayfield had collected specimens of rocks and minerals during his surveys of the Great Lakes. He had published the first account of the geology of Lake Superior and its islands.[72] Later he drew attention to banks of recent shells occur-

70. Dawson, *Fifty Years of Work in Canada,* 51.

71. Sir John William Dawson (1820–99) as a young man earned his living by teaching school, and in 1850 he was appointed superintendent of education for the common schools in Nova Scotia. In 1855 he published his *Acadian Geology* and was appointed professor of geology and principal of McGill University, Montreal, offices he retained until his retirement in 1893. He was largely responsible for the development of McGill into a leading university.

72. Bayfield, "Geology of Lake Superior."

ring along the north shore of the Gulf of St. Lawrence and to other evidence for the elevation of the whole north coast of the gulf from Quebec eastward.[73] At Seven Islands, Bayfield had found within a mile of the shore seven parallel sand ridges, each of which evidently represented a former beach. In the topsoil of the sand ridges, Bayfield had found clam shells "perfectly similar to those which abound on the present shore." Farther east on the Mingan Islands, erosion by the sea had broken the limestone coastal cliffs into columns resembling flowerpots. Back from the shore and about sixty feet above the sea were limestone cliffs and columns and beaches of shingle and gravel exactly like those along the shore. Bayfield concluded that the land had risen by successive stages. He thought its rise was connected with the earthquakes that occurred periodically along the north shore of the Gulf of St. Lawrence.[74]

After Bayfield's paper was read at the Geological Society in November 1833, Lyell had written him a note on the back of the society's official letter of thanks. In replying to the note, Bayfield had mentioned that he had begun to collect shells from elevated terraces along the St. Lawrence River. "There is much of high interest to the Geologist in those extensive provinces," wrote Bayfield, "but my duties are of so laborious & engrossing a nature that I can only catch information as it is accidentally presented to me."[75] Lyell cited Bayfield's observations in the third edition of the *Principles,* and they may have spurred him to go to Sweden during the summer of 1834 to study the recent elevation of the land there.[76]

Bayfield had taken the opportunity of the return to England of his assistant, Midshipman Augustus Bowen, to send Lyell a box of shells, including fossil shells from the Beauport formation near Quebec.[77] The box reached Lyell's house in London in December 1835, while Lyell and the Danish conchologist Dr. Heinrich Beck were in the midst of comparing Lyell's collection of fossil shells from the Suffolk Crag with living species of shells from the North Sea and Arctic Ocean. Lyell and Beck were astonished to find that Bayfield's fossil shells from Beauport were identical to fossil species

73. Bayfield, "Coral Animals in the Gulf of St. Lawrence"; Bayfield, "Geology of the North Coast of the St. Lawrence," cited by Lyell, *Principles,* 3d ed., 2:46–47. Bayfield's paper was read by George Greenough at a meeting of the Geological Society of London on 20 November 1833. It was published in abstract form in the *Proceedings* and later in full in the *Transactions.*

74. Bayfield, "Geology of the North Coast of the St. Lawrence," 90.

75. Bayfield to Lyell, 4 May 1834, Lyell MSS.

76. Wilson, *Lyell,* 399–407; Lyell, *Principles,* 3d ed., 1:327–29.

77. Bayfield to Lyell, 11 November 1835, Lyell MSS.

from Uddevalla in Sweden. Furthermore, the Beauport and Uddevalla shells were identical to living Arctic species. The Beauport and Uddevalla shells, therefore, strongly suggested that the climates of Quebec and southern Sweden formerly had been much colder and their seas as cold as those of the modern Arctic.[78]

Two years later, Bayfield sent to Lyell fossil shells from a terrace two hundred feet above the St. Lawrence River at Port Neuf, ten miles upriver from Quebec.[79] In 1839, Lyell sent Bayfield a copy of the *Elements,* which Bayfield judged "incomparably superior to any other elementary work in Geology."[80] Lyell wrote to Bayfield in April 1841 to tell him of his forthcoming visit to North America, and they arranged to meet at Pictou.[81] On Wednesday, 3 August, when he and Mary visited the *Gulnare,* Lyell and Bayfield finally met. Mary described their visit. "It is a very small vessel, but we spent a most agreeable day with Capt. Bayfield & his officers, dining & drinking tea on board & looking over Capt. Bayfield's shells & specimens which he had brought on purpose to show Charles. Then they had a great deal to compare notes upon, having corresponded so often."[82]

After the *Gulnare* sailed, Lyell continued his geological observations around Pictou. On the weekend, more guests arrived at the Pooles' house. Samuel Cunard came from Halifax with another gentleman. The settlement at Albion Mines was quite small. The coal miners were almost all Scotsmen, mainly Highlanders. The Pooles, together with the chief mining engineer, the doctor, and the storekeeper, formed a small but sociable group. On Sunday evening, the Lyells joined them at an Anglican service in the schoolhouse, there being no church.[83] After the service, Lyell went into Pictou in order to leave early Monday morning with Dawson on an expedition to the Shubenacadie River.

On the stagecoach to Truro, at one point the driver pulled in his horses and announced that he was stopping to eat some of the wild raspberries by the roadside.[84] His passengers did likewise and enjoyed the berries. The incident convinced Lyell that the indepen-

78. Lyell, *Travels,* 2:122; cf. Lyell, "Fossil and Recent Shells, Collected by Capt. Bayfield in Canada."

79. Bayfield to Lyell, 9 November 1837, Lyell MSS.

80. Bayfield to Lyell, 4 November 1839, Lyell MSS.

81. Bayfield to Lyell, 11 May 1841, Lyell MSS.

82. Mary Lyell to Caroline Lyell, 9 August 1842, Kinnordy MSS.

83. Ibid.

84. Lyell, *Travels,* 2:194.

dence of manners, so noticeable in the United States, existed also in Nova Scotia and was the result of the conditions of a new country rather than of democratic institutions. At Truro, Lyell and Dawson were joined by a Mr. Duncan, "a character who has hunted the province for mineral treasures & knows every corner," who guided them up the Shubenacadie. Early each morning, just after the tidal bore had passed up the estuary, they started up the river in a boat and were carried along by the tide. Lyell wrote of the experience:

> I saw the sudden rise of the Bay of Fundy tide day after day & had no idea it would be so fine, equal to the rapids of the St. Lawrence & continuing for hours. Sometimes when I was sitting so carelessly on a rock writing notes, Mr. Duncan would warn me that in a quarter of an hour the tide would be ten feet perpendicular & in three hours be forty feet perpendicular over the spot where I sat. By good management this tide was made a grand power for carrying our boat where we liked. One day we had to get up at 3 o'clock in the night, to be taken up 15 miles. After visiting all the places in that space by the descending tide, which flows three times as long, we were carried up again 15 miles & a half to a place where we had to sleep.[85]

The banks of the Shubenacadie contained sections through a great series of sandstone and marl strata. In the lower portion of the series were strata of white gypsum, a hundred feet thick, and limestone, which had been considered younger than the Coal formation. In 1841, in a brief examination of the formation near Windsor, Nova Scotia, William Logan collected a few of the most abundant fossils.[86] At this location the strata were greatly disturbed. They seemed to suggest that the gypsum formation was younger than the Coal. Nevertheless, when Lyell studied the strata in the cliffs of the Shubenacadie, he was confirmed in the opinion he had formed when he first saw similar strata at Windsor and in his later study of the section at South Joggins—namely, that the gypsiferous strata, far from being younger than the Coal, were in fact the lower portion of the coal-bearing formation.

Mary came from Pictou by stagecoach to rejoin Lyell at Truro. They returned together to Halifax, where on 18 August they boarded the Cunard liner *Columbia.* The voyage home was smooth and pleasant. "Reading, working, walking on deck with Charles, watch-

85. Lyell to Caroline Lyell, 25 August 1842, Kinnordy MSS.

86. Logan, "Coal-Fields of Pennsylvania and Nova Scotia."

ing him & the other gentlemen playing at shuffleboard, writing a little for Charles & above all, the *meals*," wrote Mary as they approached Liverpool, "fill up so much time that the days have slipped away pretty fast."[87] At nine o'clock in the evening on 29 August, they arrived at London by train from Liverpool.

87. Mary Lyell to Caroline Lyell, 9–28 August 1842, Kinnordy MSS.

CHAPTER 4

America in Retrospect

SOON AFTER arriving home to 16 Hart Street, Lyell and Mary went to Scotland to visit Lyell's family at Kinnordy and remained there for several weeks of rest and renewal. On their return to London in late October 1842, their first task was to find some place to store the thirty-six boxes of fossils that had arrived from America during their absence. The contents of the boxes would, Lyell wrote, "put the elasticity of a small London house to the test."[1] They also read with great interest and amusement Charles Dickens's book on his American travels.[2] Mary thought that Dickens gave a distorted account of the American people. He greatly exaggerated the amount of tobacco-chewing and spitting the juice. The habit of spitting was no more prevalent in America, she said, than in Germany. Lyell added that "there is so much of generous impulse in Dickens & so much eloquence that any want of judgment in him or rashness in pronouncing on great political & social questions is fraught with serious mischief."[3]

On 14 November, while still unpacking shells, they celebrated Lyell's forty-fifth birthday. Three days later, Charles Darwin came to discuss with Lyell the distribution of fossil mammals in North and South America. The mammalian fossils so far discovered indicated the extinction of many species, apparently without any significant change in climate or geography. In his notebook, Lyell reflected on their ignorance of the cause and effect of epidemic diseases. "If we know so little of human and canine physiology we have no right to be surprised at the dying out of species altho' climate of air & sea were similar as [is demonstrated by the] marine shells of Patagonia [which have not changed]."[4]

As he looked over his travel journals and notebooks, Lyell wondered how best to write up the results of his American tour. He had sent a number of papers to the Geological Society of London from

1. Lyell to Whewell, 29 October 1842, Whewell MSS.

2. Dickens, *American Notes.*

3. Mary Lyell to Mrs. Leonard Horner, 11 November 1842, Kinnordy MSS.

4. Lyell, Notebook 105, 89, Kinnordy MSS.

America and planned to write others on the recession of Niagara Falls, on the lake terraces at Toronto, and on the geology of Nova Scotia. At Kinnordy he had begun to make notes for a travel narrative and soon decided that he must write a book on the results of his American tour. At first he planned a book on the geology of North America, rather than a book of travels, but then found that he could not bear to omit his opinions on the scenery, people, manners, and governments of the United States and Canada. He wrote to George Ticknor:

> It may be said in a few words thought I; so I made a few notes of my voyage and tour to Niagara, Blossberg and Boston, with quotations from my journal, on the exhilarating effect of viewing the signs of such rapid progress in population, and the railways, churches and school-houses among the stumps, etc. But I know by experience the remark which this would provoke here. "It is very easy to go ahead with other people's money."[5]

He could not enter lightly into a discussion of American affairs without exciting British prejudices, recently inflamed by Pennsylvania's repudiation of its bonds, a considerable number of which had been bought by British investors.

On 14 December, Lyell read to the Geological Society a paper on the ridges or terraces around Lakes Erie and Ontario.[6] He compared the lake terraces to the *osars,* or eskers, in Sweden and thought that, like the eskers, they were not simply beaches, but banks or bars of sand. (The terraces around the Great Lakes are in fact old beaches, whereas eskers mark the beds of former rivers that flowed through glaciers.) The lake ridges were so interesting that he thought of including a discussion of them with an account of the recession of Niagara Falls to make a short, popular book of about a hundred pages. He wrote to John Murray, his London publisher, to ask his opinion of such a work.[7] Wiley and Putnam in New York had offered to print any such book for him for sale in America. The recession of Niagara fascinated Lyell as a prime example of a geological process in action. In the Niagara gorge the geological history of thousands of years could be read clearly.

In mid-December, Lyell began to study the fossil sharks' teeth from Martha's Vineyard, several species of which were identical

5. Lyell to Ticknor, 12 October 1842, printed in K. M. Lyell, *Life, Letters, and Journals,* 2:66–71.
6. Lyell, "Ridges, Elevated Beaches."
7. Lyell to Murray, 30 November 1842, Murray MSS.

with those from the faluns of Touraine. They convinced him that the strata of Martha's Vineyard must be Miocene. Sir Philip Egerton agreed to compare Lyell's American fish teeth with those in his collection of fossil fish, which had been described and named by Louis Agassiz.[8]

Frequently after their return to London, the Lyells saw William Boott, who had smoothed their way at Boston. On 11 February, they took Boott to a party at Charles Babbage's.[9] The deep ignorance of Britons about America now struck Lyell forcibly. To the American ambassador, Edward Everett, he wrote, "When I recollect how ignorant I was a few years ago of the U.S., I try to be charitable in regard to the extraordinary mistakes and prejudices which occasionally strike now with astonishment when I hear your country spoken of."[10] Lyell invited Everett to accompany him to the anniversary dinner of the Geological Society on 17 February.

In March 1843 Lyell delivered a short course of eight lectures on geology at the Marylebone Institution. His audience seemed small after the large audiences to which he had become accustomed in America, yet he had 250 paying subscribers, including many educated and even aristocratic people.[11] Gideon Mantell thought Lyell's lecture on volcanoes was an excellent discourse, with beautiful illustrations, but that his lecture on coal was not clear enough for a popular audience.[12] The success of the lectures encouraged Lyell to write up his American geology.[13] He had earned a clear 160 pounds even after having spent 50 pounds on illustrations.[14]

Late in April, Lyell also read to the Geological Society a paper on the fossil trees at Joggins, Nova Scotia.[15] From Nova Scotia he had brought back pieces of *Stigmariae* and other fossil trees and at London had them sectioned and polished to reveal the interior structure of the fossil wood. The Nova Scotia *Stigmariae* proved identical in structure to the *Stigmariae* of British coal.[16]

On 26 May, the Lyells left London for Paris on their way to the Auvergne. In Paris, Lyell and Mary attended the Théatre Français, a

8. Lyell to Egerton, 17 December 1842, Kinnordy MSS.

9. Mary Lyell to Babbage, 9 February 1843, Babbage MSS.

10. Lyell to Everett, 20 February 1843, Everett Papers.

11. Lyell to his father, 1 April 1843, Kinnordy MSS.

12. Mantell, *Journal,* 167.

13. Lyell to his father, 1 April 1843, Kinnordy MSS.

14. Lyell to Babbage, 9 April 1843, Babbage MSS.

15. Lyell, "Upright Fossil-Trees in the Coal Strata of Cumberland, Nova Scotia."

16. Lyell to Dawson, 16 May 1843 [copy], Kinnordy MSS.

Lyell Lecturing at the Marylebone Institution, March 1844. Drawing by unknown artist. In possession of the author.

lecture on astronomy by Arago, and one on geology by Élie de Beaumont. At a meeting of the Geological Society of Paris, Lyell discussed the *Stigmariae* of the American coal with Adolphe Brongniart, who was convinced that they were simply the roots of the fossil tree *Sigillaria.*[17] Lyell's French publisher, M. Langlois, gave him the cheerful news that he had sold twenty-five hundred copies of the French translation of the *Elements of Geology* and was hoping to publish a translation of the sixth edition of the *Principles.*

On 7 June, the Lyells left Paris to spend the next five weeks traveling in central France. Lyell's primary purpose was to collect fossils of the Tertiary formations, particularly those of the Miocene formations of Touraine. They traveled sometimes on one of the new railways, occasionally by steamboat on the Loire and Saône Rivers, and sometimes by hired carriage, but most often by diligence. The roads in the Auvergne were much improved since Lyell had last been there in 1828, but the French remained indifferent to time, often to Lyell's advantage. "The diligence stops to let one man write a letter," observed Lyell, "another to smoke his pipe, a third, who shall be nameless, to get out & hammer at limestone for fossils & examine an iron mine by the road."[18]

At St. Malo on Sunday, 16 July, the Lyells dined at M. de l'Abbadie's to meet his two guests, the French geologists Edouard de Verneuil and Adolphe d'Archiac. During the summers of 1840 and 1841, Verneuil had traveled in Russia with Roderick Murchison to study the Silurian, Devonian, and Carboniferous formations of that country. By 1840 Verneuil was convinced that the Silurian formations possessed the same characteristic fossils wherever they occurred, an opinion confirmed by his Russian travels. In 1841 Roderick Murchison discovered in the Russian province of Perm a new formation above the Coal. Verneuil showed that its fossils corresponded to those of the Zechstein in Germany and the Magnesian Limestone in Great Britain and that it was the youngest of the Paleozoic formations.[19]

Archiac had studied the Tertiary formations of the Aisne River valley and in 1839 published an important paper on the correlation of the Tertiary formations of northern France, Belgium, and England.[20] A member of an old but impoverished family of the French

17. Lyell to Dawson, 31 July 1843 [copy], Kinnordy MSS.

18. Lyell to Eleanor Lyell, 6 July 1843, Kinnordy MSS.

19. Édouard de Verneuil (1805–73) had studied law, but he became interested in geology through attending the lectures of Élie de Beaumont and after 1835 gave his full time to geology.

20. Archiac de Saint Simon, "La coordination des terrains tertiaires."

nobility, Archiac had served as an officer in the French cavalry for nine years; after 1832 he devoted all of his time to geology. In 1842 he collaborated with Verneuil in describing fossils from the Devonian formations of the Rhineland and continued his work with a study of the Cretaceous formations of central France.[21] Archiac and Verneuil shared Lyell's interest in using fossils to correlate the geological formations of different countries and different continents; Archiac in particular shared Lyell's interest in the use of fossils to correlate Tertiary formations. Verneuil and Archiac had come to St. Malo to examine Lyell's fossil specimens, and together they walked along the seashore. In the evening the Lyells embarked by steamboat to cross the Channel. They arrived back in Hart Street on 19 July.

In mid-August, Lyell went to Bristol to collect fossils from the Somersetshire coal formation to compare with those he had obtained from coal-bearing strata in Brittany. A few days later, Mary and her sister Frances Horner joined him. Together they took a steamboat to Cork to attend the British Association meeting, where Lyell was a vice president of the Geological Section.

At the meeting, Roderick Murchison delivered a paper on the Permian system, which corresponded to formations overlying the Coal, in Germany known as the *Rothe-todte-liegende, Kupfer-schiefer,* and *Zechstein* and in England known as the lower New Red Sandstone and Magnesian Limestone.[22] For a time Murchison had thought that the *Rothe-todte-liegende* of Germany might be an upper portion of the Carboniferous formation. His travels in Germany in 1843 convinced him that it was distinct from the Carboniferous and formed part of the Permian. The German formations contained beautifully preserved specimens of the fossil plant *Neuropteris,* found also in Permian strata, but lacked characteristic coal plants. Nevertheless, the Permian fossils showed sufficient similarity to those of older formations to include the Permian in the Paleozoic series. Previously the German formations had been considered part of the Trias, because the rocks looked alike, but their fossils now showed them to be distinct. Murchison's correlation of geological formations in England and Germany with the Permian of Russia was thus based decisively on their fossil contents.[23] Since 1835

21. Archiac de Saint Simon, "La formation cretacée des vorsants sud-ouest, nord et nord-ouest du plateau centrale de la France."

22. Murchison, "Results of a Second Geological Survey of Russia."

23. Murchison, "'Permian System' as Applied to Germany."

Murchison had distinguished three systems of Paleozoic strata by means of fossils—the Silurian and Devonian systems below the Carboniferous and now the Permian above it.

In 1841 John Phillips had published a description of fossils collected from the Silurian, Devonian, and Carboniferous rocks of southwest England. He treated them as one great Paleozoic fauna and flora that had changed gradually over the course of ancient geological time. Phillips distinguished the Paleozoic from the later Mesozoic (Triassic through Cretaceous) and Cainozoic (Tertiary) periods on the basis of the general character of the fossil life of each of the three great periods. He traced a gradual process of faunal change through successive periods of the Paleozoic, just as Lyell had found earlier for the Tertiary.[24]

After the meeting at Cork, the Lyells and Frances Horner spent the weekend with Lord Rosse at Parsonstown. Frances Horner then returned to London. The Lyells went to Belfast, where they hired a car to drive around the northern coast of Ireland to see the basaltic columns of Fairhead and the Giant's Causeway. On 12 September, they crossed the Irish Sea to Portpatrick in Scotland and took the mail coach to Stranraer to spend two days as guests of John Carrick-Moore at Corswall on Loch Ryan.[25] Carrick-Moore had sent Lyell some fossils from the sandstone strata of Loch Ryan, part of a formation extending across Scotland from which few fossils had previously been discovered.[26] Of Loch Ryan, Lyell wrote later that it was "a grand magazine of geological analogies—tidal, littoral, conchological, sedimentary etc."[27]

In mid-September the Lyells went on north to Kinnordy. After their many weeks of travel in France and Ireland during the summer, the Lyells were glad to have a long, quiet time at Kinnordy. During the next two months Lyell worked on his papers on North American geology. After Lyell's birthday on 14 November, they returned to London. On their arrival they found Lyell's clerk, George Hall, desperately ill, dying of consumption. On 17 November, Hall died "after 4 days gasping for breath with scarcely any lungs left." To

24. Phillips, *Palaeozoic Fossils,* 160.

25. John Carrick-Moore was the son of James Moore, formerly a surgeon at Glasgow, who in 1821 had assumed the name of Carrick-Moore when he succeeded to the estate of Corswal that had belonged to a relative named Carrick. James Carrick-Moore (1762–1860) was the brother of General Sir John Moore (1761–1809) of Corunna and of Admiral Sir Graham Moore (1764–1843).

26. Carrick-Moore to Lyell, 4 January 1839, Lyell MSS.

27. Lyell to Carrick-Moore, 17 September 1843, printed in K. M. Lyell, *Life, Letters, and Journals,* 2:77–78.

Gideon Mantell, who had known Hall over the years, Lyell wrote, "Dr. Chambers told me seven years ago that he did not expect him to live 2 years—He had been 28 years with me & was a most amiable & excellent servant & quite a naturalist."[28]

As winter approached, Lyell seemed surrounded with illness. At Kinnordy his mother became seriously ill. He visited Gideon Mantell who was also unwell. Lyell was trying to find an engineering apprenticeship for Mantell's son and an appointment in the East India Company for the son of his Scottish friend the Reverend John Fleming. Meanwhile, he continued to write his American travels. He was anxious to have all of his North American fossils properly identified. Edward Forbes, newly appointed curator of the Geological Society, undertook to identify the Cretaceous fossil shells from New Jersey. William Lonsdale agreed to identify the fossil corals from America and Touraine. No naturalist in the United States could identify fossil corals, and there were only two such experts in France. To take Hall's place, Lyell hired young Charles Sowerby. Lyell wrote, "He reads to me in the evening, & it seems a new thing not to have to consider a clerk's lungs, but I miss Hall in a hundred little arrangements."[29] On 26 December, further tragedy struck when Lyell's sister Maria Heathcote died at Trebarwith in Cornwall, at the age of thirty-five, a week after the birth of her first child. The Lyells went down to attend the funeral and spent ten days in Cornwall. For two days, Lyell went to Falmouth to confer with William Lonsdale and to observe the work on his American fossil corals.

During the winter of 1844 Lyell contemplated a trip to India to study the geology of the subcontinent. As he learned more of the deadly fevers of the country, he decided the excursion would not be worth the risk. Lyell's mother was still ill in Scotland, and his youngest sister, Sophy, had come down with a fever (possibly typhoid fever) after tending a sick nurse. During February she made a slow recovery. Through the winter Lyell continued to work on his American travels. He corresponded with George Ticknor about American affairs and sought information from him on American institutions and history. Edward Everett, the American ambassador, dined with the Lyells. He clearly was so happy in purely literary conversation that Lyell did not have the heart to raise American political questions. Like Ticknor a former Harvard professor, Everett impressed the British with his scholarship and polished manners.

28. Lyell to Mantell, 20 November 1843, Mantell MSS.

29. Lyell to Marianne Lyell, 7 December 1843, Kinnordy MSS.

In April, Mary's sister Frances became engaged to marry Charles Bunbury, eldest son and heir of Sir Henry Bunbury of Mildenhall and Barton Hall, Suffolk. The Bunburys were one of the great landed families of Suffolk. During the Napoleonic Wars, Sir Henry Bunbury had served with the British army, rising to the rank of major general. Charles Bunbury had traveled extensively, including a voyage to Capetown in South Africa. He was beginning to study fossil plants, particularly those of the Coal. Now in his thirties, Bunbury fell in love. "He and Frances," wrote Lyell, "are entirely, most undisguisedly & amusingly absorbed in one another, whether at home or in small or great parties."[30] Bunbury and Frances Horner were married on 30 May 1844; their wedding tour included a visit with the Lyell family at Kinnordy.

On 12 June, Lyell and Mary left for Kinnordy. Lyell took with him his American books and papers, including copies of their letters from America to their families and Mary's journal of their tour.[31] At Kinnordy he continued to write his book on America. When the Bunburys arrived, he took time out to go with Bunbury botanizing on the hills. Leonard Horner, Mary's father, also came to Kinnordy for a brief visit. Lyell spent some time at his old hobby of insect collecting, an activity that, he wrote, "does me more good, when I am tired of work, than any other pastime."[32] The Lyells were back in London on 20 August but remained there only briefly before going to visit Charles and Frances Bunbury at Mildenhall in Suffolk and then to see Lyell's old friend Thomas Spedding at Keswick in the Lake District.

In late September they were at York for the annual meeting of the British Association. Lyell delivered a lecture to twelve hundred persons on the geology of the United States and Canada. In his lecture he summarized the geological findings of his year of travel in America. He outlined the new knowledge of North American geology gained mainly from the state surveys, referring to "the faithful and conscientious manner in which the American geologists have pursued their investigations." American formations, he explained, tended to occur in broad belts extending in a northeast-southwest direction. The Cretaceous and Tertiary formations of the Atlantic coastal plain corresponded to European formations not only in the fossils represented, but also in "the proportion of the recent to the

30. Lyell to his father, 29 April 1844, Kinnordy MSS.

31. Lyell, Notebook 116, 53, Kinnordy MSS.

32. Lyell to Mantell, 20 August 1844, Mantell MSS.

extinct species of fossils" among the Tertiary formations. The inland boundary of the coastal plain was marked by waterfalls on the principal rivers, where the water tumbled over a band of granitic rocks. Further inland rose the long, parallel ridges of the Appalachians, consisting of Carboniferous, Devonian, and Silurian strata in a series of great folds. Steepest on their southeastern side, toward the west the Appalachian folds became rounded and gentler, descending to a low, level country through which the coal-bearing strata extended almost horizontally for more than 150 miles. Still farther west, the great Illinois coalfield occupied an area nearly as large as England. In contrast to Europe, the geology of America was "singularly simple," said Lyell, but both the rocks and fossils were strikingly like those of Europe. Nevertheless, on close comparison the similarity of the fossils consisted more in "the geographical representation of forms than in the absolute identity of species." In Paleozoic as in modern times, the earth was divided into distinct zoological provinces. By contrast, the coal plants were strikingly similar in Europe and America. Among 150 coal plants that Lyell had collected from the Ohio Appalachian Coal basin and the Nova Scotia Coal, Adolphe Brongniart and other botanists thought that two-thirds were identical to European coal species. Botanists knew of no modern instance of a comparable range of plant species uniform over an area four thousand miles from east to west and a thousand miles from north to south. Such uniformity of vegetation implied, said Lyell, "a peculiar uniformity of climate over a large part of the northern hemisphere" during the Carboniferous period.[33] The fossil shells and corals of American Carboniferous rocks were also closely related to those of Europe, but not with as many identical species as the land plants.

On their return to London, Lyell found waiting for him an invitation from John Amory Lowell, to come again to Boston in 1845 to deliver the Lowell lectures. He had still been considering the possibility of a geological survey of India for the East India Company but now abandoned the idea. The prospect of American travel was much more attractive. He promptly accepted the invitation.

In October 1844 the British government asked Lyell and Michael Faraday to investigate the cause of a disastrous explosion on 28 September in which ninety-five men were killed in a coal mine at Haswell in the Durham coalfield. The request came just as Lyell

33. Lyell, "Lecture on U.S. Geology, New York," MS notes [29 September 1844], Kinnordy MSS.

thought he was free to devote the winter to writing his book on America.[34] He went nevertheless. The experience was valuable, both in increasing his knowledge of coal mines and miners and in allowing him to become better acquainted with Faraday. Lyell, who had known Faraday for many years and had thought of him as shy and retiring, was impressed by Faraday's force and energy during the investigation. When they arrived at Haswell, the coroner's inquest was in progress. Lyell recalled later that a few minutes after they entered the inquest, Faraday began "to cross-examine the witnesses with as much tact, skill and self-possession as if he had been an old practitioner at the Bar."[35]

They spent two days exploring the mine where the explosion had occurred. Lyell examined the structure of the rocks and for this purpose asked the coal miners to collect for him the fossils they found most frequently. The miners were curious about the fossil plants of the coal, so Lyell told them what he knew of their characteristics and relationships to living plants. When the miners were leaving, he was about to throw away the specimens, but Faraday asked him "to wait till the men were out of sight, as he said their feelings might be hurt if they saw that nothing had proved to be worth keeping."[36] Faraday and Lyell were struck by both the intelligence of the miners and their ignorance. More than half of the pitmen were unable to write or even to sign their names as witnesses.

The miners' illiteracy was particularly serious, because on them must depend the practical steps needed to avoid future mining explosions and accidents. The explosion at Haswell was probably the result of an accumulation of firedamp in a "goaf," or area of the mine from which coal had been removed and into which the overlying rocks had partially collapsed, the firedamp arising from the residual broken coal. A fall in barometric pressure had then drawn some of the firedamp out of the goaf to create an explosive mixture in the actively worked part of the mine. To avoid such a danger, Faraday suggested that the firedamp be drawn off from goafs through tubes to the surface. However, it was essential that the miners themselves should understand the basis for the practical steps needed to avoid explosions. Faraday and Lyell recommended that the miners and their supervisors be taught the simple chemistry and physics of

34. Lyell to Egerton, 22 October 1844, Kinnordy MSS.

35. Lyell to Bence Jones, April 1868, printed in K. M. Lyell, *Life, Letters, and Journals,* 2:417–22, esp. 418.

36. Ibid., 420.

gases and air and the basic principles of hydrostatics and geology involved in the position and dislocation of strata, the intrusions of igneous rocks, and other factors that might influence the accumulation of coal gases.[37] The investigation at Haswell thus reinforced Lyell's conviction that improvement in education, particularly in scientific education, was needed at every level in Great Britain.

In October 1844 Lyell invited Charles Bunbury to study his fossil plant specimens from the coal beds of North America and, to his delight, Bunbury accepted. During his American tour, Lyell had become deeply interested in the coal plants. He was eager to determine whether the apparent identity of species between America and Europe would hold up when his American collections were subjected to detailed study. He told Bunbury, "I hope together we shall strike out some improved view of the Carboniferous era so far as related to its chief feature, the flora of its strata."[38]

During the winter of 1845, Lyell heard from Louis Agassiz in Switzerland that he was planning to make a tour of the United States, earning his way by giving public lectures. He wrote to Agassiz immediately with advice and undertook to write to John Amory Lowell to see whether Agassiz might be invited to lecture at the Lowell Institute.[39] On 1 May, Lyell was able to report that Lowell would ask Agassiz to give a six-week course of lectures at the Lowell Institute in the winter of 1846, but he cautioned Agassiz that he would have to lecture in English.

In November 1844, Lyell had written to Murray about his plans for a book based on his American tour. He planned a first volume that would be strictly a travel journal. A second volume, which might appear two or three months after the first, would contain geological observations of North America. Lyell was prepared to begin sending manuscript within a fortnight;[40] by mid-December, he was sending his American travels to the printer "as fast as I can."[41] As he wrote, his plan changed. He would incorporate the geology and travel into a single large work, generously illustrated with drawings and maps.

In May, he wrote to Murray to discuss the size, paper, and cost of printing his American travels.[42] Murray decided to publish the

37. Faraday and Lyell, *Explosion at the Haswell Collieries,* 19.

38. Lyell to Bunbury [October 1844], Kinnordy MSS.

39. Mary Lyell and Lyell to Agassiz, 28 February 1845, Agassiz MSS.

40. Lyell to Murray, 16 November 1844, Murray MSS.

41. Lyell to Egerton, 17 December 1844 [copy], Kinnordy MSS.

42. Lyell to Murray, 5 May 1845, Murray MSS.

work in two small volumes instead of one large volume. With the numerous and expensive illustrations, he would have to set the price so high that he felt that people would be dissatisfied if they received only one volume for their money. Lyell sent proof sheets to William Redfield in New York to ask him to read them and correct any errors.[43] Lyell and Murray had arranged with Wiley and Putnam of New York to bring out an American edition of Lyell's work, to be called *Travels in America,* at the same time as Murray published the London edition. On 12 June, Lyell wrote the dedication to George Ticknor, "in remembrance of the many happy days spent in your society, and in that of your family and literary friends at Boston." He wrote the preface on 14 June, completing the book. The next day Lyell complained to Murray that the printers were slow and the index still unfinished. He was hurrying to read proofs so that the corrected proofs might go off to Wiley and Putnam at New York on the next steamer.[44] On 18 June, he wrote to Ticknor that the book was imminent and he would send him a copy of the English edition. Early in July the book appeared.[45]

The frontispiece to the first volume of the *Travels* was a large "Birds-eye view of the Falls of Niagara & adjacent country, colored geologically." While visiting Niagara Falls in 1841, Lyell had gotten the idea of combining a colored sketch of the Niagara district made by Robert Bakewell Jr. with a geological representation of the rock strata as determined by James Hall. The plate illustrated strikingly Lyell's account of why the falls existed and how they had receded to create the Niagara gorge. To British readers it gave a vivid impression of the size and splendor of Niagara.

The frontispiece to the second volume was a colored geological map of the United States and Canada. Lyell listed seventeen authors whose works he had used in preparing the map, with a brief description of each work. They included the state surveys of Connecticut, Maine, Massachusetts, New Hampshire, New Jersey, New York, Pennsylvania, Rhode Island, and Virginia and David Dale Owen's account of the geology of the western states, which covered Illinois, Indiana, Ohio, Kentucky, Tennessee, and the territories of Iowa and Wisconsin. The map revealed the immense scale of American geology and the vast extent of its coal deposits. It showed, too, the rapid growth of American geology as a result of the work of many

43. Lyell to Redfield, 18 May 1845, Redfield MSS.

44. Lyell to Murray, 15 June 1845, Murray MSS.

45. Lyell to Ticknor, 18 June 1845 [copy], Kinnordy MSS.

skilled geologists. Besides the two great frontispieces, Lyell's *Travels* contained five other plates, including a map of the Niagara district and Father Louis Hennepin's view of Niagara Falls, originally printed in 1697. Lyell was anxious that his readers should see and understand the geology of Niagara Falls as he understood it. Niagara was a magnificent example of geological forces at work, now and in the recent geological past.

Among books on America by British travelers, Lyell's *Travels* was unusual in that it was the work of a naturalist. Lyell looked with an exact and curious eye at every aspect of the New World, from the fogs over the Grand Banks of Newfoundland and the brilliant sunsets over the western Atlantic to the fireflies at night in a Connecticut marsh. He delighted in America, sympathizing with the fresh and hopeful atmosphere of rapid growth and development.

Darwin found the book "very new, fresh and interesting," but he was disturbed by Lyell's discussion of Negro slavery, which he said gave him "some sleepless, most uncomfortable hours."[46] Lyell had written about slavery with tact, because he had received hospitality from many planters in Virginia, the Carolinas, and Georgia and wished to make a second tour through the South when he would again be dependent on the hospitality of the slave owners. He also thought that the exercise of righteous moral fervor over slavery by Englishmen was all too easy. Such an attitude did not take into account the size and complexity of the social problem created by slavery nor suggest a workable means of ending it. The efforts of the abolitionists in the North had aroused violent anger among white southerners, making them even more stubborn about upholding slavery. In his travels through the South, Lyell had found most Negroes cheerful, lighthearted, and apparently not overworked. At Dean Hall in South Carolina and on other plantations, he had been struck by the singing of the slaves in the evening. Yet Lyell also saw the evil of slavery and the depressing effect it exerted on the economic and social development of the South. His whole approach to the question was detached and analytical. He saw no solution in the immediate emancipation of the slaves, even if that were possible, because they would then lack a means of livelihood and would perish from want. He thought it important to educate the slaves, but the planters, in their reaction to the abolition movement, had

46. Darwin to Lyell [1 August 1845], Darwin-Lyell MSS; printed in part in Darwin, *Life and Letters*, 1:339–41, and completely in Darwin, *Correspondence*, 3:232–37.

passed laws making it a felony to teach slaves to read and write. In the *Travels,* Lyell wrote:

> The more I reflected on the condition of the slaves, and endeavoured to think on a practicable plan for hastening the period of their liberation, the more difficult the subject appeared to me, and the more I felt astonished at the confidence displayed by so many anti-slavery speakers and writers on both sides of the Atlantic. The course pursued by these agitators shows that, next to the positively wicked, the class who are usually called "well-meaning persons" are the most mischievous in society.[47]

That this was Lyell's honest opinion is indicated by a letter he wrote from Ireland to his sister Marianne in September 1843. Regarding the troubled Irish situation, Lyell thought that if he possessed absolute powers, he might be able to solve the troubles of Ireland, although with the limited powers of a British prime minister, he would not know what to do. But "if I had to put things right, & in rapid progression, in the United States south of the Potomac," he added, "I should be at a loss what to do under any circumstances."[48]

Though the letter has not survived, Lyell evidently wrote to Darwin in July or August 1845 to explain further his views of slavery. On 25 August Darwin replied, "I was delighted with your letter in which you touch on Slavery. I wish the same feelings had been apparent in your published discussion."[49] Darwin's own references to slavery in Brazil in the *Voyage of the Beagle,* also published in 1845, were, he said, "merely an explosion of feeling." Darwin clearly thought that Lyell had written about slavery with too little feeling. Lyell by temperament tended to restrain his feelings. On the subject of slavery, he considered expression of feeling not only unhelpful to clarity of thought, but actually harmful in its consequences.

Darwin was impressed by Lyell's description of the coal beds in America. The fifty-foot-thick seam of anthracite coal in the Lehigh Summit Mine in Pennsylvania, the enormous extent of the coalfields of Pennsylvania, Virginia, and Ohio, and the ten-foot-thick seam of bituminous coal exposed in the bank of the Monongahela River at Brownsville, Pennsylvania, showed the vastness of the coal

47. Lyell, *Travels,* 1:149.

48. Lyell to Marianne Lyell, 4 September 1843, Kinnordy MSS.

49. Darwin to Lyell, 25 August [1845], Darwin-Lyell MSS; printed in Darwin, *Life and Letters,* 1:341–42, and in Darwin, *Correspondence,* 3:241–43.

resources of the United States. "Nothing hardly astounded me more in your book," wrote Darwin.[50]

In his *Travels*, Lyell touched on many subjects other than geology. He admired the wisdom of John Lowell Jr., who, in leaving his bequest for the Lowell Institute, provided that neither the principal nor income from his bequest should be spent on buildings. As a result, money was available immediately to provide courses of lectures. Lowell's wisdom was in sharp contrast to that of the founders of the London University and King's College, who had expended so much on buildings that they left little money to support teaching. Lyell also praised the system of public education throughout New England, the celebration of Thanksgiving at Boston, and the kindness of his welcome among the Boston people. He was equally appreciative of the generosity of the southern planters, who gave freely of their hospitality and their horses and carriages but who refrained from consuming Lyell's time with dinners and social functions.

Lyell described the financial crisis of 1842 in the United States and the resultant distrust of banknotes. He mentioned, too, the widespread suffering that resulted when the state of Pennsylvania failed to pay the interest on its bonds, wishing to counteract the impression in England that only the foreign holders of the bonds had suffered loss. He attributed the irresponsibility of the Pennsylvania government to the consequences of universal suffrage. Lyell advanced all the reasons that an English Tory might use to argue against popular suffrage and then showed that, on closer examination, both the United States as a whole and the individual states were in essentially sound financial and economic condition. Even the new states on the western frontier were making efforts to pay the interest on their debts and gave signs of behaving responsibly in financial matters.

From a British standpoint, social life in the United States was also unusual. Lyell observed how easily he and Mary had felt at home in New York and other American cities. So many Americans traveled from the North to the South during the winter, or to Washington for the sessions of Congress or the Supreme Court, or had relatives settled in the West, that wherever the Lyells went they met people who were friends or acquaintances of those whom they had already met in other places.[51]

50. Ibid.

51. Lyell, *Travels,* 1:192.

Lyell devoted special praise to the universities of the United States, which were growing in number and whose discipline he said was very strict. Each university was connected with some church, but on the whole Lyell found that sectarian influences were less oppressive than in Great Britain. Among thirty-two professors in Harvard College, only five had been educated as clergymen. Lyell's remarks on universities in the United States gave him an opportunity to discuss also universities in Great Britain.

For more than twenty years, Lyell had been interested in the reform of the English universities. He now traced the history of Oxford and Cambridge, showing how over the preceding century and a half the college tutors, who were usually clergymen, had come to control education at both universities. When Oxford established public examinations in 1800, the examining boards consisted entirely of college tutors who conducted examinations only in subjects that they themselves taught. Since examinations for degrees were not required in the subjects taught by the professors, students gradually ceased to attend the professors' lectures. The subjects taught by the tutors included Greek and Latin literature, Christian theology, and mathematics. For lack of time, they had to be taught at an elementary level, more like that of a grammar school than a university. The tutorial faculty could neither keep abreast of nor contribute to any area of learning. Furthermore, since the tutors were clergymen, they expected to teach only a few years until they obtained a position in the church. They had little incentive to become deeply learned in any particular subject.

Lyell referred to the *Principles of University Education* by William Whewell, master of Trinity College, Cambridge, who defended classical languages and mathematics as the only subjects suitable for university education.[52] Lyell thought that the elementary study of classical languages and mathematics was hardly enough to prepare young men for a profession. Students were treated more as children than as men who needed to learn to reason independently. The system at both universities worked so well to protect the self-interest of the tutors and the church that Lyell doubted that either institution could reform itself. Only a royal commission, such as had visited the Scottish universities, could overcome the power and self-interest of the colleges at Oxford and Cambridge.[53]

Lyell's appeal for a royal commission was timely. On 10 April

52. Whewell, *University Education,* 242, 247.

53. Lyell, *Travels,* 1:226.

1845, as the *Travels* was being printed, William Christie moved in Parliament to establish a royal commission of enquiry on the English universities.[54] Lyell sent a copy of his *Travels* to Sir Robert Peel, mentioning that although the work was devoted chiefly to geology, "I have also treated of the Universities of England & the United States."[55] In July he wrote from Scotland to Leonard Horner, noting that reviews of the *Travels* had so far been very favorable, but none had mentioned his discussion of the universities. "This I believe is fortunate," commented Lyell, "as, if they had sounded the alarm, it [the book] might not have found entrance into certain houses & clubs who will now think a book of travels must be a harmless affair & stand on neutral ground."[56]

Early in August, the Lyells returned from Scotland to London and began to prepare for their second trip to America. Lyell discussed fossil elephants with Hugh Falconer, and elephants and mastodons with Richard Owen.[57] In preparation for his Lowell lectures, he was studying the recent extinctions of various species of large mammals in both Europe and the Western Hemisphere. He also enlisted artists to draw large-scale illustrations for the lectures and assembled the books he wished to take with him. On 3 September, the Lyells took the train to Liverpool, covering the 210 miles in seven hours and ten minutes, several hours less, Lyell noted, than when they had made the same journey in 1841.[58] The next day, they sailed from Liverpool in the Cunard steamship *Britannia* on their second voyage to America.

54. William Dougal Christie (1816–74), barrister, diplomat, and author, graduated from Trinity College, Cambridge, in 1838 and from 1842 to 1847 was member of Parliament for Weymouth.

55. Lyell to Peel, 2 July 1845, Peel MSS.

56. Lyell to Horner, 27 July 1845, Kinnordy MSS.

57. Lyell, Notebook 119, 79–83, Kinnordy MSS.

58. Lyell, Notebook 120, 18, Kinnordy MSS.

CHAPTER 5

Across the Atlantic Again: New England

THEIR SECOND VOYAGE to America was from the beginning among friends. The Americans on board the *Britannia* included Edward Everett and his wife, who dined with the Lyells at the captain's table. Also present was a Mr. Knight, a planter from Barbados, with whom Lyell discussed the results of the emancipation of the slaves in the West Indies and the recent importation of Hindu laborers from India. After three days of beautiful weather, the *Britannia* ran into headwinds and occasional squalls but steamed on steadily, though at reduced speed. From the deck Lyell watched porpoises play alongside the ship and on 13 September sighted his first iceberg near the Grand Banks of Newfoundland. He also visited the engine room, where he was fascinated by the indicator, which counted the number of revolutions of the engine during the voyage. On the 17th the *Britannia* docked at Halifax and the Lyells disembarked for three hours. Mary visited Mrs. Samuel Cunard; Lyell "botanized and conchologized—in the last operation he fell in the mud & a Negro woman scraped him all over."[1]

On 19 September 1845 they landed at Boston. "It was a lovely day," Mary wrote, "such bright sky as we had not seen since we left Boston & the islands are very pretty." They noticed changes since 1842, not all for the better. The new Bunker Hill monument reminded Lyell of a factory chimney. Weeks of hot weather had burnt the grass brown.

Curtis Guild, George Ticknor's eighteen-year-old nephew and a clerk in a Boston shipping house, met the Lyells at the boat and took them to George and Anna Ticknor's house on Park Street. The Ticknors themselves were away, spending the summer at Geneseo, New York, but their servants welcomed the Lyells. Although many Bostonians were still out of town during the heat of summer, a number of friends called to greet them. "I believe we felt as much returning home as the Everetts did almost," wrote Mary, "& seeing so many friendly faces."[2] That evening, they went by invitation to a hor-

1. Mary Lyell to Mrs. Leonard Horner, 24 September 1845, Kinnordy MSS.

2. Ibid.

ticultural festival in Faneuil Hall and heard speeches by Edward Everett and Daniel Webster. They were showered with invitations to dinner.

On Sunday they attended church in Brattle Square with Mr. and Mrs. Abbott Lawrence and then dined at their home. Lawrence headed the Boston mercantile firm of A. & A. Lawrence. At the time of the Lyells' visit in 1845, he had just founded the new town of Lawrence, Massachusetts, where he was building textile mills that would come to rival those of Lowell. Ten years earlier, in 1835, Lawrence had played a leading role in building a railroad from Boston westward across the Berkshire Hills to Stockbridge, Massachusetts. The Western Railroad, thus named, had opened in 1841 and, together with the Albany and West Stockbridge Railroads, completed the following year, provided a direct rail connection between Boston and Albany and a route for the transport of New England manufactures to the rapidly developing West. Public-spirited as well as enterprising, Lawrence also was currently leading a campaign to build a public water supply for the city of Boston. At dinner at the Lawrences', the Lyells met John Quincy Adams, the elderly former president of the United States, who reminisced about his days as American minister in London in 1815 and 1816 when Castlereagh was prime minister and Canning foreign secretary. That same day, Mrs. John Amory Lowell sent her carriage to bring Lyell and Mary to the Lowells' country house in Roxbury for tea, "a very sociable pleasant evening."[3]

On 24 September, they took the train to Portsmouth, on the New Hampshire coast. From the train the Lyells marveled at the profusion of goldenrod and asters in the countryside. At Portsmouth they called on John L. Hayes, a young lawyer and amateur geologist who had recently written on the role of Antarctic icebergs in transporting earth and rock.[4] During their visit, Lyell admired the separate birdhouses for the house martin and the smaller green martin in Hayes's garden. That evening at a party at Hayes's house, they met many people and enjoyed the dancing of pretty girls in white, pink, and blue party dresses and white satin shoes.[5] Among the people they met was a Mrs. Allen from Gardiner, Maine, who had made a collection of Tertiary fossil shells from deposits along the Kennebec River. She urged Lyell to visit the Kennebec and gave him a letter of

3. Ibid.

4. Hayes, "Influence of Icebergs upon Drift."

5. Mary Lyell to Mrs. Leonard Horner, 24 September 1845, Kinnordy MSS.

introduction to her husband at Gardiner. The Lyells accordingly changed their plans.

Early the next morning, Lyell went with Hayes to explore the banks of the Piscataqua River near Portsmouth. The surface geology was similar to that of Sweden and Norway, with scattered patches of clay and gravel containing fossil seashells of living Arctic species. Such deposits indicated that in recent geological times the country had been submerged beneath the sea in a climate much colder than at present.

In the afternoon, the Lyells took a train to Portland, Maine, and there boarded the steamboat *Huntress* to go to Gardiner on the Kennebec River. In bright sunshine the *Huntress* steamed out of Casco Bay into the calm, open sea and an hour or so later entered the mouth of the river. As the boat steamed up the Kennebec, among rocky islands, they admired the thriving towns and villages along the banks. At Bath, shipyards were turning oak and pine logs into ships to carry cotton from the southern United States to England. Sailing vessels loaded with baled hay were setting off for New York and even for Mobile and New Orleans. The ships that carried hay to the South would return with bales of cotton for New England mills—evidence for Lyell of the immense economic advantage created by free trade among the American states and therefore of free trade in general. Everything in Maine seemed prosperous, and in Bath whole streets of new houses were under construction.

The steamboat reached Gardiner after dark, and the Lyells went to Mr. Allen's house with their letter from Mrs. Allen. Although the family was not expecting them, Mr. Allen and his daughter gave the Lyells a warm welcome. In Mrs. Allen's collection of local fossil shells and crustacea, Lyell identified a walrus tooth similar to one he had collected on Martha's Vineyard in 1842, but other teeth he could not identify. Mr. Allen gave him specimens of the unidentified teeth to take back to London. Lyell also examined the beds of clay and sand along the banks of the Kennebec River, from which the fossils had come. Fossil shells of living species indicated that the beds had been laid down beneath the sea during geologically recent times. Every small stream draining into the Kennebec had cut its own valley through beds of clay and sand; at nearby Augusta, the strata of clay and sand were 100 feet thick, and Lyell learned that in some places they were as much as 170 feet thick. Near Gardiner there were conical hills of gravel, which Lyell attributed to the denuding action of the waves and currents of the sea on the soft sediments.

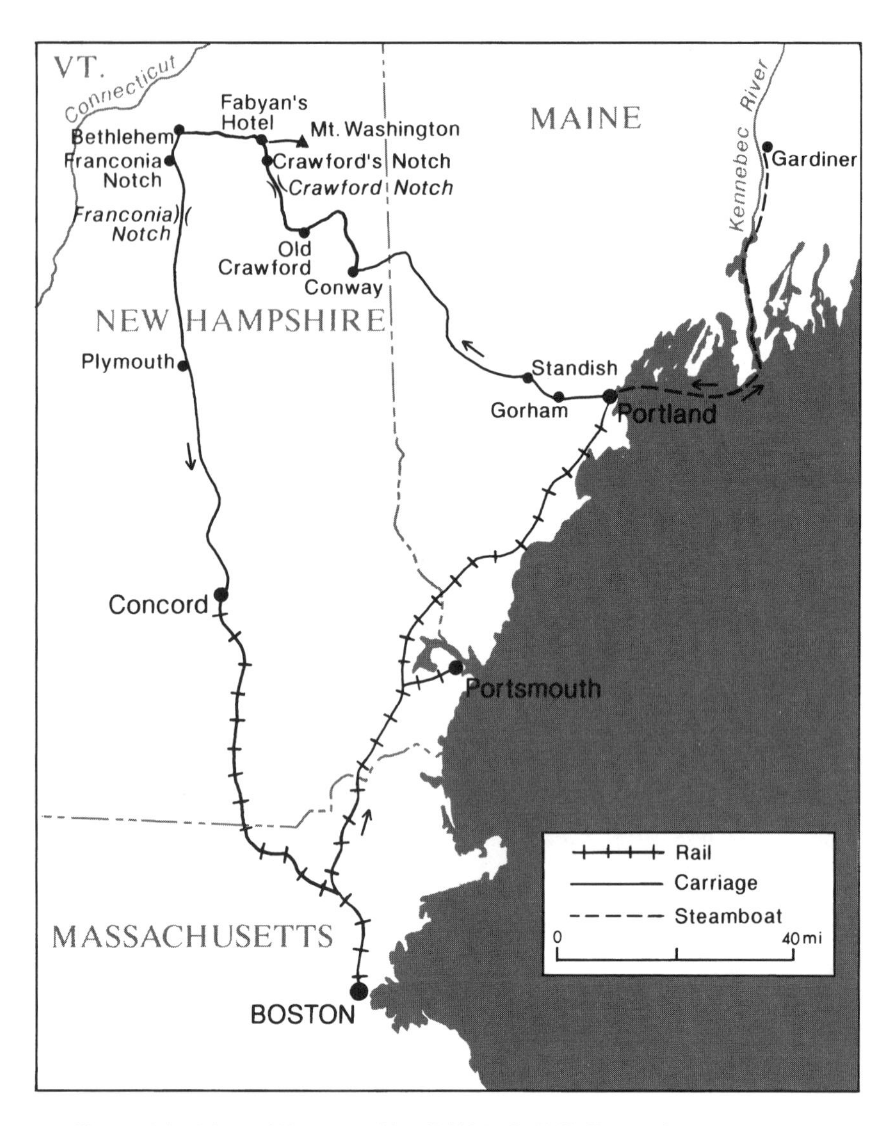

Lyell's Travels in Maine and New Hampshire, 1845. Map by Philip Heywood.

On Saturday morning the Lyells returned on the *Huntress* to Portland, where they stayed at a boardinghouse. The days were cool but clear and bright—the beautiful weather of an early New England fall. The maples along the tree-lined streets of Portland were beginning to turn red. In the gardens around the white houses with green shutters, asters, dahlias, and a few sweet peas continued to bloom.[6] On Sunday morning they attended the Unitarian church with a Mr. Davis, a lawyer and yet another friend of George Ticknor. In the afternoon Lyell hired a carriage and driver, and they drove sixteen miles to Standish, by way of Gorham. Throughout the countryside that beautiful September day, the trees were changing to the bright reds, golds, and yellows of fall. Along the roadside, Lyell noted the fading spirea mixed with sweet fern, and a profusion of goldenrod and asters. On the ground beneath fir trees, they picked some red checkerberries (wintergreen). Among the native American plants, Lyell found various European weeds, including the English selfheal, the mullein, and others. The roads were good, and Lyell was pleased to find that they did not have to pay tolls, there being only one turnpike in all of Maine.

That evening in the chill twilight, they arrived at Standish. The landlord of the small inn ushered them into the parlor warmed by a blazing wood fire. On the parlor table were serious books, on one wall a framed copy of the Declaration of Independence, and on another "a most formidable likeness of Daniel Webster." Since Lyell and Mary were the only guests at the inn, the landlord and his wife sat with them at dinner. The Lyells found the inn so pleasant that they remained two nights. On Monday, Lyell worked all morning on his lectures for Boston, but in the afternoon he and Mary went for a walk and collected plants. With Mr. Thompson, the intelligent and well-educated innkeeper, Lyell discussed the tariff question then dividing the Whigs from the Democrats. Thompson asked about the free church movement in Scotland, and how the education system in Scotland differed from that in Prussia about which he had been reading. Lyell asked about the occurrence of consumption (pulmonary tuberculosis) in New England, then a common and deadly disease. In the course of the conversation, Lyell learned that the wages of laborers in Maine were about ten dollars a month with board included and were sufficient for men to save money to go West.

6. Mary Lyell to Marianne Lyell, 27 September 1845, Kinnordy MSS.

On 30 September, the Lyells took the stagecoach from Standish to Conway, New Hampshire. Near Hiram, the road entered the White Mountains and followed the valley of the Saco River to Conway. They stayed at the Pequawket House. At midday dinner in the men's dining room, Lyell noted that the other residents possessed the big, hard hands of mechanics and working men, yet their hands were clean and the men well dressed and literate. After dinner, several of the men went into the drawing room to listen to a woman boarder play the piano. Others read. After they had left the room, Lyell, ever curious, looked to see what they had been reading. One was reading Disraeli's novel *Coningsby*, another Robert Burns's poems, and a third an article reprinted from *Fraser's Magazine* on the policy of Sir Robert Peel. He was impressed by such marks of an educated democracy in which working-class men and women had the same interests and habits as ladies and gentlemen in Britain.

A downpour kept the Lyells indoors the next morning. After dinner at one o'clock the sky cleared, and they set out on a walk through the woods that Mary described:

> We did not get back till dusk, having seen most beautiful views of the mountains, climbed fences, & I forded a stream without shoes & stockings. We picked up a small tortoise . . . with a very brilliant shell. He looked much inclined to bite. Most of the birds have migrated but there are still a great many insects & grasshoppers. Several little squirrels sit tamely by the roadside as we pass.[7]

The streams rushing down Haven Hill near Conway were like burns in the Scottish Highlands. On the hills the brilliant fall colors of the maples contrasted with the dark green of the hemlocks, and though leaves were already thick on the ground, the day was mild. Many large, angular fragments of granite on the hills reminded Lyell of the granite of Arran in Scotland. At the Pequawket House, the landlord told Lyell that in his memory the township of Conway had grown up from almost nothing. He added, "It is a healthy growth, every man his own tenant," meaning that each man owned his own house and land.[8]

On 2 October the Lyells took the stagecoach to Old Crawford. Although Lyell had seen the fall colors in Pennsylvania in 1841, he was

7. Mary Lyell to Susan Horner, 4 October 1845, Kinnordy MSS.
8. Lyell, Notebook 122, 87, Kinnordy MSS.

unprepared for their brilliance in New Hampshire. On the mountainsides Lyell was

> much struck with the grouping of the wood in zones & masses of the same colour, pale green, dark green, rich brown, yellow and crimson. In Norway the birch are often spotty when interspersed naturally amongst the evergreen firs. Here they are not so. They are intermingled & yet non-isolate. It rarely happens when art has not interfered that the foliage is not in harmony . . . . The bright sun lights up these hills finely. At each turn of the road when we saw the light falling on some hill thro' a valley opening towards the West, some Sugar Maple formed at once a conspicuous & brilliant & novel spectacle.
>
> The brawling torrents & the great river roaring like a cataract among large boulders—The tourists from N. York & Boston do well to escape from the summer heats to the cool air of these mountains, but in such Autumn weather as we are enjoying, the season of the gay & varied Autumn tints is most invigorating.[9]

That night at Crawford's Inn, Abel Crawford, "a hale old man of 79" who had lived there more than fifty years, described the changes that had occurred since the first settlement. Next morning he drove them to Crawford's Notch, stopping to show them the site of the Willey slide. On 28 August 1826 an avalanche had killed all nine members of the Willey family, although their house, protected by an outcrop of rock, had remained unharmed. In his *Principles of Geology*, Lyell had described the Willey slide as an instance of modern geological change.[10] The slide had not left grooves in the surfaces of the underlying hard rocks, like those produced by glaciers in the Alps. The grooves in the surface of rocks in northern countries, therefore, could not have been caused by avalanches or deluges of mud and stones sweeping over the land; they must instead be signs of past glaciation.

At the Notch, the Lyells stayed at the Notch House, kept by another member of the Crawford family, Thomas Crawford. Lyell asked Thomas Crawford whether in New Hampshire the rich had a fair chance to hold public office. His host replied that they had no chance at all, even with the Whig party, because people thought the

9. Ibid., 98–102.

10. Lyell, *Principles* (1830) 193–94; cf. *American Journal of Science* 15 (1829): 216.

rich would not sympathize with them; "besides they would be hunting and fishing." Crawford also thought that the people wished "to give a lift to the poorer & enterprising & say the rich man is well-off already."[11] The members of the legislature were paid, and people sought the office for the pay. Crawford was astonished that British members of Parliament received no pay. In his notebook, Lyell mused:

> Universal Suffrage
>
> The more I converse with the mechanics, laborers & small proprietors, the more I hear of the opin[ion]s of democrats and Whigs, the less probable does it appear that the capitalist & large landed proprietor will obtain an ascendancy or even obtain a reasonable share of political power. The organization of the small proprietors & laborers seems perfect & they have so much practical sense that whatever mistakes they commit whether in great public affairs or in State matters, they are not likely to go so far wrong as to endanger their power or upset the Government.[12]

On Saturday, 4 October, the Lyells went to Fabyan's Hotel below Mount Washington. While waiting for good weather to ascend the mountain, they met William Oakes, a botanist from Ipswich, Massachusetts. Oakes had been studying the flora of the White Mountains for twenty-five years and was preparing to publish a work on it.[13] Oakes identified for them some of the plants they had collected and promised to give Mary specimens of mosses from the White Mountains.

On Sunday morning the Lyells walked with Oakes through the woods to the falls of the Amoonoosuc River, where he showed them the twin flower, *Linnea borealis*, now in fruit. Lyell was amazed to find this plant so far south. He had seen it in flower in Nova Scotia in July 1842 but had encountered it first in the woods of Norway, in latitudes far to the north of those of New Hampshire. Oakes also pointed out the old man's beard, *Lichen barbatum*, on the trees and said that the top of Mount Washington was covered with lichens.

The weather finally cleared on Tuesday, 7 October, and at nine in the morning they set off on horseback with Oakes, guides, and other visitors to the hotel, forming a cavalcade of eight horses. The par-

11. Lyell, Notebook 123, 3, Kinnordy MSS.

12. Ibid., 10.

13. Unfortunately, William Oakes died in 1849 before he could complete his botany of the White Mountains. See Gray, "William Oakes."

ty rode at first through the woods, crossing and recrossing the Amoonoosuc at successive fords; they then began a much steeper ascent. "After passing through the woods for a long time," wrote Mary,

> we got to a height where the spruce & fir grew as low as the heather in Scotland & then we came to nothing but blocks of rock covered with lichen geographicus, quite a barren scene. There were a good many clouds, but they only added to the beauty as they were constantly shifting & unveiling a most beautiful view to the East of ranges of hills & many beautiful small lakes, shining like silver.[14]

Lyell was impressed by the zones revealed one after another as they climbed the mountain.

> I never saw on the Alps or Pyrenees, scarcely even on Mount Etna so beautiful a display of the zones of different plants, especially trees, in the ascent of a mountain, as in our ride to the summit of Mount Washington. After leaving the region of maple, birch, oak mingled with hemlock (a fir with leaves like our yew), white pine, balsam fir, spruce etc., the ground being covered with Lycopodium, we find one tree disappear after another, but the Spruce (Pinus nigra) I believe & balsam fir. When these two begin to diminish considerable in growth, they thicken in numbers & a great many dead ones stand among the rest, not decaying, as if some 9 or 10 months of snow preserved them. They stick up like the horns of thousands of elks & deers. A few hundred feet higher they are reduced to the size of shrubs, & then literally mat the ground like Lycopodia, so that I saw the spruce, fir & the reindeer moss, the latter often topping the tree.[15]

At the summit, they stopped to eat their lunch. Lyell looked for glacial furrows on the rocks but found none. Lyell and Oakes collected various Arctic plants and lichens. Many were familiar to Lyell. He had collected them in Glen Clova and other places in the Grampians near his home in Scotland. Oakes also recognized the

14. Mary Lyell to Caroline Lyell, 8 October 1845, Kinnordy MSS.

15. Lyell to Susan Horner, 13 October 1845, Kinnordy MSS. In July 1816, Jacob Bigelow and Francis Boott of Boston visited the mountains of New Hampshire and Vermont to collect plants for their proposed flora of the New England states, which they never completed. See Bigelow, "White Mountains of New Hampshire." In 1824, Bigelow published some of the results of their collections in his *Florula Bostoniensis.*

similarity between the floras of the White Mountains and the Scottish Highlands. He knew of the various localities in Forfarshire where Lyell's father had collected *Jungermanniae* and other plants. Lyell marveled how he and Oakes talked together of Sir William Hooker, Robert Brown, and the differences between the flora of the Scottish Highlands and that of Hampshire in southern England, of the poems of Thomas Moore, the *Pickwick Papers*, and the novels of Sir Walter Scott. It was as if Oakes were a native of Ipswich in Suffolk instead of Ipswich in Massachusetts.[16]

They returned to Fabyan's Hotel in the evening, after twelve hours in the saddle, a ride too long for Mary, who was somewhat stiff and sore the next day. Again the weather was rainy, and Oakes showed her how to dry plants so as to preserve their colors and gave her a supply of the paper he used. One evening, Fabyan played the horn so that they might hear the remarkable echo from the hills. Lyell thought of Fabyan as a gentleman and was astonished when Fabyan also blacked his shoes for him.

On 9 October, Lyell and Mary, wrapped in their cloaks "and several India rubbers Mr. Oakes lent us,"[17] rode in an open wagon twenty-five miles through heavy rain to Bethlehem and then south to Franconia Notch. On the road Lyell discussed churches and religious revivals with their driver, a freewill Baptist awakened to Christianity only five years before. They also discussed the Millerite delusion of the previous year. The Reverend William Miller had predicted that the Second Coming of Christ would occur on 23 October 1844, basing his prediction on the prophecies of Daniel. Many New Englanders who believed Miller's prediction sold their property and gave much money to his movement. They also purchased robes to wear on the day of their ascent into Heaven. At Boston, Miller's followers had built a theater in which on the appointed day they gathered to pray, dressed in their robes.[18] Fortunately, the Boston authorities had insisted that the theater be built in a substantial and permanent form, and a year later it was being used for performances by the English actors Charles and Ellen Kean. Fascinated by the accounts of Miller and his followers, Lyell pondered what it might say of the effects of popular education.

The next evening, in the hotel at Franconia Notch, a trapper sat down to dinner with the rest of the guests. As on many occasions in

16. Lyell to Caroline Lyell, 8 October 1845, Kinnordy MSS.
17. Mary Lyell to Caroline Lyell, 8–10 October 1845, Kinnordy MSS.
18. Dick, *Miller and the Advent Crisis*.

their tour through the White Mountains, Lyell was struck by such evidence of complete social equality. The New England Yankees reminded Lyell of the Scotch: "They dress in black when they mean to be smart—are tall, anxious-looking, bony, Calvinistic, hard at bargains. Why, for scarce any have come from Scotland?"[19]

On Saturday, 11 October, again traveling in the rain, the Lyells took a stagecoach down the valley of the Pemigewasset River, passing wagons drawn by oxen and loaded with Indian corn and pumpkins. At Plymouth, they stayed overnight and on Sunday attended the morning service at the Congregational church and in the afternoon at the Methodist church. In the evening, Lyell walked on the covered wooden bridge over the Pemigewasset River and talked to the keeper of the bridge, who told him tales of the winter and of early days in New Hampshire.

Monday the Lyells traveled back to Boston, where they went to the Tremont Hotel, picked up their accumulated letters at the post office, and that evening dined with the Ticknors. The next day they took the stagecoach to Plymouth, Massachusetts. As they passed out of Boston, Lyell observed fine streets of new houses in the suburb of South Cove, where three years before there had been only a marsh. Out on the blue waters of the Massachusetts Bay, they saw the white sails of ships carrying granite from quarries at Quincy to Boston. After about ten miles, the country became sandy and barren, covered with pine and fir trees.

Plymouth was a town of elm-lined avenues, beautifully situated on a fine harbor. They went to the Pilgrim House, "a small old-fashioned inn." Plymouth's heritage as the first settlement of the Pilgrims was everywhere apparent. Mary observed:

> Many of the houses are built just as you see them in English country towns with beams across the ceiling & paneled walls. Other people laugh at this, but the inhabitants are very proud of everything that shows their pure Old England descent which seems to have remained exceedingly unmixed. A bookseller there found out [who] Charles [was] & took us to one or two private houses where the family showed us chairs which had actually come over in the Mayflower & looked very genuine, also drinking cups & seals. Several relics of this sort, & pictures are preserved in Pilgrim Hall, a building where they hold fetes and balls every 22nd. of Decr. which they call Forefathers Day, which was the day they

19. Lyell, Notebook 123, 102, Kinnordy MSS.

landed & an oration is delivered in the morning. Mr. Webster gave it one year, Mr. Everett another.

The graveyard is situated on a hill where they first made a fort to defend themselves from the Indians & [contains] some old tombstones made of slate brought from England. We staid some time reading the verses, both old & modern, & you would think the whole old testament was there, Adoniram, Hiram, Eunice, Naomi, Dorcas, Ephraim, Deborah, Mercy, Rebecca, Experience, Zachaeus, Seth, Moses, Ezra, Consider, Levi, & a few, Sally, Polly, Betsy, Nancy. These were not selected but read straight from one to another.[20]

Lyell hired the landlord of the inn to drive them to Plymouth beach, a narrow spit of sand running out into the sea. While Lyell was collecting shells, the landlord drove Mary to the end of the beach. One of the horses sank into the soft sand and the other began to plunge about: "I jumped out as quick as I could," Mary wrote, "& with a good deal of trouble the landlord got his horse out, but he was much frightened & the poor horse still more so."[21] On Plymouth beach the Lyells collected eighteen species of shells, of which they recognized six as also occurring in European seas. The most noticeable shell, the very large *Mactra solidissima,* was capable of closing on the feet of birds so that they were drowned by the incoming tide. The wind was bitterly cold, and the night before the sea had frozen near the shore.

After two nights in Plymouth, they returned to Boston on 17 October to settle at the Tremont Hotel. In Boston, the late October weather was delightful—sunny days so warm that Mary walked out with no more than a scarf over her dress. The Lyells' friends, now returned to Boston, called and sent them invitations. Lyell was so occupied with preparing his lectures that they went out little, except to the Ticknors' house on Park Street, only a three-minute walk away. Lizzy Ticknor, the younger daughter, was still a child "& inclined," wrote Mary, "to have a good romp with Charles, remembering the pranks of his former visit."[22] They often dined with the Ticknors, who served dinner at half past five, in contrast to the usual Boston dinner hour of two or three o'clock. On days when Lyell gave his lecture in the afternoon, they had luncheon at Park Street and the Ticknors then accompanied them to the lecture hall.

20. Mary Lyell to Caroline Lyell, 28 October 1845, Kinnordy MSS.

21. Ibid.

22. Mary Lyell to Katherine Horner, 25 October 1845, Kinnordy MSS.

Lyell delivered the first of his Lowell lectures on Tuesday, 21 October, and repeated it the next afternoon to a second audience. Thereafter, he lectured on Tuesday and Friday evenings, repeating Tuesday's lecture on Wednesday afternoon and Friday's on Saturday afternoon. The lectures were an even greater success than those of 1841. Fourteen thousand people applied for tickets, and every evening lecture filled the Odeon Theatre to its capacity of about twenty-three hundred persons. At the afternoon lectures the attendance was smaller, the upper gallery not completely filled, but even in the afternoons many Boston men left their business or profession to attend. Experience had made Lyell a more effective lecturer than in 1841.

In his first lecture, Lyell recapitulated his general views of the uniformity of geological processes. He then discussed the extinction of species, taking as his first example the *Palaeotherium*, from the Paris Basin, originally described by Georges Cuvier. On 24 October, he discussed the mastodon, of which many large specimens had been discovered in America. Lyell showed a full-size drawing of the mastodon specimen in the British Museum, which had come from Missouri, comparing it with a specimen recently dug out of a bog in New Jersey. The British Museum specimen showed only eighteen ribs, whereas Cuvier had described nineteen and the New Jersey specimen possessed twenty. A number of mastodon skulls from the same New Jersey bog formed a series representing every stage of development, from the young calf to the full-grown mastodon. The skulls showed that the mastodon had possessed one more pair of molar teeth than Richard Owen had thought from his study of the Missouri specimen. The Boston surgeon John Collins Warren had purchased the New Jersey mastodon for Harvard University, and Dr. John Jackson was preparing to publish a description of it.[23]

From the large mammals that had so recently become extinct in both America and Europe, Lyell turned on 11 November to glaciers and glaciation.[24] The following week, he showed his audience a large map of the world marked with the provinces of distribution of land mammals, based on Buffon's concept of natural barriers to the spread of animal species, either water barriers, such as oceans, or land barriers, such as mountain ranges. Buffon's emphasis on natural barriers had been confirmed by detailed investigations of the distribution of animals in every part of the world. In South Ameri-

23. Lyell to Katherine Horner, 25 October 1845, Kinnordy MSS; cf. Jackson, "Fossil Bones of Mastodon"; Jackson, "Dentition of the Mastodon."

24. Mary Lyell to Eleanor Lyell, 12 November 1845, Kinnordy MSS.

ca, living and fossil mammals formed a distinctive set of species, including among living species the llama and armadillo and among fossils the *Megatherium.* The horse was absent from both North and South America as a result of the geographical separation of the Western Hemisphere from the Old World. Even among fossils, said Lyell, the nearest approach to a horse was a species of zebra as traced by the teeth. Similarly, the geographical isolation of Australia was reflected in its distinctive group of animal species, such as the kangaroo. In both Australia and South America, a close relationship existed between the fossil and living forms, a relationship that indicated that during the geological past natural barriers had exerted the same effects on animal distribution as they do today. Yet the number of extinct species showed that in the recent geological past at least one great change in species of animals had occurred. The *Boston Evening Transcript* reported:

> As regards the explanation of this strange law, Prof. Lyell expressed his opinion that nothing could give it but the belief that a first pair of birds or animals had been given to any particular spot favorable to their preservation and growth, evidently pointing to creation as being at many points on the earth's surface. . . . Such doctrine will doubtless be set down as rank heterodoxy by those of the old Noah's Ark belief, though it may be none the less true for all that.[25]

In his later lectures, Lyell discussed the factors that influenced climate in various parts of the world, contrasting the extreme differences between winter and summer temperatures on the eastern coast of the United States with the much smaller seasonal variations in Europe. New York, he said, had the summer of Rome and the winter of Copenhagen.[26]

The Lyells thoroughly enjoyed their six weeks in Boston. Despite considerable rain, the weather was mild and Boston social life pleasant. Many of the families they met lived in houses facing Boston Common, houses that in London would surround two fine squares. The Bostonians they met were rich but tended to live simply. Most of them did not keep horses and carriages but walked about the city. They kept few servants. The Ticknors, Lyell noted, had only four servants, whereas in London a similar household would have at least nine. The servants, too, were educated and possessed a strong sense

25. "Ocean and Land Barriers on Our Globe," *Boston Evening Transcript,* 25 November 1845.
26. "Climatology," *Boston Evening Transcript,* 2 December 1845.

of self-respect and independence. For instance, the maids would not submit to covering their hair under a cap, and George Ticknor said that his manservant always voted against him in elections as a matter of principle.

Lyell took a keen interest in every aspect of the vigorous development of America reflected in the life of Boston. He noted the growing number of Irish immigrant laborers, who lived in squalor and whose ignorance and illiteracy stood in sharp contrast to the knowledge and alertness of the native Yankee workmen. The influx of Irish was already giving rise to resentment, as reflected in nativist political movements. Industrialists like Abbott Lawrence and John Amory Lowell told Lyell that without cheap Irish labor they could not afford to dig canals or build railroads and cotton mills. One night on his way back to the Tremont Hotel after a party, Lyell dropped in on an Irish public meeting calling for repeal of the Act of Union of Ireland with Great Britain and listened to a speech on the sufferings of the Irish people. On his way out of the meeting, he met a native-born Boston workman whom he knew. The workman said that he hoped the United States might one day help Ireland secure its freedom from Great Britain. During the meeting, Lyell had his pocket picked and lost his purse; fortunately it contained only a few dollars.[27] Despite the anti-British feeling shown at the meeting, he noted that any reference to "the Queen" at Boston invariably meant Queen Victoria.[28]

The addition of immigrants to the population gave urgent importance to the system of public education in Massachusetts. At Boston Lyell visited two large public schools, one for girls and one for boys. The girls were reading Gray's *Elegy Written in a Country Church-Yard* and studying algebra. The monies spent for public and private education and the salaries of teachers were far above what was spent in England. The high level of general education among the people of New England created a demand for public lectures and supported the sale of newspapers, books, and periodicals.

At the Ticknors' the Lyells met many old friends, including William Prescott. In the summer of 1845, after his father's death, Prescott had moved to a new house at 55 Beacon Street that had room for his library and study. Early in November, John Amory Lowell invited the Lyells to dinner, and they attended two evening parties, one at the William Appletons' and the other at the Edmund

27. Lyell, *Second Visit,* 1:177–78.

28. Lyell, Notebook 125, 107, Kinnordy MSS.

Dwights'. Appleton was an enterprising young publisher working in his father's publishing business. Dwight, at the age of sixty-five, was one of the great New England cotton manufacturers. His wife, Mary, was an older sister of Anna Ticknor and, like her, a daughter of the Boston merchant Samuel Eliot. In 1845 Dwight's cotton mills, machine shops, and calico printing works employed more than three thousand workers. He also had a financial interest in the railroad across Massachusetts from Worcester to Albany, New York. But what probably made Dwight a sympathetic person for Lyell was his interest in education. As a member of the Massachusetts legislature, Dwight had helped to write the Massachusetts School Law of 1837, which established a state board of education. Moreover, in order to obtain Horace Mann as first secretary of the board, Dwight paid part of his salary.

Mary described the evening parties given by the Appletons and Dwights:

> They were much like London ones, a crowd for conversation & the best of it is, we knew about as many people & shook hands with as many in proportion as we generally do in our own circle in London, in one of Mr. Babbage's or other parties. The little differences are that the ladies go into a different room from the gentlemen to take off their cloaks, principally, as most people walk, the distances are so short, & therefore, there is more wrapping up. Instead of refreshments being handed, there is no tea, but about ten o'clock another room is thrown open, & a standing supper of oysters in various forms, raw & dressed, chicken salad, jellies, meringues & numerous pyramids of various ices & cakes of all kinds, champagne & other wine. The hours are absurdly late, evening parties never beginning till nine, while people dine at three & drink tea at six. Everybody abuses it & says it is aping Europe, yet they all go on.[29]

Despite the lateness of their parties, the Bostonians were usually up early. The Lowells were accustomed to have breakfast at seven and the Ticknors at eight.[30]

Also staying at the Tremont Hotel was their old friend William McIlvaine, who came from Philadelphia on 19 October to hear Lyell's lectures. He usually joined the Lyells for meals at the hotel. Since he knew the Ticknors and others at Boston, he was invited to

29. Mary Lyell to Eleanor Lyell, 12 November 1845, Kinnordy MSS.

30. Lyell, Notebook 125, 37, Kinnordy MSS.

the same parties. McIlvaine attended both Lyell's evening and afternoon lectures, so they saw a good deal of him until he left for Philadelphia on 11 November.

On 6 November the Lyells went for the day to the old seaport of Salem, about fourteen miles north of Boston. The railway passed salt marshes sprinkled with haycocks sitting on piles, waiting to be removed when the marsh was frozen. At Salem, the railway station proved to be an old wharf. They had no introduction to anyone at Salem, but a Boston bookseller had recommended that Lyell call at a particular bookshop. While they were in the shop, the owner introduced them to a Mr. Phillips. He in turn took them to the home of Dr. Henry Wheatland, curator of Salem's natural history museum. When they arrived, Wheatland was just sitting down to his midday dinner. He invited the Lyells to dine with him and after dinner took them around Salem. "It is a pretty, quiet looking town," wrote Mary, "with houses in gardens & trees in the streets, many very rich people. It is a few years older than Boston."[31]

At the museum, Lyell examined the jawbones and teeth of the shark *Squalus serridens* from the South Pacific, and a collection of seashells from the New England coast. In the Essex Institute, where East Indian and Javanese curiosities were preserved, Lyell examined the logbooks of some of the early sea captains who had sailed from Salem to India by the Cape of Good Hope or sailed around Cape Horn to the coast of Oregon and thence to the Sandwich Islands and China. They had made these long voyages without charts or sextants but by dead reckoning with chalk on a plank, estimating the sun's position with their hand at noon. The Lyells also visited the courthouse at Salem, where the records of the witch trials of 1692 and 1693 were preserved, including the warrants and depositions "& a small bottle of pins said by the witnesses to be taken out of their bodies & stuck into them by witches."[32] Afterward they went to Gallows Hill, where nineteen persons had been hanged as witches; from the hill they had a fine view of the sea.

Returning to Boston on the train in the late afternoon, Lyell struck up a conversation with two colored men sitting in the next seat. They were traveling to collect money for a school for escaped Negro slaves at Detroit. Lyell was impressed by the education and fluency of these men and became convinced that if Negroes were

31. Mary Lyell to Eleanor Lyell, 12 November 1845, Kinnordy MSS; Lyell to Wheatland, 16 November 1845, Wheatland MSS.

32. Ibid.

given the opportunity to receive education, they would be capable of exercising their political and civil rights and would be the intellectual equals of Europeans.

The following Sunday, Robert Winthrop escorted Mary to Trinity Church, an Episcopal church, where the service was performed "just as in a London church, with very good music & average preaching."[33] Winthrop at the age of thirty-six had been a widower since his wife's death in 1842. A lawyer and a congressman, he was descended from John Winthrop, first governor of Massachusetts Bay. Having discovered that he was distantly related to Mary, he made every effort to befriend the Lyells.

While at Boston the Lyells, accompanied by William McIlvaine and Henry Rogers, visited the Perkins Institution, or Asylum for the Blind, to see Laura Bridgman, the blind, deaf, and mute girl whom they had visited in 1841. She was now sixteen years old and much grown from when they had last seen her.[34] Dr. Samuel Howe, who had taught Laura to speak by signs and to write, accompanied them during the interview, which Mary described:

> I asked if I might take her hand & he said I might, but she would be frightened if any of the gentlemen she did not know took it. She felt all my rings, noticed the wedding ring & stroked my hand & began teaching me the deaf & dumb alphabet (single handed one) by placing my fingers all the time. Dr. Howe says there is great difficulty in teaching her, in consequence of her affective faculties having so much more exercise than her perceptive ones which have indeed scarcely any. In teaching history the account of wars makes her quite miserable, as she wants to know the reason of everything & why men kill one another.[35]

That evening Lyell attended a Whig political caucus in Faneuil Hall at which Daniel Webster and others spoke. With parties, visits, dinners, social calls, and Lyell's four weekly lectures, Mary wrote, "Time only runs away too fast in this pleasant place."[36]

Francis Calley Gray entertained the Lyells at an elaborate seven-

33. Mary Lyell to Eleanor Lyell, 12 November 1845, Kinnordy MSS.

34. Laura Dewey Bridgman (1829–89) was born at Hanover, New Hampshire. At one-and-a-half years she had scarlet fever, which left her blind and deaf. In October 1837, Dr. Samuel Gridley Howe brought her to the Perkins Institution, where he taught her to speak by signs and to write. When she was twenty-three she went home to Hanover but soon became so homesick for the Perkins Institution that she returned to live there the remainder of her life.

35. Mary Lyell to Eleanor Lyell, 12 November 1845, Kinnordy MSS.

36. Mary Lyell to Leonora Horner, 12 November 1845, Kinnordy MSS.

course dinner, also attended by Prescott, the Ticknors, and Joseph Cogswell,[37] editor of the *New York Review.* At dinner Mary "enjoyed a good chat" with Prescott. After dinner, the ladies and gentlemen rose together to have coffee in the drawing room, and Gray gave each lady a bouquet of camellias. "It was a particularly agreeable party . . . ," wrote Mary, "no one holding forth on anything like [a] dissertation, but every one young & old taking a very equal part." As the autumn days wore on, the Lyells began to feel more and more at home in their room at the Tremont Hotel. "We have a wide slab just before our window, covering a portico," wrote Mary, "& we feed pigeons with biscuit every day. They are so tame they all but eat out of our hands & I expect I shall teach them to do so before I leave."[38]

She had grown fond of Boston. One day George Ticknor took her for a walk to see the market, the new Customs House, and the commercial part of the city. Robert Winthrop invited the Lyells to dinner and showed them old letters of his ancestors from Groton in Suffolk. In the sixteenth century, Roger Winthrop had received a grant of land from King Henry VIII following the breakup of the monasteries. "There was a china dish on the table which came over in the Lady Arabella, the ship that brought out Governor Winthrop in 1630," wrote Mary, "& Mr. Winthrop's youngest child was christened out of this dish." Old portraits of members of the Winthrop family hung on the walls. Robert Winthrop also showed them a cane that his great uncle, the astronomer and mathematician Professor John Winthrop of Harvard, had presented to General George Washington during the American Revolution. After Washington's death, his family had returned the cane to the Winthrops, knowing that they would value it for its association with Washington.[39]

Lyell's lectures were very successful. His audiences held up to the end, even though he often had to compete with excited political meetings on the Oregon question held on the same nights. On 27 November, a day of steady, pelting rain, they had Thanksgiving dinner with the Ticknors, and in the evening a large family party gathered, about sixty people of all ages. "We had hunt the slipper, blind man's bluff & various dances & a supper about nine," wrote Mary.

37. Joseph Green Cogswell (1786–1871), teacher and librarian, edited and published the *New York Review* from 1839 to 1842 and during this period became a friend and adviser to John Jacob Astor. After Astor's death in 1848, Cogswell was appointed superintendent of the Astor Library, which opened in 1854.

38. Mary Lyell to Charles and Frances Bunbury, 13–14 November 1845, Kinnordy MSS.

39. Mary Lyell to Joanna Horner, 24 November 1845, Kinnordy MSS.

"Charles & I won golden opinions by our agility in running & a caper of Charles' produced a great effect. It was a very pretty sight, such a number of young cousins, near the same age & many of them very pretty."[40] George and Anna Ticknor took great trouble, she observed, to entertain the younger children.

The next evening Lyell delivered the final lecture of the series, and there were three rounds of applause after he left the platform. At the end he felt "like a bird out of a cage." On Saturday the Lyells went for a quiet dinner with the Prescotts in their new house. They also had a farewell dinner with the Ticknors the evening before they left Boston. Mary wrote, "I am half broken hearted at the thoughts of leaving Boston, were it not for the thoughts of the return in May, when we have promised to spend our time at the Ticknors' house."[41]

40. Mary Lyell to Eleanor Lyell, 25–30 November 1845, Kinnordy MSS.
41. Ibid.

CHAPTER 6

Exploring the Deep South

IT SEEMED very cold (23°F) when the train pulled out of Boston at four o'clock in the afternoon on 3 December 1845. Snow soon began to fall. Lyell read a newspaper containing President Polk's speech on the Oregon question. In 1818, the United States and Great Britain had signed a convention establishing their joint claim over the Oregon country on the Pacific coast of North America between 42° and 54°40′ north latitude. In 1844, James Polk had been elected president on a platform demanding sole United States control of the territory and threatening war to obtain it. Engrossed in the newspaper and absentmindedly standing too near the stove in the middle of the car to warm his feet, Lyell scorched his clothes and burnt a hole in his overcoat. At eight o'clock they reached Springfield and stayed overnight in a cold bedroom at the Massasoit House. The next morning they boarded the train for Hartford, Connecticut. The rails were coated with ice so that the locomotive frequently lost traction and the train proceeded slowly. At last they reached Hartford, only to find that the connecting train to New Haven was also delayed; they waited six hours for it. Finally after dark, they arrived at New Haven and went to Professor Benjamin Silliman's hospitable home.

That evening, Silliman and his son gave Lyell advice for his tour in the South and a number of letters of introduction. Everywhere Lyell traveled in America he encountered former Yale College students who had developed a love of science and natural history from listening to Silliman's lectures. The Sillimans warned the Lyells of the hazards of travel on the Mississippi River. Many steamboats blew up, were snagged, wrecked, ran aground on sandbars, or burned, but they added that such accidents usually happened only to the cheap boats.[1]

On 5 December the Lyells boarded the steamboat for New York. The trees on the great rock precipices round New Haven, coated with ice from the previous day's sleet storm, gleamed in the sunshine. "I never saw such brilliant icicles," wrote Lyell. "It was as if every rail,

1. Lyell, Notebook 126, 43, Kinnordy MSS.

bough & twig was hung with pendants of cut glass reflecting prismatic colours in the bright sun—the willows, not clear of leaves, bending very low. Even the rigging of the vessels in the harbour hung with the same icicles."[2] After a rough trip of nine hours on Long Island Sound, they reached New York City and went to the Carlton Hotel.

The growth of New York and its suburbs since 1842 was striking; the total population was now 440 thousand. New York City had acquired a splendid new waterworks, with water brought some forty miles by an aqueduct from Croton. New fountains threw up columns of water thirty feet into the air in front of City Hall in Hudson Square, a beautiful sight in the bright winter sunlight. People could pipe water even into the attics of their houses, and the fire insurance rates were reduced. The city had also become noticeably healthier, with less intestinal disease, especially among children. At opposite ends of Broadway were two new Episcopal churches, Trinity and Grace. Trinity Church, built of brown sandstone with a tower 289 feet high, resembled old Gothic churches of Lincolnshire or Northamptonshire. Also conspicuous on Broadway were wooden poles carrying the wires of the new electric telegraph, invented in 1837 by Samuel Morse. Lines of cab horses stood covered with red cloths to protect them from the winter cold; some of the horses pulling private carriages were covered with handsome blankets. In the midst of all this prosperity, pigs still wandered in the streets. To Lyell they were a symbol of American democracy—he was told that were it not for the Irish voters, the city authorities would promptly get rid of them.

President Polk's recent speech on the Oregon question had created a war panic, although Lyell thought Americans took the threat of war less seriously than did the British. In the western states there was, he wrote, "a real war party, a set of reckless, restless democrats who wish for a war for the sake of war, but it must be against England which they envy & think she treats them with disrespect." President Polk appealed to such people, but they were nonetheless a minority in the United States. "They talk of the necessity of taking possession of Mexico bye and bye," added Lyell, "and California . . . very much in the tone that we adopt in regard to the Punjab, and I suppose they are genuine chips of the old block."[3]

While in New York, Lyell called on his American publisher, Wiley and Putnam. Mr. Wiley reported that the *Travels in America* was selling well. Ordinarily, the demand for a book of travels lasted only a

2. Ibid., 44.

3. Lyell to Sophia Lyell, 11 December 1845, Kinnordy MSS.

few months, but the *Travels* would continue to sell because of the discussion of American geology. The American editions of the *Principles* and the *Elements* were also selling steadily.

In preparation for his southern tour, Lyell bought a set of Alabama fossils. He also called on Albert Koch to learn from him as much as he could about the geology of Alabama, although Lyell had no illusions about the quality of Koch's scientific work. Koch gave him the names of several men living near Clarkesville who knew sites where fossil bones had been found.

Koch had come to America in his early twenties from Saxony and settled at St. Louis, Missouri, a city with a large German community. By 1836 he had opened the St. Louis Museum, in which he displayed wax likenesses of famous people as well as a variety of stuffed birds, a live grizzly bear, and several alligators. Ever the showman, he was constantly on the lookout for strange and remarkable objects with which to attract the public; he was especially eager to secure fossil bones. In 1840 he obtained mastodon bones and reassembled them as a gigantic skeleton thirty-two feet long and fifteen feet high, naming it the *Missourium* or *Missouri leviathan.*[4] The *Missourium* had been such an attraction that Koch sold his museum and took the skeleton on tour, first in the United States and then to London, where it was exhibited at the Egyptian Hall in Piccadilly. The British Museum purchased the skeleton, and Richard Owen reconstructed it for the museum as *Mastodon americanus.*[5]

While in London, Koch had become acquainted with Lyell, who had shown him the fossil walrus head that he had found in 1842 on Martha's Vineyard.[6] Having sold the *Missourium,* Koch returned to America in 1844 and traveled about the country looking for fossils, particularly large fossil skeletons. In February 1845 in Clarke County, Alabama, he saw an enormous fossil vertebra being used as an andiron in a farmer's fireplace. Upon learning that such fossils were found frequently in the fields, he intensified his search. On the plantation of Judge Creagh at Clarkesville,[7] he learned of

4. Koch, "Remains of the Mastodon of Missouri"; Koch, *Description of the Missourium.*

5. Ernst A. Stadler, introduction to Koch, *Journey,* xvii–xxv. The Mastodon is still on display at the British Museum of Natural History.

6. Ibid., xxi; cf. Lyell, *Travels,* 1:204–5.

7. John G. Creagh (1787–1843?), a native of South Carolina, was an early settler in Alabama and established a plantation at Clarkesville about 1813. The date of Creagh's death is given as 1839 by T. M. Owen, *History of Alabama,* 3:422, but Creagh was still living in 1842 when Samuel B. Buckley visited him and dug up the skeleton of a *Zeuglodon* on Creagh's plantation. See Buckley, "On the Zeuglodon Remains of Alabama." The history of *Zeuglodon* discoveries is given in Kellogg, *Review of the Archaeoceti,* 1–10.

a fossil skeleton more than 90 feet long at the Washington Old Courthouse, across the Tombigbee River. Koch excavated it and shipped his assembled finds to New York City.[8] During the summer of 1845 he set about assembling the bones he had obtained from various sites into an immense skeleton more than 114 feet long. With great fanfare, Koch placed the gigantic skeleton on exhibit at the Apollo Rooms on Broadway in New York City, claiming it was the fossil remains of a gigantic marine reptile or sea serpent. It created a sensation. Koch conferred on Silliman the dubious honor of naming the serpent *Hydrargos sillimani.*[9] Jeffries Wyman examined the skeleton at New York and noted that it included vertebrae of more than one individual in different stages of ossification, some adult, others immature.[10] Koch certainly knew that his skeleton was a composite made up of at least five skeletons, and he may have known that the bones were those of a mammal, not a reptile.

During the 1830s, Dr. Richard Harlan of Philadelphia had described a fossil vertebra found in Louisiana. He had believed it to be from a giant lizard or crocodile and named it *Basilosaurus.*[11] In 1834 Timothy Conrad, also working at Philadelphia, received from Judge Creagh in Alabama some vertebrae and jaw fragments. The bones appeared so similar to Harlan's vertebra that Harlan wrote to Creagh requesting more specimens. Creagh had responded generously. With the additional fossils, Harlan determined that *Basilosaurus* was a very large marine animal that swam with fins or paddles and possessed a relatively long tail. Over the next several years, Judge Creagh sent additional specimens.[12] In 1838 Harlan took some of the Alabama fossil bones to England. At London he showed them to Richard Owen at the Royal College of Surgeons. When Owen made sections of the teeth, he found that they were from a mammal, evidently a gigantic extinct whale, which he named *Zeuglodon cetoides.* Owen announced this discovery to the Geological Society of London in January 1839.[13]

In 1845, Koch either was blithely unaware of the earlier work of Harlan and Owen or chose to ignore it. The communications between Harlan and Creagh, and Koch's excavations at the Creagh plantation, make it unlikely that Koch was unaware of the fossil

8. Koch, *Journey,* 99–100.
9. Koch, *Hydrargos sillimani.*
10. Wyman, "Fossil Skeleton Recently Exhibited in New York."
11. Harlan, "Fossil Bones Found in . . . Louisiana."
12. Harlan, "Remains of the 'Basilosaurus,'" 357.
13. R. Owen, "Basilosaurus of Dr. Harlan"; Harlan, "Remains of the 'Basilosaurus.'"

bones sent earlier to Harlan at Philadelphia. While Lyell was giving the Lowell lectures in October, Koch arrived at Boston to exhibit the supposed skeleton of *Hydrargos sillimani.*

At New York, Lyell went with Joseph Cogswell to call on the eighty-four-year-old Albert Gallatin, former secretary of the treasury and diplomat. They discussed the influence of maize on the development of civilizations in Mexico and Peru, the role of slavery in ancient Greece, the Oregon dispute, the age of mastodons, and other topics. He also went to see William Redfield, who mentioned that he had found impressions of raindrops and bird tracks in New Red Sandstone beds in western New Jersey. They proved that the beds originally were laid down in a horizontal position, although they were now steeply inclined.[14]

On 9 December, the Lyells crossed the Hudson River to board a train for Philadelphia. The New Jersey countryside was covered with snow, and in the railroad cuttings enormous icicles hung from the rock faces. The train passed slowly through the middle of Burlington, "a great source of convenience to the natives," Lyell thought, "and of amusement to the passengers." In the early afternoon they reached Philadelphia and went to a hotel on Chestnut Street.[15] That evening they attended a party in their honor at William McIlvaine's house; various of their Philadelphia friends were present, including Timothy Conrad and Samuel George Morton. During the next few days, Conrad and Morton provided Lyell with suggestions for their tour of the South. As at New York, the Philadelphians talked of the annexation of Texas, the Oregon question, and the possibility of war with Great Britain. In addition, they were groaning under the burden of an income tax and other taxes imposed to pay the interest on the state debt.

From Philadelphia the Lyells went on to Washington. There they visited the Smithsonian Institution, where James Dwight Dana was preparing descriptions of the rich collections of corals from the United States exploring expedition to the north Pacific. They also saw the skull of a *Megatherium* sent from Georgia by James Hamilton Couper. Robert Winthrop was also in Washington and took the Lyells to an Episcopal church Sunday morning and to meet various people. Lyell also called on the British ambassador, Richard Pakenham.[16]

14. Lyell, Notebook 126, 74–77, 85, Kinnordy MSS.

15. Lyell, *Second Visit,* 1:252.

16. Sir Richard Pakenham (1797–1868) was British ambassador to the United States from 1843 to 1847.

On Monday, Winthrop took the Lyells first to the Supreme Court to hear Daniel Webster plead a case and then to Capitol Hill, where they admired the view of the city and the river. In the House of Representatives, they heard various petitions opposing the annexation of Texas. A Mr. Rockwell of Connecticut spoke against the provision of the Texas constitution that prevented its legislature from taking any action to abolish slavery. Immediately afterward, in the Senate, they heard General Lewis Cass, the senator from Michigan,[17] extol the annexation of Texas as the great triumph of freedom. Mary wrote:

> The subject there was the preparation for arming required in the present aspect of affairs, & he made an eloquent, but wicked speech on the war side. I was much inclined to send some missile at his head from the gallery. I fancy he was speaking to Michigan which is his state & out of the way of danger in case of war & also he has a strong anti-British feeling, from his residence in France & has been irritated by some personal remarks of Lord Brougham on himself.

Senator Cass's anti-British sentiments went back to his experience in the War of 1812. After he had finished, other senators spoke, quoting from the speeches of Sir Robert Peel and Lord John Russell on the Oregon question to show that British ministers were not opposed to a peaceful settlement about Oregon.

Robert Winthrop also took the Lyells to the White House. They had a pleasant visit with Mrs. Polk, "a ladylike pleasing woman, the best Mrs. President they have had for a long time."[18] Mrs. Polk assured them that there would be no war.

Virginia

That evening, the Lyells boarded a steamboat to go down the Potomac. The next morning, they landed at Acquia Creek, where they took a train to Fredericksburg—"a grand improvement," Mary noted, "on the nine miles of stage road which we travelled twice before & which Dickens has celebrated."[19] From Fredericksburg, they

17. Lewis Cass (1782–1866) served as a colonel in the United States Army in the War of 1812. In 1813 he was promoted to major general and later was appointed governor of Michigan Territory. In 1836 Cass was appointed United States minister to France and in this position opposed a British attempt to obtain French agreement to the right of search of ships at sea.

18. Mary Lyell to Leonard Horner, 17 December 1845, Kinnordy MSS.

19. Ibid.

continued by rail to Richmond, where they were met by Augustus Gifford, manager of the coal mines at Blackheath, who took them to his house fifteen miles out of the city, a journey of three hours through the snow.

Gifford was English and a graduate of Oxford University, while Mrs. Gifford was a Virginian. During the next two days their house provided a pleasant haven for the Lyells. The servants were all colored, but free; Gifford did not wish to own a slave. A gray-haired black man named Remsen was, at the age of eighty, still an efficient servant with very pleasant manners. Mary wrote of him:

> He is really a character much to be respected, for when young he bought his own freedom, & then sold himself again to buy his wife; again redeemed himself, bought his children free twice and in his old age sold himself a fourth time to buy his grandchildren—He must, however, have had indulgent masters to enable him to put by money.[20]

At Blackheath Lyell examined the coal beds that William Barton Rogers thought were of Oolitic (i.e., Jurassic) age, rather than Carboniferous. In the Clover Hill mine, at depths between seven hundred and nine hundred feet beneath the surface, Lyell collected fossil plants and fossil fish. He found many upright Calamites in the coal, but he thought that the fossil fish would be most useful in determining the age of the beds. Lyell wanted to compare them with those he had collected from the Connecticut Sandstone during his first tour.

On 19 December the Lyells returned to Richmond for the next two days. The Supreme Court of Virginia and the legislature were in session, and Lyell was an interested observer of both. The Richmond people were very hospitable. Although they stayed at a hotel, they were entertained in private homes. At one party, the young people persuaded the Virginia attorney Charles Carter Lee to mimic a conversation between two Negro servants,[21] one the slave of Governor Taswell and the other a grocer's slave, the point being the importance assumed by servants according to their master's social

20. Mary Lyell to Sophia Lyell, 21 December 1845, Kinnordy MSS.

21. Charles Carter Lee (1798–1871) was born at Stratford, Westmoreland County, Virginia, and educated at Harvard College. He entered the practice of law, first at Washington, then in Virginia. He next spent several years in Mississippi before returning to Virginia to resume his practice. He was the son of Henry Lee (1756–1818), better known as "Light-Horse Harry" Lee; his elder brother was Robert E. Lee.

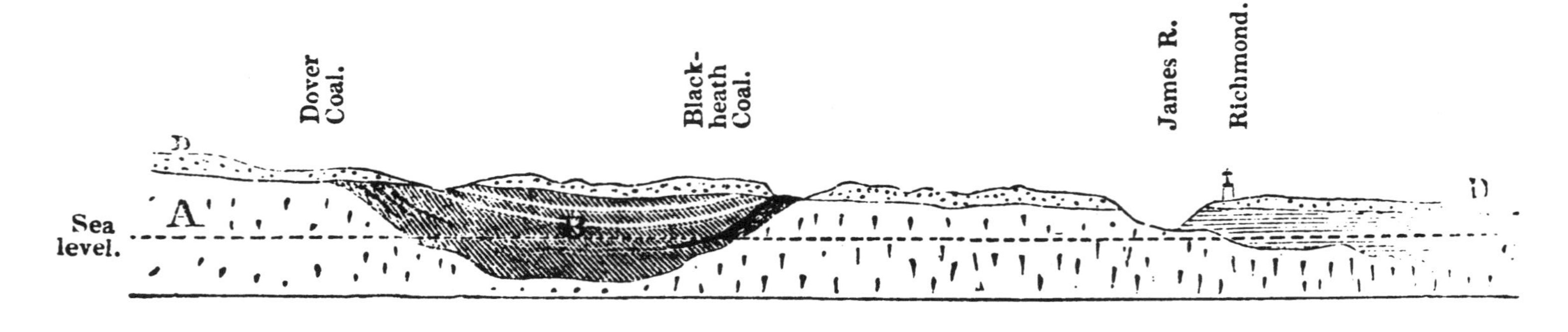

Section Showing the Geological Position of the James River or East Virginia Coal-Field. Lyell's *Manual* (1851), p. 284.

position. Lee told Lyell that Negroes were more truly Virginians than many whites, having been in the state for many generations. They were, he said, "patient, generous, forgiving, grateful, kind, cheerful—religious [and] superstitious."[22] He liked them very much. During the 1820s, Lee had spent several years in the new state of Mississippi, an experience that he drew on for another skit, which Mary described in a letter to her sister-in-law:

> Mr. Lee preached a sermon in a Mississippi mosquito swamp in the character of a New England itinerant preacher, "On the patience of Job," the "gallinippers" tormenting him the whole time & compelling him to slap his hands & cheeks. It was great fun much heightened by the delight of the children—The New England accent is as foreign to the Virginian as to us.[23]

At Richmond, Lyell also met Dr. Jeffries Wyman,[24] the New England physician he considered "the best comparative anatomist in the U.S." Wyman was then professor of anatomy in the Medical College at Richmond.

After attending Sunday morning service at an Episcopal church, the Lyells took the train to Petersburg. They called on Michael Tuomey to see his fossil collection. He was not at home, but his wife showed them some of her husband's fossils. On Monday afternoon, 22 December, they left Petersburg by train. South of town the locomotive struck a cow on the tracks, the first incident on a monotonous and jolting trip. All night and the next day, they traveled across Virginia and North Carolina. The train proceeded over the rough roadbed at no more than twelve or fourteen miles an hour and often, because of ice on the rails, even more slowly. On the single track, they sometimes had to back up for miles to a siding to let another train pass, and frequently they stopped for water and fuel. The cars were uncomfortable but at least kept warm with a stove. In North Carolina they stopped for supper and breakfast. "Being hungry & having a good digestion & not eating a great deal I find the heavy bread & greasy fried pork does not disagree with me at all," wrote Mary, "while I hear other travellers complaining bitterly."[25]

22. Lyell, Notebook 127, 92, Kinnordy MSS.

23. Mary Lyell to Sophia Lyell, 21 December 1845, Kinnordy MSS.

24. Lyell, Notebook 127, 92, Kinnordy MSS. Jeffries Wyman (1814–74) was, from 1843 to 1847, professor of anatomy and physiology in the medical school of Hampden-Sydney College at Richmond, Virginia. In 1847 he became professor of comparative anatomy at Harvard College.

25. Mary Lyell and Lyell to Frances Lyell, 29 December 1845, Kinnordy MSS.

Neither the railroad line nor the stations and inns along the way had improved since they had traveled the same route in 1842, but the line was soon to be replaced by another to pass farther inland, through Raleigh.

When they reached Wilmington, North Carolina, late Tuesday afternoon, the steamboat for Charleston, South Carolina, had already left, but the Lyells were not sorry to have to stop. Obtaining a quiet room at a hotel, they enjoyed a night's rest after their long train journey. They had left the snow behind, although it was still an unusually cold winter in the South, the coldest since 1835. On the outskirts of Wilmington, the live oaks, festooned with Spanish moss, were green, as was the holly, bright with its red berries. The next morning, Lyell collected fossils from the cliffs along the Cape Fear River.

South Carolina

The Lyells spent the day before Christmas on the steamboat to Charleston. At Smithville, at the mouth of the Cape Fear River, Lyell went ashore and walked through the village, admiring the many large live oaks, the hollies, and the red oaks with large leaves. Reaching a seaside marsh, he saw flocks of birds whose call reminded him of the European peewit.[26]

Early on Christmas morning they landed at Charleston, looking pretty with its white houses set against the green of the live oaks, orange trees, and Pride of India trees covered with yellow berries. After settling into the Charleston Hotel, they attended Christmas service at St. Philip's Episcopal Church, "adorned with multitudes of evergreen, & stuck among them large white paper flowers, very good imitations of magnolias, asters & chrysanthemums." Few people were in the church, since many Charlestonians had gone to their plantations for the holiday season. During the service, the congregation was disturbed by Negroes setting off firecrackers in the street outside. The celebration reminded Lyell of fireworks and even the firing of muskets within a church in Sicily at Christmastime in 1828. After the service the Lyells remained to take communion. "A good proportion of coloured communicants attended after the whites," wrote Mary, "some quite black, but one or two as light as Joanna [her sister] & with hardly a trace of the African features. It does seem very hard & cruel that this badge of serfdom should never be effaced."[27]

26. Lyell, *Second Visit,* 1:292.

27. Mary Lyell to Frances Lyell, 29 December 1845, Kinnordy MSS.

On returning to their hotel room, they found cake and tumblers of eggnog, courtesy of the landlord.

That evening the Lyells called on Dr. John Bachman, pastor of St. John's Lutheran Church. Although a minister by profession, Bachman, then fifty-five, had been interested in natural history since his boyhood in Duchess County, New York. In October 1831, Jean Jacques Audubon had spent a month at Bachman's house in Charleston, a visit that had marked the beginning of years of collaboration with Bachman. At the time of the Lyells' visit, Bachman had just finished the first volume of their great joint work on the mammals of North America.[28] He showed the Lyells two of his series of colored plates, telling them anecdotes about each. Lyell was struck by the large number of North American species of squirrels, including three species of flying squirrels. Two days later, Lyell again called on Bachman, and they had a lively discussion about the geographical distribution of North American mammals and birds. In their new work, Audubon and Bachman described two hundred species of North American mammals, of which only seventy-six had been described before. Bachman said that among ten genera of birds and ten of mammals peculiar to North America, each was represented by distinct species on opposite sides of the Rocky Mountains. The distribution was consistent with the doctrine of specific centers, that is, that each species arose at only one spot on the earth. "But," Lyell wrote later, "the limitation of peculiar generic types in certain geographical areas, now observed in so many parts of the globe, points to some other and higher law governing the creation of species itself, which in the present state of science is inscrutable to us, and may perhaps remain a mystery forever."[29] Lyell seems to have had in mind Darwin's account of Galapagos species in the 1845 edition of his *Journal of Researches*, where Darwin referred to "that mystery of mysteries—the first appearance of new beings on this earth."[30]

Lyell also called on Dr. Lewis R. Gibbes,[31] a naturalist and professor of mathematics in the College of Charleston, who showed

28. Audubon and Bachman, *Quadrupeds of North America.*

29. Lyell, *Second Visit,* 1:304.

30. Darwin, *Journal of Researches,* 378.

31. Lewis R. Gibbes, M.D. (1793–1894), was born at Charleston and educated at South Carolina College. In 1836 he graduated with a medical degree from the Medical College of Charleston and traveled to Paris, where he studied at the École de Médecine and the Jardin des Plantes. In 1838 he was appointed professor of mathematics in the College of Charleston. LaBorde, *History of the South Carolina College,* 196–200.

him his collection of South Carolina fossils.[32] From Gibbes, Lyell borrowed a copy of Edmund Ruffin's agricultural survey of South Carolina, published in 1843, in which Ruffin described some of the geological features of the state.[33] On a second visit with Gibbes, Lyell went over his collection that included Eocene shells from various localities. Gibbes gave him shells of twenty-three species.[34]

One evening the Lyells accompanied Charles Shepard and his wife, who were also staying at the Charleston Hotel, to the house of Mitchell King, the eminent Charleston lawyer.[35] Among the guests that evening was James Louis Petigru,[36] also a lawyer, and famous for his wit as well as his knowledge of the law. The Kings served their guests tea and refreshments in comfortable, book-lined rooms; the chief topic of conversation was the threat of war with Britain over the Oregon boundary, which those present opposed. Petigru told Lyell, "We have a set of demagogues in the country who trade in the article called 'hatred to England' as so much political capital, just as a Southern merchant trades in cotton, or a Canadian one in lumber."[37] The chief support of such American demagogues was, however, the British press, which was tireless in its misrepresentation of everything American.

Lyell recorded some of his impressions in his notebook. In contrast to the rapid development of Boston and New York, the Lyells found Charleston little changed since 1842. There was no influx of immigrants; none of the ships in Charleston harbor belonged to Charleston people. Slavery was an obstacle to progress in South Carolina. An even more serious factor was the absence during six months of the year of many rich people, who went north to escape the heat and the malarial fevers of the Low Country. Charleston itself was healthy, with a mortality rate little higher than that of Boston, but from May to October its weather was oppressively hot.[38]

32. Lyell to Gibbes, 26 December 1845, Gibbes MSS.

33. Ruffin, *Agricultural Survey of South Carolina.*

34. Lyell, Notebook 128, 19, Kinnordy MSS.

35. Mitchell King (1783–1862) was born at Crail, Scotland. After various adventures, he came to Charleston in 1805.

36. James Louis Petigru (1789–1863) was elected attorney general of South Carolina in 1822 but resigned in 1830 to oppose nullification. In 1860 he opposed Secession, and although he was a Unionist he was so highly respected that he lived undisturbed in Charleston through the Civil War. Carson, *Petigru.*

37. Lyell, *Second Visit,* 1:298–99.

38. Lyell, Notebook 128, 34, Kinnordy MSS.

Georgia

In the bright sunshine of the morning of 28 December, the Lyells sailed from Charleston in their old friend, the steamboat *General Clinch*. As they steamed out of the harbor, they admired the beauty of the islands. Looking back, they were surprised to see a cloud of smoke hanging over the city, caused by the burning of coal during the winter. After passing Edisto Point, the steamboat ran aground at the entrance to St. Helena's Sound, and they waited for several hours for the incoming tide to float them off. They then steamed westward along the Beaufort River. The islands to seaward and the mainland were low and sandy, covered with live oaks, palms, and palmettoes, or fan palms. Occasionally, they passed cotton plantations and great salt marshes. In the clear, calm water of the channel, many wild ducks were diving, and on the beaches low tide revealed beds of oysters. The timbers of the wharf at Beaufort were covered with oysters, just as in England wharves were covered with barnacles. Negroes in their Sunday clothes stood on the wharf to watch the steamboat come in. Beaufort was a pretty village and, in the summer, home to many planters who wanted to escape the fevers on their plantations. Each house was surrounded by live oaks and orange trees laden with oranges. South of Beaufort the steamboat passed an old Spanish fort. As the evening drew on, they enjoyed the brilliant colors of the sunset, and at eleven o'clock that night they reached Savannah. At the Pulaski House hotel, they found a welcoming letter from James Hamilton Couper, who had invited them to visit Hopeton, his plantation on the Altamaha River.

At the hotel they were served bananas and pineapples for dessert, reflecting the closeness of the West Indies, and rice cakes at breakfast. The rice crop that year had been abundant.[39] New buildings were going up in the city and camellias were in flower in the gardens. Lyell paid calls on various Savannah naturalists. One morning he went with Dr. John LeConte, Captain Barton Alexander, and William Hodgson on an excursion to Skiddaway Island.[40] LeConte

39. Mary Lyell to Frances Lyell, 29 December 1845, Kinnordy MSS.

40. John LeConte, M.D. (1818–91), in later life professor of physics at the University of California, was born in Liberty County, Georgia, of a Huguenot family. He was educated at Franklin College, later the University of Georgia, and studied medicine at the College of Physicians and Surgeons, New York City, from which he graduated with a medical degree in 1841. From 1841 to 1846 he practiced medicine at Savannah. Barton Stone Alexander (1819–78), a native of Nicholas County, Kentucky, entered West Point in 1838 and in 1842 was commissioned in the United States Army Corps of Engineers. From 1842 until the Civil War, he was occupied in the construction of coastal

was a young Savannah physician deeply interested in science, Alexander a tall young Kentuckian attached to the Army Corps of Engineers, and Hodgson a leading Savannah businessman. In 1842, shortly after Lyell's previous visit to Savannah, Hodgson had discovered on Skiddaway Island a remarkable group of fossil bones. In a low cliff at the northwest end of the island, he had dug out from a bed of dark peaty mud three skeletons of *Megatherium,* together with the bones of *Mylodon, Elephas primigenius, Mastodon giganteus,* and a species of fossil ox.[41] Below the peaty mud was a bed of sand containing many fossil seashells belonging to species still living along the coast.

They rode to Skiddaway Island through a forest of tall yellow pines with palmettoes growing beneath them. On the island itself, the vegetation was remarkably tropical. Several species of fan palm grew under the tall cabbage palm, a species that Lyell had not seen before. This tree grew slowly, reaching a height of forty feet. The horses started up many birds. LeConte identified for Lyell ten species, including the Virginian partridge, crow, marsh hawk, white heron, bald-headed eagle, summer duck, and meadowlark. Alexander told Lyell of having seen in 1844 an eagle with its foot caught by an oyster at low tide. The eagle would surely have been drowned by the incoming tide had Alexander not released it from the oyster's grip. LeConte and Alexander also described some of the habits of the alligators that were common in the swamps near the mouth of the Savannah River. LeConte had recently carried out a number of experiments on a decapitated alligator, showing that the animal's headless body remained motionless when undisturbed, but if it were pricked or pinched on the sides, a hind leg or foreleg would scratch the spot.[42] Lyell was strongly impressed by such examples of actions, now known as reflex actions.[43]

On Skiddaway Island Lyell collected twenty-one species of fossil shells from the bed of sand below the peaty bed containing the fossil bones. The shells, all belonging to living species, showed that the fossil bones in the overlying bed were geologically very recent, somewhat more recent than the fossil bones of elephant and rhinoceros found in the Thames Valley in England. They were con-

fortifications and various government buildings. During the Civil War he helped to construct the defenses of Washington. William Brown Hodgson (1800–71) was a Savannah businessman.

41. Lyell, *Second Visit,* 1:313–14; cf. Hodgson, *Megatherium.*

42. LeConte, "Experiments."

43. Lyell, *Second Visit,* 1:317–19.

temporary with the fossil elephant *Elephas primigenius* (the mammoth) of Europe and the mastodon of North America. "The modern or Post Pliocene series of changes before & after the Megatherium in this low country are very interesting," Lyell wrote to his father-in-law, Leonard Horner. "Tell Darwin I have quite a counterpart to his Patagonian steps, or successive cliffs cut out of the tertiary."[44] Lyell's interest in the fossil bones of Skiddaway Island stimulated Hodgson to publish a full account of them with an anatomical description of the bones by Dr. John Habersham of Savannah and an account of the geology of the Georgia coast by James Hamilton Couper.[45] Like Darwin's fossil animals from the plains of Patagonia, the fossil animals of Skiddaway Island were much larger than their living representatives or related species. Both groups of fossil animals were associated with fossil seashells of living species. Thus in both North and South America, species of large mammals had recently become extinct.

On 31 December in warm, bright sunshine, the Lyells boarded a steamboat to go 125 miles farther south to Darien. "We had a beautiful inland passage from Savannah to Darien," wrote Mary, "among narrow channels between the islands & main land, saw the palm tree growing on the islands, which looked very beautiful & tropical."[46] On the boat Lyell had lively discussions with some planters returning home and several businessmen from the North on the Oregon question, the possibility of war with Britain, and the subject of slavery. The planters were worried about the effect that a war with Britain would have on the cotton trade. They had also read accounts of speeches made at a recent antislavery meeting in London. Such speeches made the planters indignant, because they had also read reports of the poverty, ignorance, and social stagnation of English agricultural laborers and the sufferings of women and children working in English coal mines.

After dark that night the Lyells landed at Darien. Five Negroes offered to carry their luggage the short distance to the inn. Since they had only four pieces of luggage, the fifth Negro led the procession, having first assumed an air of command over the party and informed Lyell, "If you not ready, I will hesitate for half an hour." It was a mild night with a clear sky as they walked under great live oaks hanging with Spanish moss. At the inn, Lyell paid the four porters.

44. Lyell to Leonard Horner, 9 January 1846, Kinnordy MSS.

45. Hodgson, *Megatherium*.

46. Mary Lyell to Frances Bunbury, 4 January 1846, Kinnordy MSS.

The man who had led the way laughingly requested payment for his services as "pilot," and Lyell, amused, paid him.[47]

On the morning of New Year's Day 1846, the Lyells were standing on the bank of the Altamaha River as James Hamilton Couper came down from Hopeton to meet them in a long dugout canoe rowed by six slaves singing heartily. After the Lyells had read a bundle of letters from England that Couper had brought them, they embarked with him in the canoe. The banks of the Altamaha were lined with canes and cypress trees, leafless during the winter season. Occasional laurel, myrtle, or magnolia trees created patches of green. In some places they could see tall pines rising beyond the marshes. Fifteen miles upriver, they entered a canal that took them through low rice fields to Hopeton. As the canoe passed along the canal, great flights of red-winged blackbirds rose up from the rice fields.

On landing they walked up to Hopeton plantation house, on higher ground overlooking the rice fields and the river. Hopeton was one of the great rice plantations of Georgia, and Couper a leading southern agriculturalist. He was another of Silliman's many former students. After attending Silliman's classes and graduating from Yale College in 1814, Couper had traveled to Holland to study Dutch techniques of canal building and water control. On his return to Georgia, he became manager of Hopeton plantation, which belonged jointly to his father, John Couper, and Couper's partner, James Hamilton.[48] At that time, only two hundred acres of swamp had been cleared and diked, but Couper cleared and diked additional land until he had eighteen hundred acres under cultivation.

After first trying to grow long staple cotton and finding that the soil of the drained swamps was too rich for cotton to mature properly, Couper began growing sugar cane. In 1829, with eight hundred acres in sugar cane, he built a sugar mill driven by a Boulton and Watt steam engine imported from England. As more economical sugar plantations developed in Louisiana and Mississippi, Couper changed gradually from sugar cane to rice culture. In 1832 he had 500 acres in rice, 330 acres in sugar cane, and 170 acres in cotton. He also grew corn, sweet potatoes, and cowpeas to feed his people,

47. Ibid.

48. John Couper (1759–1850) was born at Lochwinnoch, Renfrewshire, Scotland, and in 1775 came to St. Augustine, Florida, in company with his friend James Hamilton. In 1780, in partnership with Hamilton, Couper opened a store at Savannah, Georgia. Couper and Hamilton together bought two thousand acres on the Altamaha River and began the process of clearing swamp and upland to form Hopeton plantation. See Wylly, *Annals . . . of Glynn County, Georgia,* 13, 15.

rotating the crops from field to field.[49] At his father's plantation on St. Simon's Island, Couper experimented with the growing of olive trees. He pioneered in the extraction of oil from cottonseed, establishing extraction mills at Natchez, Mississippi, and Mobile, Alabama. Couper was not only a scientific planter, but an industrial entrepreneur on a considerable scale. In addition, he was deeply interested in geology and natural history and managed his time systematically in order to have leisure for reading and scientific studies.

The Lyells found Hopeton a thoroughly sympathetic household, partly because the Coupers were Scottish. Mrs. Couper was the daughter of Captain Alexander Wylly, an officer in the British Army during the American Revolution and later a planter on St. Simon's Island. Both she and Mary were related to a Mrs. Wardrobe, a Scotswoman and former resident of Savannah. The day they arrived, Mrs. Couper asked Mary whether she was possibly the Mary Horner of whom her Scots cousins, the Baillies, had talked so often when describing their games and charades. The household included the Coupers' eight children, their tutor from the North, and John Couper, then eighty-seven. After seventy years in Georgia, old Mr. Couper still spoke with a broad Scotch accent. "He comes from Lochwinnoch [Renfrewshire], where his Father was Minister," wrote Mary, "& one of his brothers was a tutor in MacDonald of Castle Semple's family, & another Professor of Astronomy at Glasgow. He was so delighted to find I have been at Lochwinnoch & Barr, but sometimes he forgets my memory does not go back as far as his."[50]

Each day Lyell went with Couper, either in a carriage or on horseback, to places where fossil bones had been found. Meanwhile, Mrs. Couper took Mary to see the rice mill and the cotton gin and to visit the hospital for the slaves. Malaria was endemic in the Low Country, and thus there were always some slaves in the hospital, especially during the winter months.

The Coupers lived simply, without show, but their house was richly provided with books and prints. Mary described life at Hopeton to her sister Frances:

> In short they are particularly sensible people who manage to overcome many of the difficulties of their situation. It is a curious & interesting life for me to see, so different from any other.—This

49. On Couper's crops and acreage at this time, see Philips, *Life and Labor in the Old South,* 264.
50. Mary Lyell to Frances Bunbury, 4 January 1846, Kinnordy MSS.

> is considered the best managed plantation in the whole southern country so of course I am seeing it most favorably. This is the flattest country possible on the borders of a swamp, on a branch of the Altamaha river, with abundance of live oaks, magnolias & bays of various kinds. The birds are quite beautiful, of the most brilliant plumage & they have bears, alligators etc., though we have not yet seen any.
>
> . . . [The negroes] are very black on this plantation, but some have good countenances. They look very cheerful, & some of them hold out their hands to me to shake. At this moment I see a grand game of romps going on among the children, rolling on the grass & screaming with joy. They have no work to do hardly till they are twelve years old, except to help look after the babies.[51]

Couper was a liberal-minded, highly educated man with whom the Lyells could talk frankly about the whole troublesome and complicated subject of slavery. Mrs. Couper was equally candid and sympathetic. Couper was willing to discuss every aspect of slavery both on his own plantations and as he had seen it on others in Georgia, Alabama, and Mississippi. There were about five hundred slaves on Hopeton, but a large proportion were children or old people. Lyell found that adult slaves might be required to work eight hours a day or might be given a specific task to complete in a day. When they were assigned taskwork, they usually completed their assignment in five hours and had the rest of the day free. They worked under the direction of black foremen, or drivers. Compared to northern factory workers, who commonly worked twelve hours a day, slaves were far from overworked. While the mothers were at work, older women looked after the young children.

Although Hopeton was a particularly well-run plantation, it was at the same time representative of Georgia plantations. The slaves lived in family units, each family in its own cottage. The cottages, thought Lyell, compared well in comfort with those of farm laborers in Scotland. The Negroes were better fed and better dressed than most of the working classes in Britain and had more leisure. In their leisure time, some of the Negroes made dugout canoes from cypress trees and sold them for four dollars each. On the darker side, an evil of slavery arose from sexual relationships between white men and black women. The children born of such unions were a

51. Ibid.

source of great unhappiness to the wives and mothers of the men involved. "The planters do not like the white mixtures," noted Lyell, "by association less agreable [*sic*] to them than the blacks—the [slave] women boast openly of their mulatto children; to them it is an honor; they claim support."[52]

On one occasion Lyell was struck by the sense of humor of one of the slaves on Hopeton who, observing a great blue heron with its stately walk, remarked, "Impudent bird! Mocking my Masser!"—an incident that hints at the dignified walk and manner of Couper and the affection and respect that his slaves had for him.[53]

Lyell was also impressed by their peaceable character. Among the slaves on Hopeton there had been only about six cases of assault and battery. Petty thievery from each other was common, however, and Lyell noted large wooden padlocks on the doors of the bedrooms in the Negro cabins, intended to protect their belongings.

> I have looked into the statistics of crimes & punishment on this estate, for thirty years, & wish that an average English parish could return as favorable results. Besides being clothed & very well fed, they are by nature most peaceful, so that they hardly ever fight, & the contrast of some Irish labourers who came here to dig a canal would really be laughable, if it were not such a serious evil.[54]

Slave traders were rare in the Georgia Low Country. The planters were particular to whom they sold slaves. It was a disgrace for a slave to be sold to a trader, and usually a trader could buy only the worst characters. Couper told Lyell, "I regard it as a great trial to have the responsibility of owning slaves." He did not allow any of his children to slap a slave, although many southern families permitted such punishments.[55]

Couper told Lyell that the condition of the slaves was worse in Alabama than in Georgia, while in Mississippi they were still worse off than in Alabama. Northern men who came to the South to make money, growing cotton or sugar cane, were hard taskmasters. They worked hard themselves and worked their slaves still harder. Under such conditions the slaves did not have as many children. A Mississippi planter had told Couper that among a hundred slaves on his plantation, he had ninety workers, whereas on Hopeton only half

52. Lyell, Notebook 129, 66, Kinnordy MSS.

53. Ibid., 106.

54. Ibid., 72.

55. Ibid., 10, 14.

the slaves were workers, the rest being children. On Hopeton Couper had noted that whenever the slaves had to work harder than usual as a result of moving to a new settlement, fewer children were born. In Georgia, Couper told Lyell, public opinion opposed the harsh treatment of slaves and a grand jury had even indicted a planter for mistreatment of his slaves. Couper deliberately overlooked the fact that many of his slaves were being taught to read by other slaves, but he could not establish a school for them because Georgia law prohibited the teaching of slaves to read. The law dated from colonial times, but it had fallen into abeyance until the rise of the abolition movement had caused it to be enforced.[56]

Each year on the first of May, the Coupers moved for the summer to St. Simon's Island, by the sea, to avoid the malarial fevers on Hopeton. During the summer, Couper came once a week to Hopeton during the day to check on the plantation, but he always left before nightfall. Mrs. Couper had never seen the rice crop when it was ripe. During the summer, the slaves began work at six o'clock in the morning and stopped work at midday. After a sleep during the afternoon they spent the evening gossiping, singing, and listening to sermons. Although the Negroes were better able than the whites to withstand the fevers of the Low Country rice plantations in summer, Lyell noted, they were only comparatively more healthy. The Negroes in the interior uplands of Georgia were healthier than those in the Low Country. They increased in number more rapidly than those on the rice grounds at Hopeton.[57] The upland country, Lyell thought, would be one of the healthiest countries for white men once it was cleared.

In 1814, during the war with Britain, British naval ships under Admiral Cockburn landed on St. Simon's Island and offered to carry any slaves to freedom in Canada. About half of John Couper's slaves at Cannon's Point went, but the others remained. His black headman, African Tom, a native of Timbuktoo, had first experienced slavery in the British West Indies and was convinced that the British were worse masters than the Americans. He influenced many of the slaves to remain. John Couper had purchased Tom in about 1800 in the Bahama Islands and soon came to respect his intelligence, honesty, and interest in the plantation. A Muslim, Tom owned a copy of the Koran; he read but could not write Arabic. By the time of the Lyells' visit to Hopeton in 1846, Tom was seventy-five and had

56. Ibid., 22–23.
57. Ibid., 78.

served for many years as headman on the plantation. James Hamilton Couper was accustomed to leave him in charge of the plantation, occasionally for months at a time, and found him entirely capable of managing it.[58]

With his Scottish family background, Yale education, European travel, and lifelong experience as a slave owner, Couper had developed an unusual vision of the process of historic change in the South. As settlement spread westward through Georgia and Alabama, accompanied by the growth of new towns and the extension of stagecoach service, steamboats, and now railway lines, Couper saw the white people of the South achieving a more ordered and civilized life. He thought that in the United States as a rule the level of civilization of a state was in proportion to its age. The older, longer-settled states of the eastern seaboard were more civilized than the new western states. Of this theory, Couper wrote to Lyell:

> It is to be qualified by referring to the character of the population from which the emigration takes place, and to the density of the first settlements. If the emigrants go from an old and refined community they carry with them a tone of character, which, although lowered by the circumstances of a new country, is yet much higher than if they came from one of greater ignorance. It may thus happen that a very new country populated from our intelligent states will exhibit higher attributes than an older one settled from a rude one. Intelligent persons moving and settling together in large masses counteracts in a great measure the tendency to a retrocession in civilization. In looking into the subject you must keep in view the character of the communities from which the emigrants come, as well as the age of the settlements.

As an example, Couper asked Lyell to compare the new states of Ohio, Indiana, and Illinois with such older states as Massachusetts and Connecticut. In the South, Couper saw the Negroes advancing rapidly in civilization, even though in slavery. A quarter of the slaves on Hopeton had been born in Africa in primitive societies, yet they had learned to become carpenters and bricklayers, and to operate a steam engine. Many had converted to Christianity and were developing a rich religious life. As both whites and blacks rose to higher levels of civilization, Couper foresaw that the barriers between the races would melt away.

58. Wightman and Cate, *Coastal Georgia,* 153.

> Amalgamation is an idea abhorrent to the whites at present, because it is forbidden by every association. When time shall have brought the races nearer to an equality in character and intellect, then associations which are founded on realities now, will cease, and with them the prejudice against a mixture of colour. This must be the work of time. The human mind cannot be wrenched suddenly into views and tastes, merely to meet the wishes of the purest philanthropy, when the realities of a degraded character are opposed to it. The first step is to elevate character, and in proportion as this is done prejudice will disappear.[59]

At Hopeton, Couper strove to act in conformity with his vision of the future of the white and black races. Lyell was deeply impressed by both his aims and accomplishments. Near the end of his visit, Lyell wrote:

> One fourth of the 500 negroes now collected round this house were born in Africa, & to bring them up or elevate them to the grade of the lowest Irish is a step far beyond training the said Irish into the average American labourer's standard. When I see how much has been done in so short a time, I begin to be more hopeful than in my last tour, for unless the fanatical party of the North force on a collision, the next generation will be just as much beyond the blacks we see, as they are above the Africans & the treatment of them is necessarily regulated by the position & intellectual & moral condition to which they have attained. Doubtless the newer states are much worse; they are what this state has been & the annexation of Texas, which is the work of the Abolitionists, who would not support Clay, increases the mischief & retards the hour of emancipation.[60]

When the slaves had been set free in the British West Indies in 1835, the British Parliament had compensated the slave owners for the loss of their property. In the United States, the abolitionists opposed any compensation to slave owners for the loss of their slaves. There were still other implications to emancipation. Lyell wrote:

> If [emancipation were] sudden & before the Negroes have been gradually improved, it will lead to extermination as in the case of the Indians. If I were a Negro I sh[oul]d rejoice in the Annexat[io]n [of Texas] for the spreading of the race into Mexico may

59. Couper to Lyell, 15 February 1846, Kinnordy MSS.

60. Lyell to Leonard Horner, 9 January 1846, Kinnordy MSS.

> save it. As a philanthropist interested for the Negroes I sh[oul]d wish it. As a lover of civiliz[atio]n, I sh[oul]d lament it. It defers the period of the perfection of the U.S.; it requires ages more to incorporate so many millions of blacks as must now be mixed up with the Europeans. Had the Northerners not been checked, they might have caused by a speedy emancipat[io]n, the extinction of the Negroes. They w[oul]d have suffered, but perhaps the U.S. been sooner civilized.[61]

In Georgia the Negroes seemed to Lyell to be advancing rapidly in their level of education and civilization as a result of their close association with white people in the master-slave relationship. Couper told Lyell that on St. Simon's Island the Negroes were noticeably more advanced on the smaller estates, where they had greater contact with their white owners. Lyell observed that the speech of the household slaves was better than that of the slaves who worked in the fields.

> Education
>
> . . . does not consist in reading & writing. Let 100 black men from Africa work with & converse with white as in upper Georgia; the children run in & out of the house playing on terms of great equality with white children, often petted by the whites. Let them as mechanics learn to be blacksmiths, carpenters, masons, work in the field, kitchen, nursery & domestic service & with the kind of quality wh[ich] brings the master & servant much nearer to each other in the families of the middle classes & what state of things can be more favourable for education, what schools, or preaching, or catechizing or books could in 2 or 3 genera[tion]s raise so many millions as fast as such a system. What Government is rich or influential enough to procure such a body of white men to improve as many Africans in as short a time.[62]

When Negroes became free, they were immediately subject to racial discrimination, especially by exclusion from skilled trades. In Georgia, the state legislature had recently passed a law prohibiting free Negroes from entering into contracts for building and other work. Lyell thought that this unfair and retrograde law could not "retard the great onward progress of emancipation."[63] The law was

61. Lyell, Notebook 129, 75, Kinnordy MSS.
62. Ibid., 77.
63. Ibid., 79.

itself evidence that the Negroes were getting on and proving their skill as carpenters, masons, and mechanics.

From what he saw in Georgia, Lyell was inclined to be optimistic about the future of blacks and whites in the South. He wrote:

> Negroes
>
> Unless the affair is precipitated, unless by war or insurrection, by a collision forced on by interference, a sudden end is put to the black man & they suffer the fate of the Indian, they will amalgamate—The black will rise & the whites will be divested of a great part of their present prejudice. The approximation of both will lessen the distance, the one rising in intelligence, the other in associat[ion]s favourable to the Negro. . . .
>
> It is not the present races, the ignorant, half African foulah, superstitious or Mahometan slaves & the excited, speculating northern adventurers, it is not these two extremes that may amalgamate, but an elevated black, an unprejudiced white.
>
> Why should colour militate as a permanent bar between the races? In the case of the Indian, Randolph boasted the dash of Indian blood & other Virginians are proud of it. We do not feel to the Hindoos, on account of mere colour, an insurmountable aversion—The mixture of white blood, the advance in intellect of the black will go hand in hand with the moral feeling of the white man.
>
> In the upper part of Ga. & S.Ca. the Negro goes to the same meeting house, worships with the white, his cottage is nearly as good & for such a climate, comfortable, he ploughs with him, has as much meat. . . . We must not take a Millerite view. The northern Abolitionists are as impatient as if the world was coming to an end—Time is essential—Did any 3 millions of men ever make a greater progress in one generation than the blacks of 1800 as compared to those of 1845?

The agitation of the abolitionists was trying to many southerners. Lyell thought that they required "not merely judgement, but temper, in bearing with the insults of the northern men."[64]

Fascinated by the religious life of the Negroes, Lyell noted the similarity between their "brotherly dance," in which a group joined hands and danced in a circle, alternating right hand and left, and the dances of the Shakers. When the Baptists and Methodists gained

64. Ibid., 81, 85, 95.

influence among the Negroes, they forbade dancing as sinful but allowed the brotherly dance as an outlet for their love of music and movement. "There were formerly 20 violin players on Mr. Couper's plantation," noted Lyell, "now scarcely a fiddler." In an effort to meet the expectations of the Negroes, Bishop Stephen Elliot of the Episcopal church had performed the ceremony of baptism by total immersion, just as the Baptists did. Even so, the Episcopal church was ill adapted to the feelings and level of education of either whites or blacks under the rough conditions of frontier settlement. "The Negroes require excitement," Lyell concluded. "They are mesmerized by a man shouting out with a loud voice Oh Lord! Oh Lord! till all join in the exclamations, the Preacher himself being excited, or appearing to be so."[65]

As Lyell reflected on the future of the Negro slaves, their present sufferings appeared to result not so much from the necessities of the cultivation of cotton or sugar cane as from the level of civilization among their white masters. If slavery could be modified gradually, as the planters became more civilized, Lyell thought that no human system could be devised that would be "capable so rapidly of educating & improving the Negro as slavery."[66] In contrast, the rapid extension of slavery into new states thrust the Negroes into the power of uncivilized and unprincipled white adventurers, eager to make fortunes quickly at any cost.

Lyell made many excursions with Couper. Every facility of Hopeton—horses, carriages, boats, and servants—was made available to him so readily and spontaneously as to make him feel completely at home. Near Hopeton he saw an example "of that natural rotation of crops"[67] that we now call ecological succession. An area of pine forest had been cut down, but instead of young pine trees springing up to replace those cut, a crop of young oak trees had appeared. Evidently, acorns had been buried in the ground under the pines by squirrels but could not germinate so long as the ground was shaded by the canopy of pines. Lyell was puzzled why the seeds of pines did not also germinate in the soil of the clearing.

On 5 January, Couper took the Lyells on a two-day expedition fifteen miles down the Altamaha River to his father's plantation at Cannon's Point, at the northeast end of St. Simon's Island. Many years before, the elder Couper had planted lemon trees, olive trees,

65. Lyell, Notebook 130, 7, 9, Kinnordy MSS.

66. Ibid., 11–12.

67. Lyell, *Second Visit,* 1:330.

and date palms around his handsome two and a half story plantation house, famous for its hospitality.[68] He was known also for the excellent fruits and vegetables grown at Cannon's Point.[69] The live oaks on St. Simon's Island were particularly magnificent, one of them seventy-three feet in height, with its branches spreading over an area sixty-three feet in diameter. Lyell went with Couper to Long Island (now called Sea Island), lying between St. Simon's and the open sea. About four miles long and a half mile wide, Long Island consisted of parallel rows of sand dunes, sometimes as high as fifty feet, separated by areas of salt marsh. On the beach facing the ocean, Lyell collected twenty-nine species of seashells, all identical to the species of fossil shells he had collected on Skiddaway Island.

In all his observations, Lyell was trying to form a clearer idea of the recent geological history of North America. At Cannon's Point a number of Indian mounds, formed of oyster shells and sometimes as high as ten feet, covered several acres of ground. The shell mounds must have been accumulated over a long period of time, yet the Indians who had made them had disappeared before the time of European settlement. They showed that the coastal islands had existed as they now were for a long time, perhaps thousands of years, yet the period of the Indian mounds was only the youngest of several periods in the recent geological history of North America. In his notebook, Lyell made a tentative list of the successive periods:

1 - Drift Period—& the Arctic fauna of Quebec
2 - period of No. 1 step, marine
3 - Megatherium Period, & freshwater shells
4 - Goat Isld. [Niagara Falls] oysters?
5 - Indian Mounds [St. Simon's Island][70]

In the well-stocked library at Cannon's Point, Lyell found such works as Audubon's *Birds of America,* Michaux's *Forest Trees,* and other books on natural history.[71] He took with him to read at Hopeton a copy of William Bartram's *Travels.* Of particular interest to Lyell was Bartram's description of terraces extending parallel to the coast from North Carolina to Florida, each marked by different forest trees and other botanical characteristics.[72] They were, Lyell

68. Wightman and Cate, *Coastal Georgia,* 55.
69. Kemble, *Journal.*
70. Lyell, Notebook 129, 36, Kinnordy MSS.
71. Audubon, *Birds of America;* Michaux, *North American Sylva.*
72. Bartram, *Travels.*

thought, similar to the terraces that Charles Darwin had described as extending for hundreds of miles along the coast of Patagonia.[73] The first terrace, about twenty miles inland from the Georgia coast, was very low and consisted of sand and clay containing living species of seashells. The second terrace, also about twenty miles wide, rose abruptly about seventy feet above the first. The third terrace rose another seventy feet or so to a tableland. The first terrace, near the coast, appeared to have been formed during a period of coastal subsidence and was comparable to the Pampean deposit described by Darwin in South America. In the salt marshes along the coast were stumps of cypress trees buried two or three feet beneath the surface, indicating that the salt marsh had once been a freshwater swamp and was later inundated by the sea. The terraces of Georgia are now considered by geologists to have been formed by rises of sea level during successive interglacial periods and the salt marshes to have formed after the last retreat of the glaciers. In 1838–39, when the Brunswick Canal was dug to connect the Altamaha River to the Turtle River, the workers found various fossil bones. They included bones of the *Megatherium,* mastodon, elephant, horse, and a large extinct tortoise, intermingled with the stumps and roots of cypress and other trees in a bed of alluvial clay. The clay was covered by the more recent deposit characteristic of the salt marsh as a whole.[74] Couper had collected and preserved the bones, sending some to the Academy of Natural Sciences of Philadelphia and others to the National Museum at Washington.

On their return trip by canoe up the Altamaha to Hopeton, the Lyells and Couper saw a large alligator, nine feet long, basking in the sun on the bank. As they approached, it entered the water. That evening, Lyell had much conversation with old Mr. Couper, whose knowledge and memories of Georgia extended back to before the American Revolution. John Couper described the great hurricanes that had struck the Georgia coast, the worst being those of 1756, 1804, and 1824. In 1756 the whole of St. Simon's Island had been submerged by the rising sea. He also described nests of alligators he had seen and said that formerly the alligators on the St. John's River in Florida had no fear of man, whereas now they dreaded him and were especially fearful of steamers.[75] As a measure of how fast trees grew in the South, John Couper told Lyell that live oaks plant-

73. Lyell, Notebook 128, 102, Kinnordy MSS.

74. Hodgson, *Megatherium,* 31–37.

75. Lyell, Notebook 129, 62–64, Kinnordy MSS.

ed on St. Simon's Island in 1803 now, only forty-two years later, had trunks thirty-eight inches in diameter.

The Lyells' delightful visit at Hopeton came to an end on Saturday, 10 January. James Hamilton Couper took them down the Altamaha River to Darien, stopping on the way at Butler's Island so that they might walk over the plantation, with its beautiful orange trees and neatly built Negro houses, where in 1838–39 Fanny Kemble had lived so unhappily as Mrs. Pierce Butler.[76] At eight o'clock that evening they boarded the steamboat and arrived at Savannah on Sunday morning at seven. After breakfast at the Pulaski House, they went to church. In the afternoon, Lyell attended a Negro Baptist church, where he joined a congregation of about seven hundred people to listen to their preacher. The preacher was, Lyell observed,

> a greyheaded venerable looking man with a fine sonorous voice. Only 3 or 4 inaccuracies of language such as "God pity us in our affections" in the whole extempore discourse which was positively above an average dissenter sermon in an English village. None of the canting tone of the Methodists & most of it very practical about the future state of rewards & punishments to which "rich & poor, white & black" were alike destined. He got through some flights about the gloom of the valley of the shadow of death very well & without any inflated language & compared the intervention of the Deity in favor of a Christian "left for a time to his own wisdom, but yielding to temptation & being in danger of falling" to an old eagle having its nest on the summit of a tree taking its newly fledged young & dropping it in the air & seeing it unable to fly & so with the rapidity of lightning darting down before it reaches the earth & bearing it back uninjured to its nest. The singing was very good, the church well ventilated, pews comfortable.[77]

At eleven o'clock that night the Lyells boarded the train for Macon, Georgia, to begin their travels farther westward. The journey of 191 miles on the Savannah and Macon Railroad (later the Georgia Central Railroad) took about twelve hours. Halfway, Lyell left the train to travel the rest of the distance on a railway handcar, accompanied by a railroad engineer and two slaves, so that he might stop to examine the cuttings along the right-of-way. Mary, escorted by a Mr. Reynolds, the chief engineer and director of the railroad,

76. Kemble, *Journal*.

77. Mary Lyell and Lyell to Frances Lyell, 29 December 1845–11 January 1846, Kinnordy MSS.

arrived in Macon about eleven o'clock Tuesday morning. She went to a hotel to wait for Lyell, who would join her in two days.

Lyell was continuing his study of the strata of the coastal plain in Georgia begun in 1842 along the Savannah River. In order to have the easiest grades, the railway line followed the valley of the Ogeechee River much of the way to Macon, but in so doing it tended to avoid the series of abrupt terraces separating the various levels of the coastal plain. Thus for geological work, Lyell had to make side trips away from the railway line. From the Burrstone formation at Parramore's Hill, ninety-five miles out of Savannah, he collected seashells and corals that confirmed his earlier conclusion that the Burrstone was Eocene. Lyell and his party stayed overnight at Parramore's Hill on the plantation of a Mr. Gray. The night was so cold that the next morning there was ice on the puddles in the road. At breakfast, despite a blazing wood fire in the fireplace, Lyell was thoroughly chilled by the breeze blowing through the open windows and doors. He asked to close the window, only to learn that there was no glass in it. He then put on his overcoat, but the women of Gray's family went about the house bare necked, with light dresses, and seemed indifferent to the cold.

This jaunt along the railway in a handcar through the pine woods, with frequent stops and an excursion in a hired gig to visit plantations away from the railway line, gave Lyell many opportunities to talk to slaves. He was impressed by their intelligence and ability. On one plantation, the head Negro had been there under two or three different white owners. Some years before, he was responsible for the digging of a well on the plantation. He told Lyell about the various layers of sand, clay, and limestone they had dug through and the fish teeth they had found. Curious about such fossils, the Negroes asked Lyell how seashells, sharks' teeth, sea urchins, and corals came to be buried so deeply in the earth at such a height above the sea. They had thought of Noah's Flood as a cause but could not understand how the Flood could have deposited fossils so deeply within the earth. Many Negroes on the upland plantations were remarkably responsible and competent. On Gray's plantation, the household was managed by an intelligent and good-looking young Negro woman. When Lyell asked Gray why she was not married, Gray told him that she thought "no country raised nigger good enough for her." Gray did not see how she could find a suitable husband in the country, although she could easily do so in a city like Savannah. Gray also said that when his own children were learning the alphabet and to read, he allowed them to teach the black children of the same age, even though such teaching was against the law. The black children learned quickly.[78]

Meanwhile at Macon various ladies called on Mary at the hotel. They took her for drives to see the town, which was only twenty-five years old and prettily situated at the falls on the Ocmulgee River. Mr. Reynolds escorted her to meals in the hotel dining room. The day she arrived, a young northern man died of consumption in the hotel, alone, with no friends or relatives near him. Lucy, the mulatto chambermaid in the hotel, asked Mary to attend his funeral, which she did, one of fifty or sixty people to attend the service. She also went with a number of others in three carriages to the burial service at the cemetery on Rose Hill, overlooking the falls, sharing in the kindness of the southerners toward the dead young stranger.

Lyell reached Macon on 17 January. He and Mary then set out for Milledgeville, the capital of Georgia. Their route took them twenty-five miles back along the railroad to Gordon, where they boarded a stagecoach to travel the rest of the way. Across the pine barrens, the road was sandy, but through the swamps they had a jolting ride over a corduroy road consisting of logs laid close together across the roadbed.

Lyell wished to go to Milledgeville to see the Reverend John Cotting, a New England clergyman who had come to Georgia in 1835. He was now the geologist for the state.[79] At the statehouse, Cotting showed Lyell fossils collected from Georgia formations. Near Milledgeville Lyell observed a remarkable gully that had developed since the first clearing of the land. He noted that such hard rocks as gneiss and mica schist had disintegrated into soft, friable earth to a great depth, without otherwise being disturbed. In the sides of the gully, layers of disintegrated rock were visible. The gully was then 300 yards long and 50 feet deep and varied in width from 20 to 180 feet. It was growing steadily, having not even existed twenty years before. Such rapid erosion in the short time since the clearing of the forest suggested to Lyell that the land had been heavily forested since it first appeared above the sea. Otherwise, the Georgia countryside would have been deeply fissured by ravines and valleys instead of flat and uniform, as for the most part it was.

78. Lyell, Notebook 130, 32, 40, Kinnordy MSS.

79. John Ruggles Cotting (1784–1868) was born in Acton, Massachusetts, and in 1835 moved to Augusta, Georgia. He served as state geologist of Georgia for two years. Merrill, *Contributions,* 693.

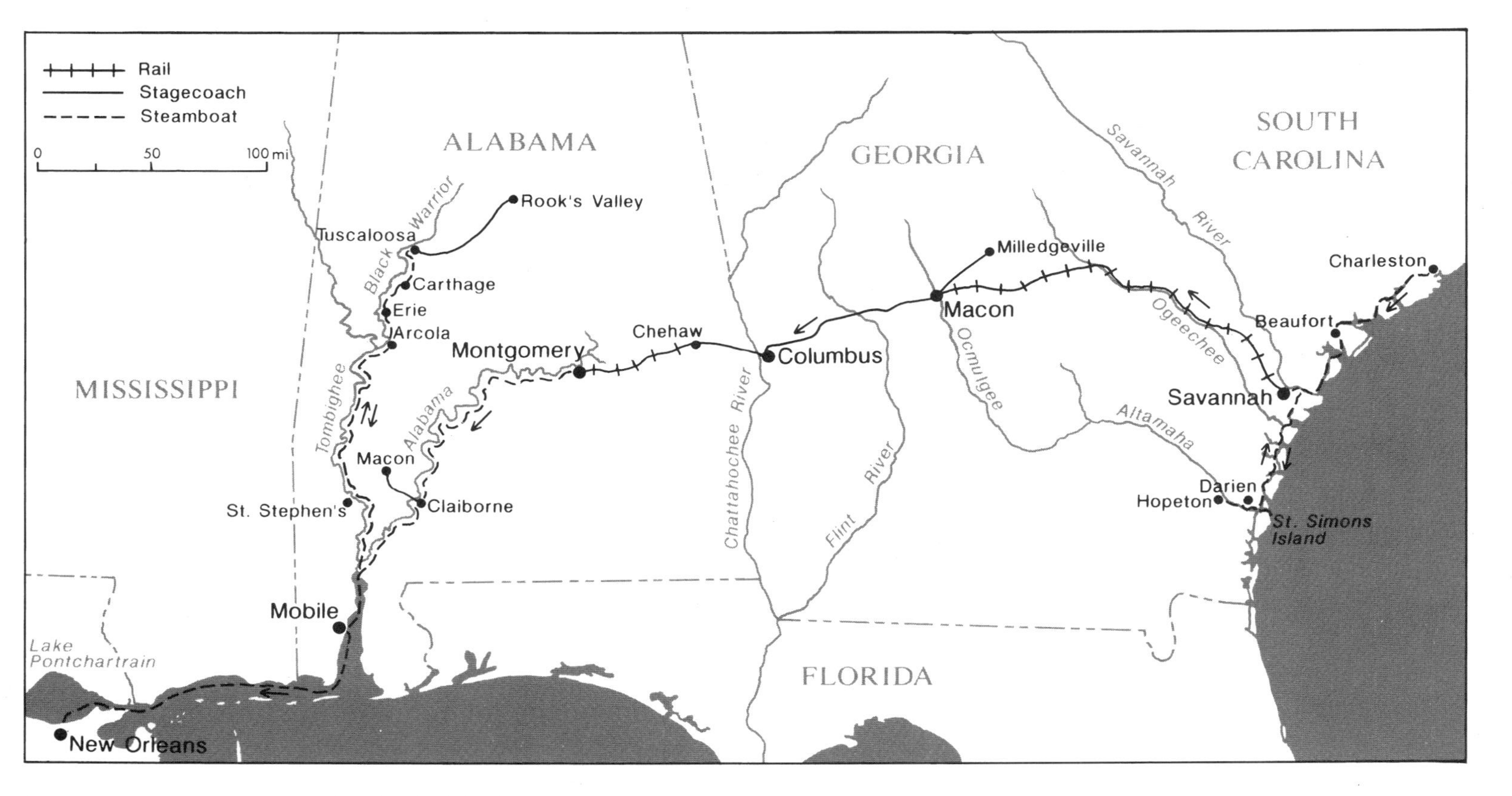

Lyell's Tour in Georgia and Alabama, 1846. Map by Philip Heywood.

At Milledgeville the hotel was not crowded, because the legislature was not in session. The landlady, a Mrs. Huson, "was just like an old Englishwoman very proud of her management of her house & table which were in excellent order." The colored maid apologized for their room, recently occupied by four legislators, who, she said, had made "a hogpen" of it.[80] According to Cotting, some legislators could not even read. He told of one, who upon receiving from the governor a card bearing an invitation to dinner, rose from his seat and, holding the card upside down, said, "Mr. Speaker. I oppose this motion; I am against such a bill."[81]

The governor, George Walker Crawford, was not in Milledgeville, but Mrs. Crawford called on Mary at the hotel and took her in a carriage to see the governor's mansion. She impressed Mary "as a very ladylike refined person." When Mrs. Crawford paid a second visit to Mary at Huson House, Mrs. Huson joined them as if they were all guests in her house. She talked of how busy she had been since the end of the legislative session, "putting down 60 hogs" and making soap from the lard combined with potash leached from hardwood ashes. She asked Mary how she made soap in England and was amazed to learn that Mary bought her soap in a shop. Mrs. Huson was also amused that Mary was keeping her hands warm in a muff, an article she had not seen since she was a girl fifty years before in Baltimore. The contrasts between life in the pine woods of Georgia and in distant London struck Lyell as so comic that in his notebook he started to sketch a scene for a play: "Their days in Milledgeville—the Huson House by Dr. Hammerman," with scene 1 set in the drawing room of the hotel and Mrs. Huson and Mary on stage.[82]

At the Presbyterian church on Sunday morning, Lyell enjoyed the service, which included a hymn, prayer, scripture reading, another hymn, sermon, a closing hymn, and a blessing.[83] The minister astonished him by announcing that there would be no service that evening, because he was preaching at the Methodist church.

On their return trip to Macon on Monday, one of their fellow passengers on the stagecoach was a young man who the day before had shot an Irishman in a saloon and was avoiding the law. He had not killed the man but clearly had intended to do so. Mary wrote:

> There are a great many of these reckless young men who have not been brought up to steady habits of industry, & having negroes

80. Mary Lyell to Susan Horner, 29 January 1846, Kinnordy MSS.
81. Lyell, Notebook 130, 104, Kinnordy MSS.
82. Lyell, Notebook 131, 39, Kinnordy MSS.
83. Ibid., 16, 22.

> to work for them, look on themselves as a superior class & pass their time in grog shops where they get into quarrels, & having the shocking habit of carrying arms about with them, of course frays often ensue.[84]

On the morning of 21 January, the Lyells boarded a stagecoach for Columbus, Georgia, ninety-six miles away on the Chattahoochee River. After many heavy rains, the road, never very good, was deeply rutted by wagons hauling cotton to the railhead at Macon. The stagecoach passed through Knoxville and after dark stopped for supper at a small roadside inn. Twice the stage nearly upset, and frequently the driver asked the men to get out and walk while he drove over a fallen tree or through a deep puddle. At such places, Mary was thrown against the roof and sides of the coach by the tremendous jolts. Late at night and in pouring rain, the stage was driven onto a flatboat to cross the Flint River. Slaves on either bank held pine torches, casting a red glare over the scene. They had to drive through several smaller streams, swollen with rain. The coachman worried that they would find some too deep to cross and have to camp out. Finally, at dawn on 22 January, the stagecoach entered Columbus. The first thing the Lyells saw in Columbus was a line of cheerful, well-dressed Negroes, including men, women, and boys, who were a gang of slaves, probably from Virginia, going to market to be sold.

After twenty-two hours on the road, the Lyells went to a hotel and slept. The next day Lyell went with some local naturalists to Snake's Shoals, a short distance down the Chattahoochee River, to see the strata laid bare in its banks. The Chattahoochee was running high. Lyell found that he should have come in November, when the water was low, in order to see the geology of the riverbanks—the difference between high and low water being as much as sixty to seventy feet. The Chattahoochee had cut its valley "like a ditch, a deep ditch" through the Tertiary strata into the underlying Cretaceous formation along the crest of a low anticlinal ridge.[85] From Dr. Samuel Boykin, to whom he had a letter of introduction, Lyell obtained representative fossils from the Chattahoochee beds. He also went with a Mr. Pond to the Upotoy Creek, where he again found Cretaceous strata underlying the Eocene formation that extended across Georgia.

Just as Macon was located at the falls of the Ocmulgee River, and Augusta at the falls of the Savannah River, Columbus was situated at

84. Mary Lyell to Susan Horner, 19 January 1846, Kinnordy MSS.

85. Lyell, Notebook 131, 44, Kinnordy MSS.

the falls of the Chattahoochee River. In each case, the location of the city marked the boundary between the granite upland region and the Cretaceous or Tertiary formations of the coastal plain. A cotton mill in Columbus used the water power of the falls to make cotton cloth for clothing the Negroes. All the workers at the mill were white, and the owners of the mill were trying to make cotton spinning and weaving respectable occupations for white men. Mr. Pond, a New Englander living in Columbus, complained to Lyell that he could not bring up his children as carpenters or blacksmiths or other skilled tradesmen, "because a planter prefers a negro at the same charge, he can command him, he has not to consider how he orders him—to measure his words."[86]

ALABAMA

After two days at Columbus, the Lyells boarded a stagecoach drawn by six horses to travel the fifty-five miles to Chehaw (Giaupee) in Alabama. The road was much better than that between Macon and Columbus, passing mostly through a rolling country of pine woods. The wind in the tall pines sounded like waves breaking on a distant seashore. In places the forest was of oak and hickory. The relationships between whites and blacks visible along the road endlessly fascinated Lyell. He observed "a set of children at a small house by road side, alternate black & white, shouting loudly at our coach & six—the same mixture of fun, impudence & mischief & bravado as in England—all on equality."[87] At midday the coach stopped for dinner at a small house by the road, where they were served "wild turkey, partridge pie and venison steaks, together with the unusual treat of a large jug of fresh milk, which one very seldom meets with hereabouts."[88]

At Chehaw they slept in a room "with such spaces between the boards on three sides," Mary wrote, "that we could see through them into the open air."[89] Lyell counted the annual rings on a number of fresh pine stumps near the inn, which he determined were 120 to 320 years old, the stump of the oldest tree being four feet in diameter. When the existing forests were cut, no such great trees would be seen by future generations, because of the length of time they took to grow.

86. Ibid., 52.
87. Ibid., 58.
88. Mary Lyell to Susan Horner, 29 January 1846, Kinnordy MSS.
89. Ibid.

After a conversation with the landlord, Lyell made further notes on slavery:

> Some planters at Chehaw have this year, owing to the drought, so deficient a cotton crop that, after supporting their negroes & giving them medical assistance, they are considerable losers & have to borrow. Next year they will probably recover, but the land here only supports its population of negroes by the white superintendance & economy. The landlord never stints his negroes in food; they eat the same as the family. In many of the houses here the total number of whites, adults and children, exceeds the blacks, the latter with the white children who are very fond of them. Cotton employs young as well as old negroes & by bringing them earlier to work & supplying work all the year round checks the acquirement of those indolent habits with which the black race have most to contend. Slaves who come from Virginia and parts of the north where there is no cotton are more indolent. . . .
>
> The landlord has all his blacksmith's work done by a rich white owner of negro blacksmiths, the white in return favoring him in his business as carrier. This method of employing artizans is against white competition. Nevertheless, at Montgomery New England artizans beginning with nothing have grown rich as superior workmen outdoing the negroes as upper mechanics.[90]

At Chehaw, the Lyells boarded a train to Montgomery. The train was crowded, and a drunken man put his feet up on the seat cushions and began to sing. After the conductor twice asked him to put his feet down, the drunken man put them up a third time. The conductor then stopped the train and ordered the drunk to get off. He was left by the side of the line. "It was all done so quietly," Mary observed, "even the tipsy man made no noise. The conductors of railways & clerks of steamboats are allowed great authority. All the passengers feel it is for their own advantage. At meals the ladies are summoned first & not a gentleman sits down till the ladies or women are seated." On the train newsboys were selling apples and biscuits as well as newspapers and novels. Lyell was impressed by the sensible tone of the newspapers, even on such inflammatory questions as Oregon.

"I never saw anything so merry as the negroes are," wrote Mary. "At all the railway stations we stopped at, it was Sunday & they came

90. Lyell, Notebook 131, 66–68, Kinnordy MSS.

down in their best clothes laughing like children at some jokes of their own."[91] At one station they saw a runaway slave in handcuffs, the first Negro Lyell had seen in irons since 1842, when he saw the two captured runaway slaves landed in Kentucky from a steamboat on the Ohio River.

The sight of the handcuffed slave caused a New Englander, who had traveled with the Lyells on the stage from Columbus to Chehaw, to speak of the suffering of the Negroes in Alabama, Mississippi, and Louisiana. He described the cruelty of the overseers, their opposition to the education of the Negroes, or even their conversion to Christianity. The New Englander spoke so intelligently and with so much authority that Lyell began to dread what he might see of slavery as he went on through the South. However, as their conversation continued, Lyell began to detect a tendency toward exaggeration and prejudice in the New Englander, so he led him to speak of the condition of the slaves in South Carolina and Georgia. There, the New Englander assured Lyell, he had seen the same suffering and cruelty as farther west. Lyell knew that the Negroes on the plantations he had visited in South Carolina and Georgia were well fed and housed and not overworked. Consequently, he became doubtful of the accuracy of the New Englander's statements. Shortly afterward, he learned from the railway conductor that the New Englander had come to the South as an agent for a northern commercial house to press claims against various firms and planters bankrupted by the financial crisis of 1839–40. The failure to obtain payment of bad debts had given him a jaundiced view of the South. Frequently he had to deal with planters who had borrowed heavily to buy slaves or land and then attempted to escape their debts. He tended to see the evils of slavery through the magnifying lens of his financial losses and had, wrote Lyell,

> imbibed a strong anti-negro feeling, which he endeavoured to conceal from himself, under the cloak of a love of freedom and progress. While he was inveighing against the cruelty of slavery, he had evidently discovered no remedy for the mischief but one (namely, war with England over Oregon in which England would declare the slaves free and thereby promote a slave insurrection), the hope of which he confessedly cherished, for he was ready to precipitate measures which would cause the Africans to suffer the

91. Mary Lyell to Susan Horner, 29 January–7 February 1846, Kinnordy MSS.

fate which the aboriginal Indians have experienced throughout the Union.[92]

When the Lyells reached Montgomery, "a very pretty, decent looking place" on the Alabama River, hardly more than twenty-five years old,[93] Lyell went immediately to Jackson's Ferry three miles upriver. There he found beds of loose gravel thirty feet thick, alternating with beds of red clay and sand containing Cretaceous fossil shells. The gravels, clays, and sands, therefore, were Cretaceous and much older than they looked.

At Montgomery the next morning, Lyell saw an auctioneer in the marketplace selling slaves; as he passed, a Negro was sold for four hundred dollars. The following day in the same place another auctioneer was selling horses. The public display of the slave trade was, Lyell learned, disturbing to some Montgomery citizens, who thought it might be carried on more quietly. "It w[oul]d be more decent in a room," noted Lyell, "yet in Scotland at term time you go to the market place & find all the people in their best clothing to be hired."[94] At nine o'clock each evening, a great bell in the marketplace was rung for the curfew imposed after a slave insurrection in Southampton, Virginia, in 1835. Any Negro on the streets without a pass after that time was liable to arrest. Lyell had a letter of introduction to a Mr. Knapp, an Episcopal clergyman, who told Lyell that he had recently established a Sunday school for Negroes in Montgomery and hoped to have Negro congregations represented at the triennial conventions of the Episcopal church.

Lyell wished to follow the junction of the coastal plain formations with the granitic upland to Tuscaloosa, about a hundred miles northwest of Montgomery. However, because the roads were bad and the inns doubtful, he was advised to go to Tuscaloosa by river, first down the Alabama River to Mobile and then up the Tombigbee River and its tributary the Black Warrior to Tuscaloosa, thereby making a detour of eight hundred miles to travel one hundred. At ten o'clock on the morning of 28 January, he and Mary boarded the steamboat *Amaranth.* Upon boarding, they had to walk over bales of cotton covering the lower deck, but on the upper deck they found airy and comfortable cabins. Each cabin opened on its outer side to

92. Lyell, *Second Visit,* 2:30–40.

93. Mary Lyell to Susan Horner, 29 January–7 February 1846, Kinnordy MSS.

94. Lyell, Notebook 131, 82, Kinnordy MSS.

the deck and on its interior side to a long central parlor, divided into a ladies' cabin and gentlemen's cabin. In charge of the cabins was a young German stewardess, with whom Mary was soon speaking in German. The boat was not ready to leave, but after Lyell spent some time looking at the geology of the river bluff, they settled themselves on board in their cabin to read and write. In midafternoon the steamboat left, but Captain Bragdon started upriver rather than down, intending to pick up more cotton at a landing above Montgomery. Only when he heard that another steamboat had already gone for the cotton did he turn the *Amaranth* to head downriver. Despite its unpredictable schedule, the *Amaranth* was most comfortable. Lyell and Mary found it delightful to sit on their private deck outside their cabin while they went downriver, passing canebrakes and cypress swamps similar to those along the Savannah River.

The casual attitude toward time was actually to Lyell's advantage. At each landing where the steamboat stopped, Captain Bragdon, himself an amateur geologist, gave Lyell time to collect fossils from the bluffs, advised him where fossils were to be found, and sometimes came to help him collect them. For some distance below Montgomery, the bluffs consisted of red clay loam above a bed of gravel about thirty feet thick, unlike any Cretaceous bed that Lyell had seen before.[95] Below Cahawba, the rocks in the bluffs were thinly bedded marly limestones, like the Lias formation in England, but containing Cretaceous fossils. The same fossils occurred in the bluffs at Selma and at Prairie Bluff, from which Lyell gathered a particularly rich collection.

Planning to get off the steamer at Claiborne late that night, the Lyells did not go to bed. After dark there was a sudden loud crash and a shower of broken glass on the floor. When Lyell rushed on deck, he learned that the steamboat had run into the branches of the trees on the bank. The riverbanks were lined by tall cypresses with branches overhanging the river. When the river was low, a steamboat could pass beneath them, but with the water high, the pilot had to steer between the branches extending from either bank. In coming swiftly around a bend in the dark, he had miscalculated and run into the branches on one bank. The pilot reversed the engines and men with axes soon cut the *Amaranth* free.

At eleven o'clock they reached Claiborne, a typical cotton land-

95. Lyell, Notebook 132, 13, Kinnordy MSS.

Cotton Landing on the Alabama River, Showing Staircase and Cotton Slide. At Claiborne the staircase and cotton slide were unroofed. From *Ballou's Pictorial Drawing-Room Companion* (1855). Courtesy of the Minnesota Historical Society.

ing, with wooden steps leading up the bluff, 150 feet high. Beside the steps was a smooth inclined plane down which bales of cotton were slid to the steamboat. When they landed, Captain Bragdon gave Mary his arm to walk up the long, wooden staircase, "while in the other hand," she wrote, "he carried a pine torch blazing which cast a red light all round on the trees, some large magnolias with their glossy leaves."[96] A group of slaves followed, carrying the Lyells' baggage. At the top, facing a level green, was a small inn that proved to be much more comfortable than they had been led to expect.

Claiborne was then a village of about four hundred people, with straight, wide, but unpaved streets, planted with flowering trees. During the night it rained, but next morning was sunny and quite warm. The Lyells spent all day collecting fossils among the magnolia trees and fan palms on the Claiborne bluff. The bluff consisted of beds of limestone about 150 feet thick, overlain at the top by a bed of red clay about 20 feet thick. In the evening, as the shadows lengthened and the sun was setting in red and gold splendor over the hills across the river, they returned to the inn. On its broad piazza, enjoying the evening sunshine, they found a family who had sold their plantation and were moving to Texas with forty of their slaves. The Alabamians were leaving partly because they feared the heavy taxes, which they thought would be needed to pay Alabama's state debts. Some children had caught a robin, intoxicated from eating the berries of the Pride of India trees in front of the inn. Mary bought it from them in exchange for some candy. In their room the bird soon recovered and flew out the window.

On Saturday, 31 January, Lyell became discouraged working at geology with no one to help him. He set off across the Alabama River to Woodlands plantation with a letter of introduction to the planter, Frederick Blount. A lawyer and native of North Carolina, Blount had earlier worked for James Hamilton Couper's brother at Mobile, where the latter operated a mill for extracting oil from cottonseed. Lyell intended to return from Woodlands on Sunday, but Blount was so willing to help him that together they planned an expedition of several days to various geological sites in Clarke County.

Lyell sent his driver back to Claiborne with a note to Mary, explaining his change of plans, and an invitation from Mrs. Blount for Mary to come to Woodlands to stay with her. Mary preferred to re-

96. Mary Lyell to Susan Horner, 29 January–7 February 1846, Kinnordy MSS.

Claiborne Bluff in Winter, 22 December 1934. Photograph by Q. B. Schenk, in Wheeler, *Conrad.*

main at Claiborne over Sunday to do a little more collecting along the bluff. She also attended service at the Methodist church, where she heard "a very fair sermon from a young man who was evidently well educated by his language." On Monday morning she went to Woodlands in a carriage hired from the inn, crossing the Alabama River on a flatboat. "It is a place quite in the pine woods," she wrote, "with however a good house & a very pretty garden making, as things very frequently are in this part of the world." Mrs. Blount welcomed her pleasantly, but during the afternoon the Blounts' youngest child, a little girl of five, came down with a sharp attack of fever. That night a doctor came to see her. Mary wrote:

> In a new country one dispenses with many forms, for though I have not yet known Mrs. Blount 24 hours, I go in and out of the room to see if I can help about the child, quite as a matter of course. Two little coloured girls, very intelligent children, are continually in waiting & seem to do all the work & would wait on me continually if I wanted them. I have a charming luxurious bedroom looking on the piazza.[97]

Meanwhile, Lyell and Blount drove in Blount's carriage sixteen miles through thick forest, broken only by an occasional clearing, via Suggsville to Macon (now Grove Hill). Blount was a judge and had business at Macon, the county seat of Clarke County. Blount pointed out to Lyell the place where a bloody battle had occurred between the Chocktaw and Chickasaw Indians and told him of the feats of General Andrew Jackson in organizing the settlers to fight the Indians in the early days. He also introduced Lyell to his friend William Pickett, who had helped Albert Koch dig out the skeleton of the *Zeuglodon cetoides,* or fossil whale, near Clarkesville. On Monday Lyell went with Pickett to visit the *Zeuglodon* site—or rather, sites. The various portions of the *Zeuglodon* skeleton that Koch had exhibited in New York, Lyell learned, had been found at various places: the head and teeth in a hillside on Judge Creagh's plantation four miles southwest of Clarkesville and the main group of vertebrae at another spot fifteen miles away.

When Lyell and Pickett went for their horses to start out from Macon on Monday morning, they found they could not get them because the stable boy was up for sale, so they attended the auction. First an old Negro man was sold for $150, then a seventeen-year-old

97. Mary Lyell to Marianne Lyell, 3 February 1846, Kinnordy MSS.

boy for $535, and finally their stable boy for $675. The stable boy was bought by the livery stable owner to whom he had been hired for three years. When the boy came with them to get their horses, Pickett said to him, "I would not have given so much for you by $100." Lyell asked the lad whether he was satisfied with his new master, and the slave replied, "Yes, I have been with him three years so he knows what I am worth." Lyell observed, "The bearing of most of the negro servants or work people is not more humble or subdued than English labourers, on the whole much on a par."[98]

That night Lyell and Pickett stayed at Doyle's Inn in Clarkesville. Lyell was startled to find that he could not have a room to himself and more alarmed to learn that he was expected to share a bed. The landlord gave him a bed to himself, but in the same room Pickett shared a bed with two other men. Astonished though he was, Lyell accepted the need for roughing it philosophically. Blount had told Lyell that once when he was running for the Alabama senate, he had shared a room in which three married couples slept in beds while he slept on the floor along with three children and two grown-up girls. Pickett related that during the Texas war he had not changed his clothes for six months and was clearly proud of it.[99] At Doyle's Inn, Lyell also felt the lack of soap, water, and towels in the room. In the morning, the innkeeper was suffering from a hangover and was taking more whiskey to brace himself for the day. He invited Lyell to join him and thought Lyell slightly rude when he refused. Similarly the innkeeper's neighbor, a German planter, was missing and was believed to be drunk somewhere in the woods. Such heavy drinking shocked Lyell. Although he witnessed no cruelty to Negroes, he could believe that drunken slave owners might be capable of almost any kind of violence.

The principal geological purpose of Lyell's expedition to the *Zeuglodon* sites was to determine the age of the rocks containing the *Zeuglodon* bones in relation to the age of the Nummulitic Limestone of Alabama. The Nummulitic Limestone was a soft, cream-colored stone, fifty to one hundred feet thick, lying immediately above the beds containing the *Zeuglodon* bones. The *Nummulites* for which the rock was named were small disclike bodies (*Orbitoides mantelli*) so numerous as to give the rock a granular texture. Previously the Nummulitic Limestone of Alabama had been thought to be Creta-

98. Lyell, Notebook 132, 86, Kinnordy MSS.
99. Ibid.

ceous or possibly intermediate between Cretaceous and Eocene. Lyell found that the Nummulitic beds in Clarke and Washington Counties lay above the Eocene strata of the Claiborne bluff and contained typical Eocene fossils. Therefore, the Nummulitic beds were not only Eocene, but well up in the Eocene series.[100] Lyell thought they were probably of the same age as the Burrstone formation of Georgia, but he could not be certain until he compared fossils from the two formations.

When the Nummulitic Limestone of Alabama was considered a Cretaceous formation, the fossils from it had naturally been considered Cretaceous fossils. Many of the fossils occurred also in Eocene formations in other parts of the United States. The presence of the same fossils in formations thought to be Cretaceous and Eocene led to the idea that the Nummulitic Limestone might be intermediate between the Cretaceous and the Eocene. When the Nummulitic Limestone became Eocene, all its fossils became Eocene rather than Cretaceous, and by this one step almost all the fossils thought to be common to Cretaceous and Tertiary formations in the United States were eliminated.[101]

On 4 February, Lyell and Blount returned to Woodlands, where they found Mrs. Blount and Mary in alarm over the sickness of the Blounts' little girl. Two doctors were there and the Blounts had little hope that the child would live.[102] Mary sat up half one night with the child to allow Mrs. Blount to get some sleep. Although she sympathized deeply with the Blounts, Mary formed decided opinions about the way in which the Blounts were bringing up their children.

> I never saw the misfortune of spoiling children more exemplified. The poor little girl was quite ungovernable & medicine having to be given every two hours was such a business. All sorts of deceit practised, which the child was quite sharp enough to see through, & each time we had to force it down her throat, with kicking, fighting & shrieks enough to frighten one. I really never saw anything like the way they allow children to do as they please & eat as they please & they are a set of sickly, miserable-looking, odious creatures in general. I hardly ever notice a child, but give

100. The formation was later classified as Oligocene by Timothy A. Conrad. See Heilprin, *Tertiary Geology,* 2.

101. Lyell, "[Portion of a Letter to B. Silliman, Jr.]"; cf. Lyell, "Newer Deposits of the Southern States," esp. 408–9.

102. But apparently she did. The Blounts had three children living in 1850. See United States Census, 1850, Clarke County, Alabama.

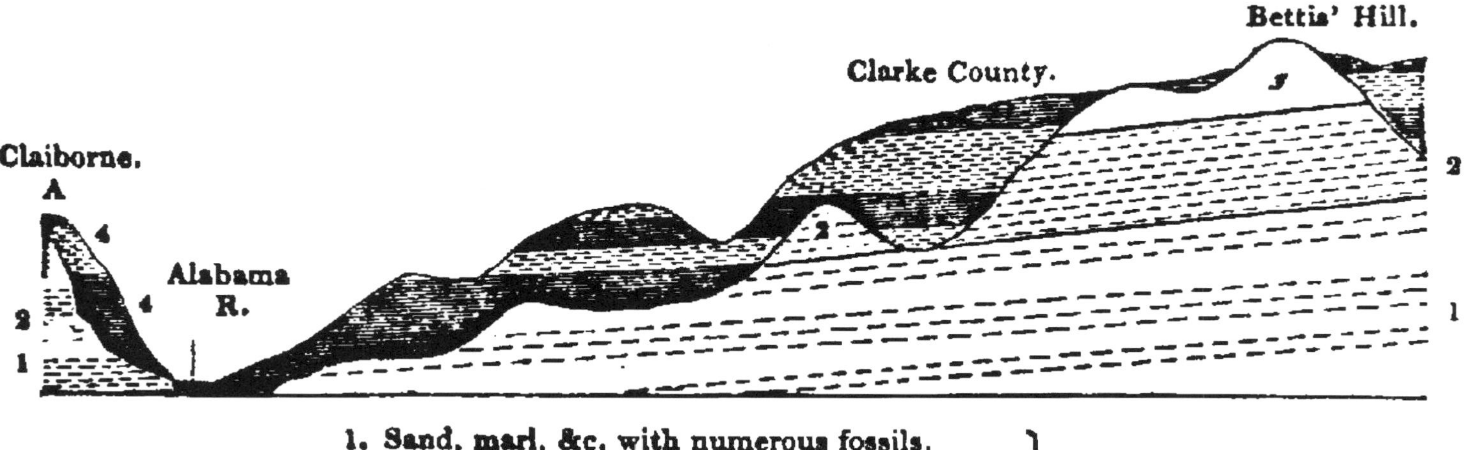

Geological Section, Clarke County, Alabama. Lyell, "On the Newer Deposits of the Southern States" (1846).

> all my attentions to dogs & cats, who are at least grateful. There are certain exceptions; for instance Mr. [James] H. Couper's large family, where there are all reasonable freedoms allowed, but I do not think one of his children would dare to disobey him. But then he is a man of very commanding character. The poor white children are the worst. There is a great talk always of whipping, but no execution. The little negroes are twice as good & pleasant, probably from their being somewhat disciplined.[103]

At Woodlands, Lyell wrote a letter to Benjamin Silliman Jr., describing the geological results of his tour in Clarke and Washington Counties, including his demonstration of the Eocene age of the Nummulitic Limestone and the fraudulent character of Koch's *Hydrargos.* Next morning, the Lyells returned to Claiborne, where they spent the rest of the day collecting fossils on the Claiborne bluff. The first steamboat to come by was going upriver and pulled into Claiborne just long enough to ask whether there were any passengers. There were none, and when asked for news, the captain shouted "Cotton up one eighth—no war." By torchlight at ten o'clock that night (Thursday, 5 February), the *Amaranth,* in which they had come down from Montgomery, picked up the Lyells to go down-river. Next morning, in the midst of a heavy rain, they arrived at Mobile, where they drove through muddy streets to the Mansion House, a large and comfortable hotel. In Mobile, the cotton brought downriver by steamboat was traded, stored, and shipped abroad, while the steamboats carried manufactured goods back upriver to supply Alabama plantations. That evening the weather cleared, and the Lyells went for a drive to deliver some letters of introduction.

On 6 February, Lyell boarded a steamboat to go up the Tombigbee River to St. Stephen's, where he wished to examine the geology of the bluff. Mary remained at Mobile two more days and then started upriver on another steamboat, which Lyell boarded at St. Stephen's. Together they continued upriver to Tuscaloosa, located at the falls on the Black Warrior River. The falls marked the boundary between the horizontal Cretaceous strata and the older, harder, more inclined Carboniferous rocks.

Lyell came to Tuscaloosa with a letter of introduction to Professor Richard Brumby at the University of Alabama. Founded in 1831 with a large endowment of public land, by 1846 the University of Ala-

103. Mary Lyell to Marianne Lyell, 3–7 February 1846, Kinnordy MSS.

bama possessed a handsome group of buildings and a small but brilliant faculty. In 1837, the Reverend Basil Manly had come from South Carolina to be its second president, bringing with him a clear idea of the need for learning among young men growing up in the backwoods of Alabama. In reorganizing the faculty, Manly appointed Samuel M. Stafford, a fellow South Carolinian, professor of ancient languages and Frederick A. P. Barnard,[104] a graduate of Yale College and former student of Silliman's, professor of mathematics and natural philosophy. He appointed Richard Brumby professor of chemistry, mineralogy, and geology. A native of South Carolina, Brumby had been educated at South Carolina College and trained in the law. Although without training in the sciences he was to teach, Brumby had the use of a well-equipped chemistry laboratory. Full of enthusiasm, he developed the laboratory and collections of minerals and fossils, which he used in his teaching. He welcomed Lyell with the eagerness of a hardworking but solitary teacher, far removed from the great centers of science.[105] The day after the Lyells arrived, Lyell and Brumby set out on an expedition to the Alabama coalfields. Mary stayed with Mrs. Brumby.

When Lyell prepared the geological map of the eastern United States and Canada for his *Travels in America,* he had not known of the existence of a coalfield in Alabama. He had marked the southwestern limit of the Appalachian Coalfield at the big bend of the Tennessee River, a hundred miles north of Tuscaloosa. However, while traveling in the South he had learned that the city of Mobile was supplied with coal brought down the Tombigbee River from Tuscaloosa. He wondered whether the Tuscaloosa coal was, like the coalfield at Richmond, Virginia, younger than the Carboniferous, and came to Tuscaloosa to find out. Brumby had already studied the extent and structure of the coalfield to the north of Tuscaloosa and could guide Lyell to the most interesting sites. He said that the Alabama coalfield was Carboniferous and was a southwestern extension of the Appalachian Coalfield, about ninety miles in length. The coal-bearing strata outcropped along the sides of an anticline, of which the Rooks Valley formed the central axis.

With a team of horses and a buggy, Brumby and Lyell drove over hills of sandstone and shale, heavily forested with long-needled

104. Frederick Augustus Porter Barnard (1809–99) was in 1856 appointed president of Columbia College at New York City, where he became a leader in the development of graduate education and university education for women.

105. LaBorde, *History of the South Carolina College,* 389–94.

pines, following the valley of the Black Warrior River. Near the river, they visited several open-pit coal mines, where Lyell found *Calamites, Sphenopteris, Neuropteris,* and other characteristic Carboniferous fossil plants, showing that the Alabama Coal was a true Carboniferous formation. About thirty miles northeast of Tuscaloosa, they entered the Rooks Valley. The beds of limestone and ironstone contained large deposits of hematite suitable for iron ore, suggesting to Lyell that this valley would become a great iron-making center.[106] As a result of the winter rains, the roads were extremely muddy and Brumby and Lyell could not cover more than a few miles a day. At times they had to take down a section of snake fence to drive through a field in order to avoid a particularly bad mud hole. On 12 February they returned to Tuscaloosa.

The Brumbys lived in a comfortable, three-story brick house, one of four professors' houses next to the Lyceum, which housed the principal lecture rooms and laboratories. Modeled on the campus of the University of Virginia, the Alabama campus was laid out around a grassy quadrangle. Along the sides of the quadrangle were residence halls for the students, while in the center was a round, domed building containing an assembly hall and the university library.[107]

Mrs. Brumby was a pleasant woman, chiefly concerned with the care of her seven children. She made Mary feel quite at home. Next door lived Professor Stafford and his wife.[108] Mrs. Stafford was an attractive, intelligent northern woman with strong intellectual interests and Mary spent much time with her. The professors and their families formed a community somewhat separate from the town of Tuscaloosa. Brumby had eight Negro house slaves, who lived apart in three small, two-room houses, where they went when their day's work was finished. To her sister, Mary wrote:

> A great deal of visiting & gaiety goes on at night in these houses, for the negroes are great sitters up. But it decidedly gives them a feeling of independence as they like to adorn these houses & have little treats & suppers there. Perhaps you will wonder where their money comes from, but they have many little perquisites, for instance one of the servants mends the students' shoes at leisure

106. Elyton in the Rooks Valley is now Birmingham, Alabama, which for many years had great iron and steel plants. See Lyell, *Second Visit,* 2:81.

107. Sellers, *History of the University of Alabama,* 32–33; cf. Wolfe, *University of Alabama,* 9–13.

108. Maria B. Stafford became principal of the Tuscaloosa Female Institute in 1856 on the resignation of her husband from the University of Alabama on account of ill health.

> moments, another servant, belonging to another person, is a barber & cuts hair & is paid by everybody, even his own master etc. so they generally have a little pocket money.[109]

Brumby told Lyell that when he calculated the interest on the money invested in a slave and the annual cost of his support, a slave was less expensive than a servant in England, but since two Negroes were required to do the work of one white servant, Negro labor was really much more expensive than free white labor. Brumby said that some free colored men in Alabama were prosperous merchants. For example, in Tuscaloosa one man, Solomon Perteet, was still thriving with his store, even after losing seventeen thousand dollars by having acted as security for the debt of another.

Brumby also told Lyell that among the planters there was resistance to the separation of families when slaves were sold. Near Tuscaloosa, when a Negro man and woman were put up for sale, the woman cried and begged that their daughter be sold with them. The auctioneer at first refused, but when after fifteen minutes he had received no bids for them, he included the girl and immediately received a bid of twelve hundred dollars, the family ultimately selling for seventeen hundred dollars. In another instance, when a Negro man was sold separately from his son, no one would bid for the son; finally, the person who had purchased the father bought the son for a very low price.[110]

On Sunday morning the Lyells attended service at the Episcopal church, where they heard the Bishop of Alabama preach. In the afternoon they accompanied Professor Brumby and his family to the Presbyterian church. The minister spoke on the need for greater honesty in business affairs and mentioned the high proportion of bankruptcies among traders and merchants in Alabama. On Monday evening, Barnard showed the Lyells the new astronomical observatory, equipped with large reflecting telescopes, chronometers, and other astronomical instruments from Europe.[111] Looking through a telescope, they saw the constellation Orion, more conspicuous in the clear sky of Alabama than in England, because it was farther above the horizon. The Great Bear was in a different position, Castor and Pollux were bright, and Sirius was especially bright. They also had a good view of the satellites of Jupiter. Lyell later mused:

109. Mary Lyell to Katherine Horner, 15 February 1846, Kinnordy MSS.

110. Lyell, Notebook 133, 63, 90, Kinnordy MSS.

111. W. G. Clark, *History of Education in Alabama,* 1–83.

> When looking at double stars with Mr. Barnard in the University with instruments from London's first makers & hearing discussions on astronomy, we ought to remember how recently the Indians were there, that their trail is hardly effaced from the woods, then expect a large equatorial [telescope] or a duplicate of L[or]d Rosse's telescope.[112]

Although Alabama had been settled little more than twenty years, many Alabama planters were already selling their land and moving with their slaves to Texas. The planters were unwilling to stop the movement of Negroes from state to state in the internal slave trade, Lyell observed, "because when they make money they like to invest in negroes, land being so cheap that they can get as much of that as they can manage." Many plantations in Alabama were owned by rich Carolina planters, and many successful Alabama planters had bought additional plantations in Mississippi. The abundance of land and the steady expansion of plantation agriculture kept up the demand for slaves to work on the new plantations. At auctions in Tuscaloosa the price of slaves was rising. Nevertheless, there were also new developments in the economy of the South. At Tuscaloosa the owner of a new steam-powered cotton factory that was under construction intended to try to use white labor and mechanics. In cotton factories in the North, white labor was much cheaper and more productive than Negro labor.[113]

On 17 February, the Lyells again boarded a steamboat to go downriver to Mobile. Eighteen miles below Tuscaloosa the boat stopped to load cotton at Carthage, where Lyell examined the beds of gravel, sand, and clay in the bluff. At first glance these great beds of unconsolidated sediments looked like very recent deposits; in fact, they belonged to the lower Cretaceous period. Successive stops made the journey downriver slow, but they allowed Lyell to study the changing geology of the bluffs. Below Erie, the bluffs consisted of a blue, marly limestone containing many hippurites and other Cretaceous fossils. At Erie and Arcola, artesian wells had been drilled to great depths (five hundred to a thousand feet) through the blue limestone to water-bearing beds of Cretaceous gravel beneath. One such well at Erie poured forth 350 gallons of water per minute with sufficient pressure to rise forty feet above the surface. At Arcola, a planter gave Lyell several Cretaceous fossils and some irregular

112. Lyell, Notebook 133, 99, 110, Kinnordy MSS.
113. Ibid., 99, 101.

tubular bodies. Lyell recognized them as identical to irregular glassy, tubular bodies found in sand dunes in Cumberland, England, formed by the passage of lightning through the sand.

In the small hours of the morning of 20 February (which, Lyell recalled, was the anniversary of the Geological Society of London), the steamboat reached the bluffs of Eocene limestone at St. Stephen's. The captain stopped to take on wood so that Lyell might examine the bluffs. While the wood was being loaded, he gave Lyell a boat and two slaves with pine torches, which provided enough light for Lyell to examine the bluff from one end to the other and to collect many fossils. The bluff was composed of the same Nummulitic Limestone that Lyell had seen forming knolls covered with red cedar (*Juniperus*) in Clarke County. From the top of the bluff, he looked down on the steamboat. Slaves were loading firewood by torchlight, while the slave who was tending the rowboat at the foot of the bluff used his torch to set fire to the Spanish moss hanging from a tree. Although glad of the good nature of the captain in stopping at St. Stephen's, Lyell felt slightly guilty at being the cause of a further delay to his fellow passengers. On Alabama steamboats, Lyell observed, passengers were "mere bales of cotton like other freight."[114]

At Mobile, Lyell went with the Reverend Dr. William Thomas Hamilton,[115] minister of the Presbyterian church, to visit Mobile Bay. Driving first to the lighthouse, they climbed to the top of the tower, from which they viewed the bay to the south and the city of Mobile to the north. From the lighthouse they went to the shore of the bay. The bay, surrounded by pine forests, was smooth as glass. Along the shore lay much driftwood, some of it drifted eastward from the mouth of the Mississippi River. Lyell tasted the water of the bay and found it fresh. For several days before, the wind had blown from the north and the Alabama River was constantly pouring in fresh water. On the beach Lyell and Dr. Hamilton dug up some living shells of *Gnathodon (Rangia cuneata)*, a species accustomed to live in brackish water. They also obtained a species of *Cyrena* and a *Neritina*. As they returned to Mobile, Lyell saw along the road large banks of *Gnathodon* shells, appearing to mark old lines of beach. The city itself was built on a bed of *Gnathodon* shells. Lyell also called on C. S. Hale, a preacher originally from New England, who had

114. Ibid., 122.

115. Reverend William Thomas Hamilton (1796–1884) in 1834 became minister of the Government Street Presbyterian Church at Mobile, a post he held until 1854.

made a large collection of Alabama fossils. Hale told Lyell that he had discovered large banks of *Gnathodon* shells as far as twenty miles inland. They were so far above the high-tide level that he thought they demonstrated the elevation of the land. He gave Lyell a map on which he had marked the locations of the shell banks.

While coming down the Tombigbee River on the steamboat, Lyell had written a paper on the Tuscaloosa coalfield that he now sent to Benjamin Silliman Jr. to appear in the *American Journal of Science.*[116] Lyell also sent an earlier version of the paper written at Tuscaloosa to his father-in-law, Leonard Horner, to be read at the Geological Society of London.[117]

At Mobile, Lyell talked with the physician Josiah Nott. A native of Columbia, South Carolina, Nott had studied medicine at New York, Philadelphia, and Paris before settling at Mobile in 1835.[118] Besides writing on medical subjects, he had begun to write articles on ethnology, dealing with the differences between the Negro and white races, which he contended were separate species. In Nott's view, the mulatto was a hybrid—weaker, shorter-lived, and less healthy than members of either of the parent races. Nott considered mulattoes intermediate in intelligence between blacks and whites, a sign, he thought, of their hybrid status. He argued that mulatto women were particularly delicate, subject to many diseases, and of low fertility. He concluded that if whites and blacks were allowed to intermarry, such marriages would lead to the extinction of both races. He mentioned also that mulattoes made bad slaves and that when insurrections occurred mulattoes were always involved in them. Despite their alleged greater susceptibility to disease, Nott noted that during the severe epidemics of yellow fever at Mobile between 1837 and 1842, he "did not see a single individual attacked with this disease, who was in the remotest degree allied to the Negro race."[119] In 1844 Nott had published a small work on the natural history of the white and black races.[120] At the time of Lyell's visit, he had just written a long reply to a critical review of the book in the *Southern Quarterly Review.*[121]

Lyell was doubtful of many of Nott's assertions. He had seen that

116. Lyell, "Coal Field of Tuscaloosa, Alabama."
117. Lyell, "Coal Fields of Alabama."
118. Josiah Clark Nott (1804–73); see Horsman, *Nott.*
119. Nott, "Mulatto a Hybrid," 254.
120. Nott, *Two Lectures.*
121. Nott, "Unity of the Human Race."

matings of white men with Negro women occurred frequently throughout the South and wondered why such unions should be considered unnatural. He knew that mulatto women were frequently reduced to prostitution and wondered whether prostitution might not be a factor in their lack of fertility. He was skeptical of Nott's contention that Negroes lived longer than mulattoes. Negroes' birth dates were not usually recorded by registry of their baptism. Lyell speculated that their supposed longevity was a result of uncertainty about their true ages. Finally, if mulattoes did not die of yellow fever as whites did, Lyell wondered what did they die of to reduce their longevity so much?[122] From what he had seen of the Negroes in Georgia and Alabama, Lyell thought that their social condition was improving. As they learned various trades and mechanical skills, they were gaining in intelligence. He believed that such mental changes were accompanied by changes in facial appearance and physique. Nott, for his part, thought Lyell's ideas absurd. He wrote to Samuel George Morton at Philadelphia:

> Mr. Lyell left here today—He has read my articles & is very full of the subject. I like him very much,—he is an honest seeker after truth, takes very liberal views of everything in this country, but is travelling so fast that he is taking up many false impressions & like all travellers I fear will write a good deal of trash—
>
> He believes the negroes might in a few generations be brought up to the whites in intelligence, he acknowledges the brain to be the organ of intellect & thinks the "physique" would be developed with the "morale." I gave him my notions on the subject & he seemed to me much staggered by them.[123]

By contrast, Lyell recorded in his notebook that if Nott were right in thinking the mulatto intermediate between black and white and that the function of the brain depended on its material structure, then as the Negro improved his brain by learning skilled trades (as Lyell had seen happening in Georgia and Alabama), he might be able to transmit such improved intelligence to his children. After what he had seen in the South, Lyell could not accept Nott's deep pessimism about Negroes. Nevertheless, as a scientist he took seriously Nott's view that if the human race formed one species, an immense amount of time had been required for the development of

122. Lyell, Notebook 134, 13, Kinnordy MSS.
123. Nott to Morton, 23 February 1846, Morton MSS.

the various races; ancient Egyptian reliefs showed Negroes as much different from Egyptians as they were in modern times.[124]

After his conversation with Nott, Lyell had another talk about slavery with Dr. Hamilton, the Presbyterian minister. In December 1844, Hamilton had delivered a sermon in which he attempted to develop a biblical justification for slavery, based on the reciprocal obligations of masters and slaves. When the sermon was printed in 1845, it brought down upon him the wrath of the abolitionists.[125] Hamilton told Lyell that the abolitionists had called him "a hoary ruffian, a proud hypocrite, a bloodthirsty villain. If you can prove that slavery is approved by God in the Bible, you prove God to have out-sataned Satan, to out-juggernaut Juggernaut, to out-herod Herod."[126]

On Monday evening the Lyells sailed from Mobile in the steamboat *James L. Day.* In contrast to the vibration of the powerful high-pressure engines of the river steamboats, the *James L. Day,* driven by low-pressure engines, was so quiet that Lyell thought the voyage like one in a sailing vessel. On the boat were a number of Alabama planters moving to Texas with their Negro slaves. One planter told Lyell that he had been "eaten out of Alabama" by his Negroes. When Lyell and Mary awoke the next morning, the boat had left the sea and, having passed through a deep, narrow channel, was steaming across the broad waters of Lake Pontchartrain. Lyell recognized Lake Pontchartrain as a great lagoon, cut off from both the sea and the Mississippi River by the development of the Mississippi Delta.[127] Six miles north of New Orleans they landed and traveled into the city in railway cars.

124. Lyell, Notebook 134, 15–17, Kinnordy MSS.
125. Hamilton, *Duties of Masters and Slaves.*
126. Lyell, Notebook 134, 19, Kinnordy MSS.
127. Skinner, "Lyell in Louisiana," 244.

CHAPTER 7

The Mississippi Valley

LYELL WAS ANXIOUS to see the Mississippi River, having read descriptions of it and referred to it in the *Principles*. He wished particularly to study its delta—clearly a product of the river over time—to learn what it told of the age of the river. In addition, he wanted to trace the Tertiary and Cretaceous formations of the gulf coastal plain into the valley, as they might be revealed in bluffs along the Mississippi River.

NEW ORLEANS AND THE DELTA

After many weeks among the new towns and backwoods of Georgia and Alabama, of swaying in stagecoaches over rough roads and living in simple, sometimes rough, country hotels, Charles and Mary found New Orleans fascinating and exciting. An old city, it immediately reminded them of France, where they had traveled so often. New Orleans offered the comforts and luxuries of a European city together with a strangeness of atmosphere that resulted from the mingling of French, Spanish, and English influences and the presence of slave and free, black and white, and all intermediate shades of color. It was the chief port for the expanding West and the burgeoning agriculture and trade of the Mississippi Valley.

New Orleans contained two distinct municipalities, American and French, separated by Canal Street. The French Quarter was a French city, in which the buildings resembled those in France and in the streets and shops people spoke French. The Lyells went first to the St. Charles Hotel, a handsome building at the end of the Rue Royale. It was full so they went to another large hotel in the French Quarter, the St. Louis, where they obtained rooms furnished with muslin curtains and scarlet drapes. After settling in, they walked to the levee along the Mississippi River, where they counted thirty-four large river steamboats, each with two tall black smokestacks. The steamboats took five days to go upriver to St. Louis, and the fare was twenty dollars; the trip to Natchez took a day and a half and cost eight dollars. From the upper deck of one steamboat that they boarded, the Mississippi did not look as wide as they had anticipated. Lyell thought it about as broad as the Thames at London. The large wharves along the levee were piled with bales of cotton. As he walked along Canal Street, Lyell reflected that he was "tired of everlasting cotton."[1]

As they walked back to the hotel, the streets were filling with revelers. It was Shrove Tuesday (24 February), the last day of Carnival. "All the French population were dressed in masks, on horseback or in carriages, dressed up & in large waggons, in the most grotesque dresses flinging flour over one another & the passerby," Mary noted. "It was so perfectly unlike anything I ever saw in this old World where there is so much work & so little play." Some of the revelers were dressed as Indians, with feathers on their heads. Among the onlookers, themselves contributing to the strangeness of the scene, were "American, French, Creoles, Negroes and Quadroons with their straight black hair being only two shades removed from white."[2]

At the St. Louis Hotel, the Lyells found Richard Henry Wilde, whom they had met in 1842 at Augusta, Georgia. He now joined them at meals. Wilde, a widower, had moved to New Orleans in 1843, where he was practicing law. He told them that everything was expensive in New Orleans. The city was really an active center of business for only seven months of the year, from December through June. From July through November, the heat and, even more, the mosquitoes together with the constant threat of yellow fever drove many people from the city. Business came almost to a standstill. The population of New Orleans was between 130,000 and 140,000 people. The English-speaking portion was growing more rapidly than the French-speaking portion. The number of French-speaking Louisianans who returned to France with fortunes nearly equaled the number of new immigrants from France. Among the Americans who had migrated from the eastern seaboard to Louisiana, Mississippi, and Arkansas, Wilde observed, "the sea coast people had kept to their fish & oysters, the pine barren crackers to their light wood & the oak & hickory men to their red soil."[3] Each settler kept to the kind of country and way of life with which he was familiar. No books, not even the Law Reports, were printed in Louisiana, because labor was so expensive. The combination of the short working year with the threat of yellow fever kept printers away. Yellow fever rarely at-

1. Lyell, Notebook 134, 32, Kinnordy MSS.

2. Lyell to Caroline Lyell, 25 February 1846, Kinnordy MSS.

3. Lyell, Notebook 134, 37, Kinnordy MSS.

tacked natives of the city, usually occurring only among newcomers. People who survived an attack were safe thereafter, but many died from the first attack.[4] Wilde said that the belief of the ancient Romans that malaria was caused by invisible animalcules had recently been taken up by several medical men to explain the cause of yellow fever. The fact that the previous summer, although exceedingly hot, had been free of yellow fever had overturned all recent theories.

Also staying at the St. Louis Hotel was Mrs. Charles Stewart, the wife of Commodore Charles Stewart of the United States Navy, famous as commander of the U.S.S. *Constitution* in 1813–14.[5] A sister of Mrs. Allen of Gardiner, Maine, who had been so hospitable to the Lyells the previous September, Mrs. Stewart was also acquainted with the Ticknors and other Boston friends of the Lyells. She had traveled to Europe, where her daughter had married a son of Sir Henry Parnell. She remarked that in England the higher the rank of people in society, the less prejudice they seemed to feel against Americans. Most of the guests at the St. Louis were Creoles, that is, native-born Louisianans of French descent. The Creoles, Mary wrote, "form the best society of the place & have till very lately kept quite distinct from the American population, but are now beginning to intermarry & so will no doubt lose their peculiarities."[6] The Creoles spoke French, and the atmosphere of the hotel continually reminded the Lyells of France.

On Wednesday, Wilde took the Lyells on a walk through the French Quarter. In the Place d'Armes they entered the cathedral, poor and bare in the interior. People were on their knees praying, a sight that seemed strange on a weekday after they had been traveling so long through Protestant communities. They also saw the old Spanish government house, in 1846 a private house, and went into the State House, where the legislature was in session. The members were making speeches, some in French, others in English, on various proposals to move the capital from New Orleans to some smaller center upriver. One member favored Baton Rouge and another, Donaldsonville. Wilde explained that the law in Louisiana

4. On 10 September 1847, a year and a half after Lyell's visit, Wilde died of yellow fever at New Orleans.

5. Charles Stewart (1778–1869) was born at Philadelphia and entered the United States Navy, where he saw extensive service in the Mediterranean in the wars against the Barbary pirates. In 1813 he married Delia Tudor and the same year was given command of the U.S.S. *Constitution*, with which in 1814 he captured two British warships off the Madeira Islands.

6. Mary Lyell to Leonora Horner, 26 February 1846, Kinnordy MSS.

was governed chiefly by the *Code Napoléon*. The old French civil law, introduced when Louisiana was founded, had been altered to conform with the code, except in those portions that had already been changed when Louisiana was under Spanish rule. Now that Louisiana was one of the United States, the laws were gradually being altered to resemble the Common Law of England.

One evening, the Lyells attended the French opera for a short time. Mary observed:

> It is a very pretty little theatre & I hardly ever saw so many handsome faces. The Creoles are considered remarkably handsome & it is quite true. The original race came from Normandy, but they have improved here, for it is so fine a style of face. They dress with true Parisian taste, but their manner of dressing their hair is exquisite, a great variety, each fit for a Grecian statue. Very, very long hair they must have, with such bright coloured ribbon, or a single flower, or pins tuck in. Then they are not so thin as American women are in general.
>
> We had a little French vaudeville, some medium dancing & an act of the Freichutz. That was all we staid for. The theatre is very pretty, a gallery for the Negroes pretty well filled & another gallery & boxes for the quadroons.[7]

On 27 February, Lyell went with Dr. William M. Carpenter, a young New Orleans physician and naturalist, to Balize, 110 miles downriver from New Orleans at the tip of the Mississippi Delta, where the great river poured into the Gulf of Mexico. Dr. Carpenter's knowledge of botany and geology, especially the geology of Louisiana, combined with a lively intelligence and keen sense of humor, made him a particularly valuable companion and guide. As a boy at St. Francisville, Louisiana, Carpenter had assisted John James Audubon to collect and mount birds.[8] After a period at the United States Military Academy at West Point,[9] he had studied medicine in New Orleans. In 1838, he discovered fossil teeth and the jaw of a mastodon and a fossil horse's tooth in West Feliciana Parish.[10] He

7. Ibid.

8. Herrick, *Audubon,* 1:322–24, 345–46; Cock, "Carpenter."

9. United States Military Academy Archives, West Point, New York. Information on Carpenter's attendance at West Point was kindly provided by Suzanne Christoff, Assistant Archivist, USMA Archives. Fossier, "Medical Education in New Orleans," 338. Carpenter is listed by Fossier as M. M. Carpenter, an obvious misreading of Carpenter's signature, because his *W* looked very much like an *M*.

10. Carpenter, "Fossils Found in Louisiana."

described a forest of fossil trees at Port Hudson and made the first discovery in North America of the fossil tooth of a tapir.[11] In 1842, he had just begun to practice medicine.[12] Just a few weeks before Lyell's visit, the purported fossil remains of a man eighteen feet high were brought to New Orleans from Tennessee for public exhibition. The owner of the exhibit invited Carpenter to examine it and give his opinion. The skeleton was supported in an upright position by a vertical beam of timber, with missing bones replaced with wooden parts. Carpenter saw at once that it was the skeleton of a young mastodon, but "with its half human, half beast-like look, and its great hooked incisive teeth," wrote Carpenter, "it certainly must have conveyed to the ignorant spectator a most horrible idea of a hideous, diabolical giant, of which he no doubt dreamed for months."[13] Along with the mastodon, Carpenter described the fossil bones of ox, tapir, elephant, and other indeterminate animals, recently brought from Texas.

While Lyell and Carpenter were on their expedition to the mouth of the Mississippi, Richard Wilde showed Mary the sights of New Orleans. She also went with Mrs. Stewart and another lady to a reception for Samuel Goodrich, the author of Peter Parley's tales.[14] The house was crowded with adults and children. Mary met the governor and lieutenant governor of Louisiana, various prominent persons, and the guest of honor. On Sunday morning she accompanied a New Orleans lawyer, Lucius C. Duncan,[15] and his wife twice to Christ Church, where they heard sermons by the Reverend Francis Lister Hawks, a most eloquent preacher.[16]

When Lyell and Carpenter left New Orleans Friday evening, they expected to reach Balize by midnight, but the steamboat had gone little more than halfway when it had to drop anchor in a thick fog,

11. Carpenter, "Miscellaneous Notices"; Carpenter, "Bitumenization of Wood"; Carpenter, "Geological Survey of Louisiana."

12. Duffy, *Tulane University Medical Center,* 6–7; Duffy, ed., *Rudolph Matas History of Medicine in Louisiana,* 2:274.

13. Carpenter, "Fossil Bones Recently Brought to New Orleans."

14. Samuel Griswold Goodrich (1793–1860), a native of Connecticut, in 1827 published *The Tales of Peter Parley about America,* the first of a long series of Peter Parley books for children.

15. Lucius Campbell Duncan (1802–55) was born in Kentucky but in 1812 came with his family to New Orleans, where he later studied law and became eminent in the legal profession. Obituary notices, *New Orleans Bee,* 11 August 1855.

16. Mary Lyell to Leonora Horner, 26 February–7 March 1846, Kinnordy MSS. Francis Lister Hawks (1798–1866) in 1844 was appointed rector of Christ Church, New Orleans, and first president of the University of Louisiana. A distinguished clergyman and outstanding preacher, Hawks was also a historian and writer.

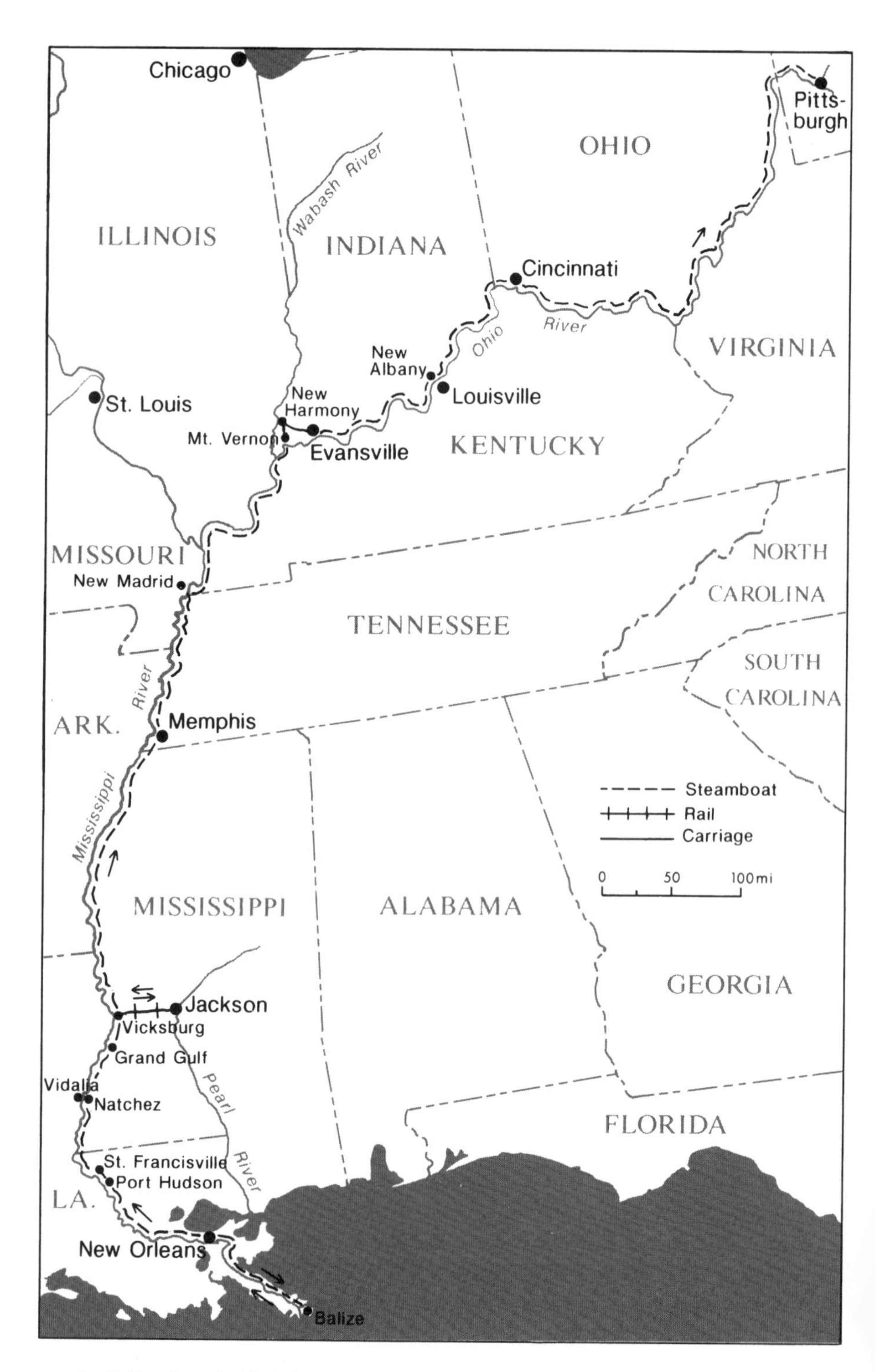

Lyell's Travels on the Mississippi and Ohio Rivers, 1846. Map by Philip Heywood.

where it remained all night. During the evening there was a thunderstorm with heavy rain. The reason for the fog was that the water of the Mississippi, flowing from the north, was quite cold, while the surrounding air was warm. The old pilot on the boat told Lyell that when you followed the current of the Mississippi out into the Gulf of Mexico, the fog still hung over the band of Mississippi water, while the air above the surrounding sea remained perfectly clear.

The next morning, the fog soon cleared enough for the steamboat to proceed. Whenever the boat stopped, Lyell and Carpenter landed to study the trees and plants along the bank. From the deck, they saw occasional flocks of great blue herons and sandhill cranes. As they descended the river, the forest gradually thinned. At Fort Jackson they caught their first glimpse of the Gulf of Mexico; thereafter, the trees disappeared except for a few small willows. In the marshes on either side were flocks of white whooping cranes and many other birds. Gulls followed the steamboat, eating apples thrown into the water for them.

At Bayou Liere, Lyell was struck by the plentiful palmettoes and fascinated by the fiddler crabs, or land crabs. When he attempted to approach them, they fled into their holes. On the bank, he picked up pieces of pumice that he was told had come from the Rocky Mountains, evidence that volcanic regions must exist in the western part of the Mississippi Basin. Lyell was astonished to find no land shells in the Mississippi mud. Along the banks of the Rhine and other European rivers, land shells were common. Below Bayou Liere, the bank on either side became merely a narrow strip of land separating the channel of the Mississippi from the waters of the Gulf of Mexico. Along the banks, willow trees grew in rows so regular they seemed to have been planted, but in fact they had grown naturally. The youngest trees, only a few feet high, grew on the most recently formed land next to the river. Behind them was a second row, about eight feet tall, and further up the bank was a third row, about twenty-five feet in height. The trees grew where the river was depositing new soil; on the opposite bank, where the river was cutting into the land, the trees were being undermined and swept away.

Near the mouth of the Mississippi, the rows of willows disappeared to leave only a strip of reed and grass-covered silt, or even bare sand, to separate the river channel from the sea. At Balize (a Spanish name meaning the Beacon), a mere cluster of houses near the tip of the delta, Lyell and Carpenter landed. They went up to the lookout, a high wooden platform where the pilots scanned the sea for incoming ships. From the lookout Lyell and Carpenter surveyed the little town of some seventy houses, surrounded by reeds, basking in the evening light. The only sound to be heard was the cry of the marsh hen. Out to sea, to the left lay the North-East Pass lighthouse and far away on the right stood the South Point lighthouse. To the north towards New Orleans, they could see scarcely more land than to seaward—"so small," observed Lyell, "was the space occupied by the narrow banks of the great river."[17] To the west lay Bird Island, covered with trees. The oldest pilot at Balize, George Linton, told Lyell that Bird Island was very high, as much as three or four feet above the sea.[18] On the single boardwalk along the bayou, several girls were walking dogs.

Sunday morning Lyell and Carpenter hired a rowboat to explore the channels among the mud banks. At half past eight in the morning, the temperature of the air was 64 °F, but the temperature of the river was only 48 °F. Lyell wished to learn from the pilots how rapidly the delta was extending into the Gulf of Mexico, but he found such information difficult to obtain. Carpenter had brought with him a copy of Pierre Charlevoix's map of the Mississippi Delta based on observations made in 1722.[19] Lyell and Carpenter were astonished at how accurate it still was and, therefore, how little change had occurred in the delta in more than a century. When Lyell and Carpenter showed Charlevoix's work to the pilots, they said that the maps might have been drawn only the year before, except that bars had developed across the entrance of every bayou that was no longer a discharge point of the Mississippi. In 1846 the main mouth of the Mississippi was the South-West Pass. When Lyell compared the present position of the South-West Pass with where it had been in 1722 according to Charlevoix, he found that it had advanced three miles, or, he estimated, about one mile in thirty years. By contrast, the North-East Pass had not advanced at all since 1722. George Linton told Lyell that until 1770 the Balize Pass had been the principal mouth of the Mississippi. Then the South-East became the main pass, followed by the North-East, and then the South-West.[20]

The rate of growth of the delta was difficult to estimate. Although

17. Lyell, Notebook 135, 28–30, Kinnordy MSS.

18. Bird Island has since disappeared, swept away by hurricanes.

19. Charlevoix, *Histoire et description générale de la Nouvelle France.*

20. Lyell, Notebook 135, 38–42, Kinnordy MSS. Lyell's estimate was approximate. If Charlevoix compiled the data for his maps in 1722, the rate of advance would have been three miles in 124 years, or about one mile in 41 years.

Lyell's Sketch of the Balize, 1846. Lyell, Notebook 135, 28–29, Kinnordy MSS.

Linton said that during the preceding twenty-four years, each of the passes had advanced a mile, yet each time the main channel of the river changed to a new pass, the sea began to eat away at the promontories of the old passes, the tide coming up the old channels and carrying soil out to sea. Both the pilots and the engineers to whom Lyell talked later at New Orleans agreed that the combined actions of the river and the tides tended to maintain the depths of the various channels fairly constantly over long periods of time. When Lyell took into account the relatively constant configuration of the delta, he realized that it was growing as a whole much more slowly than the extension of the South-West Pass since 1722 would indicate. Furthermore, the Mississippi carried much sediment for miles out to sea and distributed it over a broad area of the Gulf of Mexico. The delta had formed slowly, over an immense period of time, yet its size was a measure of the age of the Mississippi River.

The delta fascinated Lyell. It was a richly varied example of immense geological forces endlessly at work, with the result of thousands of years of that work manifest in the very existence of the delta, with its great sweep of marshes and many channels, bayous, and islands. When Lyell and Carpenter started back upriver, Lyell observed in his notebook how his own attitudes had been changed by his visit to the delta:

> High Land
>
> After being at the Balize I was surprised on landing above Ft. Jackson at my altered conception in so short a time of "high land." I have been accustomed to regard land not liable to be inundated as high. You see knee-deep and ankle-deep land, houses built on raised marshes, communicating with their outhouses by bridges—they point to Deer Island and call it "very high land"—"How high?" "At least 4 feet," appealing to the pilots. "Ay it's all that," was the affirmative seconding the assertion.

Linton was concerned that with four hundred people now living at Balize there might not be enough boats to evacuate them in case of a hurricane. Nevertheless, hurricanes were rare and the Mississippi, even during its greatest floods, rose but little in the delta. Lyell observed, "One must remain a day or two & be accustomed to the vast extent of the swamps—their security and permanence—in order to know what high & low land means in a delta."[21]

21. Ibid., 52, 54.

On their way upriver, the steamboat stopped at a sugar plantation about twenty miles below New Orleans to load barrels of sugar. The hoops had come off some of the barrels and needed to be replaced, so the steamboat had to wait some time. While waiting, Lyell and Carpenter walked at leisure through the sugar mill to examine the steam engine that operated the press to extract the juice from the cane. In contrast to the French Creole planters who used horses to operate their presses, the American sugar planters were investing large amounts in the more powerful and efficient steam engines. The mill also had large vats in which the sugar cane juice was boiled down. While the steamboat was stopped, the Creoles, with no thought of time, calmly played cards in the saloon. By contrast, the English-speaking Americans fumed as they waited hour after hour in sight of New Orleans. With their greater drive and efficiency, the Americans were steadily displacing the Creoles from the best land in Louisiana.

As they ascended the Mississippi, Lyell was struck by the remarkable straightness of the channel. For 150 miles from the mouth of the delta it lacked any of the great loops and bends that occurred so frequently above New Orleans. Below New Orleans there was only one such bend, the English Turn, a few miles downriver from the city. Below the English Turn, the river was shaping its bed below sea level and, therefore, was influenced by forces different from those operating above sea level. Above the English Turn, Lyell landed to examine the site of the Battle of New Orleans in 1815, when United States forces under Andrew Jackson had defeated a British force. He saw the swamp through which the British troops had dragged their cannon while under fire from the American forces on higher ground.

On Tuesday evening, 3 March, Lyell returned from his five-day excursion with Carpenter. At New Orleans during the next few days, he sought all the information he could find in old maps and charts of historic changes in the Mississippi Delta. From Dr. John Riddell, professor of chemistry at the Medical College in New Orleans, Lyell learned that the average proportion of sediment in Mississippi River water was 1/1,245 by weight, or about 1/3,000 by volume.[22] From the calculations of Riddell, Carpenter, and Caleb Forshey, the Louisiana state engineer, Lyell learned the average width, depth,

22. John Leonard Riddell (1807–65) in 1836 received a medical degree from the Cincinnati Medical College and moved to New Orleans to become professor of chemistry in the Medical College of Louisiana. Riess, *Riddell.*

and velocity of the Mississippi River and estimated both the volume of water and the quantity of sediment brought down each year. He also calculated the volume of the Mississippi Delta, assuming that its average depth would be about equal to the average depth of the Gulf of Mexico between the tip of Florida and Balize, or about a hundred fathoms. From complex calculations, Lyell decided that the accumulation of the delta must have required at least 67,500 years. Another 33,500 years or more would have been required for the accumulation of the alluvial plain above New Orleans. Even these figures, he thought, were probably much too low. A considerable portion of Mississippi River water sediment was not deposited in the delta but was carried far out into the Gulf of Mexico, possibly even swept by the Gulf Stream into the Atlantic Ocean. The immense time required for the accumulation of the delta fitted with everything that Lyell had learned of the slowness of the growth of the delta in modern times and the remarkable constancy of its geography, despite the ceaseless work of the river.[23] Yet the most striking feature was that although the delta demonstrated that the Mississippi had been pouring its waters into the Gulf of Mexico for more than 100,000 years, the river had carved its valley out of modern Tertiary formations. Like Mount Etna in Sicily, the Mississippi Delta showed how very old were even the youngest geological formations. To his father-in-law Lyell wrote that the delta was "as grand a field as I expected & almost expands even my conceptions of time within the modern era."[24]

On Wednesday, Henry Clay came to see the Lyells at the St. Louis Hotel and urged them to visit him in Kentucky. They also watched the annual procession of the firemen with their ribbon-bedecked fire engines. The well-dressed firemen looked much healthier than the people in the pine woods of Alabama and Georgia, who were continually weakened by recurrent attacks of malarial fevers, diseases relatively rare at New Orleans.[25]

On Sunday morning, 8 March, Lyell and Mary rose early to go to the market before breakfast. As they walked through the foggy streets, Lyell noted that the fog was white, not yellow as in London. Yet it was just as thick. From the market they could barely see through the fog the masts of ships in the river. The market delighted them. Heaps of Spanish moss brought in from the country for

23. Lyell, "Delta and Alluvial Deposits of the Mississippi."

24. Lyell to Leonard Horner, 5 March 1846, Kinnordy MSS.

25. Mary Lyell to Caroline Lyell, 25 February–7 March 1846, Kinnordy MSS.

stuffing mattresses lay alongside palmetto brooms, and every kind of meat and fish. Tropical fruits from the West Indies abounded—bananas, excellent pineapples for twenty-five cents, and an immense variety of nuts and vegetables, including golden cobs of Indian corn. In some stalls they could buy hot coffee in white china cups, reminding them of Paris.

The people in the market were as varied and interesting as their wares. Two Indian women, wrapped in blankets, were selling roots. The Lyells saw more mixture of races than they had ever seen before.[26] Many of the French Creoles were darker in complexion than some of the quadroons or octoroons. They heard both French and Spanish spoken and a patois that seemed to intermingle the two languages. A Negro speaking English mingled French with English idioms. The crowd was cheerful and noisy, the Negroes particularly merry. "The slaves have Sunday to themselves," wrote Mary, "& come in to sell their little produce in the morning & it was such fun to see them & hear them—a perfect Babel, laughing & jumping & gesticulating, most of them speaking French."[27] In the midst of such a colorful gathering, Lyell found it almost startling to see a young white man and woman, walking arm in arm, evidently recently arrived from the North, and standing out by contrast from the crowd.

After attending Christ Church to hear Dr. Hawkes, the Lyells went in the afternoon for a drive along the shell road outside the city. They took with them the British actors Charles and Ellen Kean, who were then playing at New Orleans.[28] The flat country and well-kept country houses alongside the canal reminded them of Holland. The trees were coming out in leaf and the grass was now green. For months they had seen nothing but brown grass. During the drive, Ellen Kean got out of the carriage to gather wildflowers by the roadside. On Monday evening a Mrs. Lippincott, a northerner wintering in New Orleans, invited them to join her at the theater, where the Keans were giving a benefit performance of *The Gamester*. The play was extraordinarily good, Ellen Kean's acting particularly effective. "After the curtain dropped," wrote Mary, "not one of us spoke a word for some minutes." The theater was "crowded to the

26. Lyell, Notebook 136, 15, Kinnordy MSS.

27. Mary Lyell to Eleanor Lyell, 9 March 1846, Kinnordy MSS.

28. Charles Kean (1811?–68) was the son of the famous actor Edmund Kean (1787–1833) and was himself a well-known actor of the English stage. Ellen Kean (1805–80) was born Ellen Tree and was a distinguished actress.

very roof," and the Keans "were called for & most loudly cheered at the end."[29]

Voyage up the Mississippi River

The next morning Lyell left New Orleans to go upriver on the steamboat *Rainbow.* He wished to stop at Port Hudson, about 160 miles up the Mississippi, to examine the bluff described by William Bartram in his *Travels* and where Dr. Carpenter had observed the conversion of a submerged forest of fossil trees into coal.[30] Port Hudson was a small place. He would have to take his chances of finding a place to stay, so Mary remained in New Orleans. On Tuesday evening, Richard Wilde would escort her to the steamboat *Magnolia* to go directly to Natchez, Mississippi, to await Lyell's arrival.

From the deck of the *Rainbow,* Lyell admired the dome of the St. Charles Hotel, from which the day before he and Mary had looked out over New Orleans, and the tower of St. Patrick's Cathedral. For a long time, the landmarks of the city remained in sight as the river bent round, the *Rainbow* at times steaming directly toward instead of away from them. On the riverbanks, many trees were still bare, only a few beginning to show an aura of green. They met a number of steamboats and schooners loaded with cotton going downriver. At four-thirty in the afternoon the captain estimated that since they had left the wharf at New Orleans, they had met ten thousand bales of cotton. Each bale was worth $35 so they had seen cotton passing downriver to a total value of $350,000—an immense sum in 1846. They also met a flatboat from near Pittsburgh, floating downriver with the current, carrying a cargo of farm produce destined for New Orleans. During the afternoon a heavy downpour of rain was followed by fog. Immediately, the *Rainbow* stopped and dropped anchor to avoid collision in the fog with flatboats or other craft coming downriver. As they resumed going upriver, the levee along either bank gradually rose in height. Only about four feet high at New Orleans, sixty-five miles upriver the levee had risen so high that only the tops of the trees were visible in the swamps beyond. When the

29. Mary Lyell to Eleanor Lyell, 9 March 1846, Kinnordy MSS.

30. After describing the multicolored strata of clay, marl, and chalk that made up most of the bluff at what later became Port Hudson, Bartram wrote: "The lowest stratum next the water is exactly of the same black mud or rich soil as the adjacent low Cypress swamps, above and below the bluff; and here in the cliffs we see vast stumps of Cypress and other trees, which at this day grow in these low, wet swamps, and which range on a level with them. These stumps are sound, stand upright, and seem to be rotted off about two or three feet above the spread of their roots; their trunks, limbs, etc. lie in all directions about them." Bartram, *Travels,* 433.

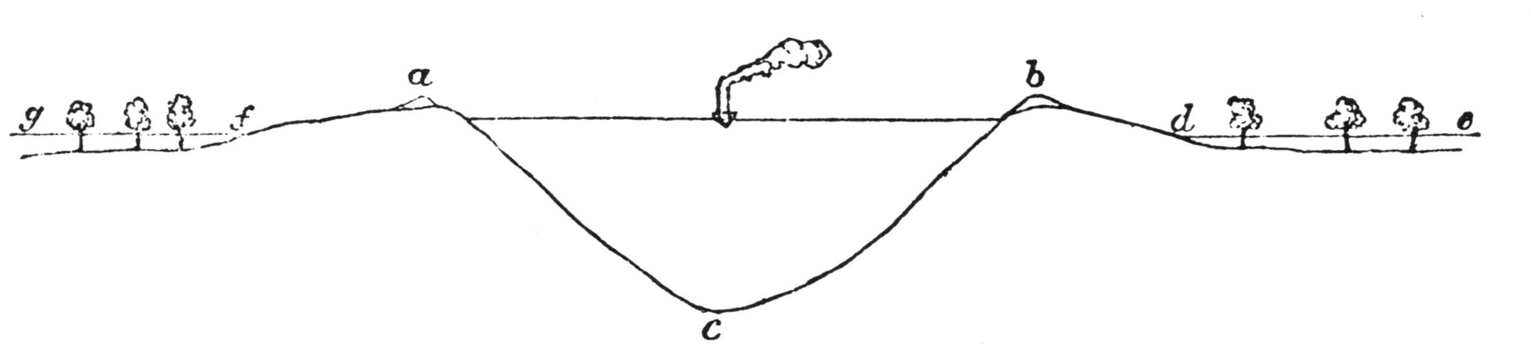

Section of Channel, Bank, Levees (a & b) and Swamps of Mississippi River. Lyell's *Second Visit* (1849), 2:170.

Mississippi was in flood, a break in the levee allowed the water to pour into the swamps, often sweeping flatboats through the opening to become marooned helplessly in the swamps.

Early on the morning of 11 March, the *Rainbow* landed Lyell at Port Hudson. He had carefully studied Bartram's description of the Port Hudson bluff in 1777, although by 1846 the bluff was several hundred feet east of where it had been in Bartram's time. He had also read Carpenter's account of the bluff as it was in 1838. Its striking feature was a bed containing the stumps of trees, logs, and other pieces of wood, all partially converted to coal, visible at about the high-water level of the Mississippi. Some of the logs were converted completely into coal at one end, while at the other end they remained wood, although wood so soft that it might be crushed between the fingers. Among the logs, Carpenter had recognized the water oak and a species of pine, but not the cypress characteristic of the modern Mississippi swamps. Further down, about at the level of low water on the Mississippi and beneath a bed of clay twelve feet thick, Carpenter discovered a second bed containing sticks, leaves, and the remains of stumps. In this lower bed he found fossil swamp-hickory nuts abundant, while the logs included cypress, swamp hickory, a species of cottonwood, and other trees characteristic of the modern swamps of the Mississippi Delta. The fossil cypress stumps showed the knees characteristic of cypress roots when the trees grow on ground subject to flooding. Cypress characteristically occupied the flooded lowlands of the Mississippi. When sufficient silt had been deposited to raise a portion of the flood plain above the level of the river at low water, cottonwood and swamp hickory succeeded cypress.[31]

When William Bartram observed fossil cypress stumps in the cliff at Port Hudson in 1777, he wondered how they had come to be covered with the strata of earth, nearly one hundred feet thick, forming the bluff. The same question fascinated Lyell. At Port Hudson on 11 March, the water was so high that the bed containing the cypress stumps was twelve feet underwater. The captain of the *Rainbow* had told Lyell that during the previous summer the stumps could be seen in the bluff as plainly as ever, and at Port Hudson several persons confirmed the observation. Lyell examined a ravine where a small stream had cut through the beds of the bluff to a depth of sixty feet. He found that the deposits were such as would

31. Carpenter, "Bitumenization of Wood."

be formed in a riverbed or in the swamps on either side of a great river. In a canoe he followed the bluff downriver for about a mile below Port Hudson. At one place, some men were digging in the bluff to obtain sand for brick making. They had just discovered a large buried tree, blackened but not yet transformed into lignite. In its trunk Lyell counted 220 rings. Nearby were two smaller fossil logs, all the logs lying together as if they had been driftwood buried in sand. The men told him that one of the logs was oak, another hickory, and the third sassafras. During the previous eight years, the river had cut back the bluff at Port Hudson as much as two hundred feet. The men described a great collapse of the bluff a few years before, when about three acres of land fell into the river.[32]

The Port Hudson bluff showed that the Mississippi Delta had undergone both gradual subsidence and reelevation through a vertical distance of perhaps 150 feet. Lyell reasoned that the subsidence must have been slow enough to allow the river to deposit layers of sediment at an equivalent rate so as to prevent the delta from being flooded by the sea. If the present delta should sink suddenly, even by as much as ten or twelve feet, it would be covered by the sea.

After making his observations at Port Hudson, Lyell went downriver in the canoe to Fontania plantation on the east bank of the river. Dr. Carpenter had given him a letter of introduction to the planter, Mr. Faulkner. Faulkner took Lyell to visit Lake Solitude, an old oxbow bend of the Mississippi now cut off from the river. A passenger on the *Rainbow* had told Lyell of a remarkable floating island in the lake. Such islands were formed by tangled masses of driftwood floated into the lake from the Mississippi. Becoming covered with silt at flood time, shrubs and trees soon began to grow on them. At first the island floated freely about the lake, but as the cypress trees growing on it became larger, their roots extended down through the water to anchor the raft in the mud of the bottom.

Lake Solitude possessed a wealth of wildlife, its waters swarming with fish and hundreds of alligators. Lyell saw several alligators lying like black logs in the water, only their noses showing above the surface; one was about five feet long. The trees were not yet in leaf, and on the bare branches of willow trees along the shore, numbers of white cranes perched. He saw also a great blue heron, a woodpecker, and many red-winged blackbirds. Cormorants and water rails swam in the water, and towards evening the bullfrogs began to

32. Lyell, *Second Visit,* 2:182.

croak. Faulkner, a deadly accurate marksman, brought down a specimen of every bird that came within range of his gun so that Lyell could examine its plumage. They took two hen ricebirds, starlings, a woodpecker, a sapsucker, a small sparrow, and a large speckled owl.[33] Faulkner also shot down a cluster of flowers and fruit of the red maple for Lyell to examine. On the east side of the lake, they climbed the bluff, a southern extension of the Port Hudson bluff, and from its top gained a wide view of the great plain of the Mississippi Delta.

That evening, Lyell wished to take the next steamer bound upriver for Natchez, and Faulkner told his slave to waken Lyell if a steamboat should approach. About an hour after midnight, the slave called him. In bright moonlight Lyell walked down to the riverbank, where the slave had a signal fire. Around Lake Solitude the owls and other night birds were calling and many frogs were peeping, while in the distance could be heard the puffing of a steamboat. When they hailed the boat as it came near, the reply was "*La Belle Creole,* bound for Bayou Sara"—far short of Natchez. As Lyell turned to walk back to the house, standing on the bluff amid great magnolia trees, he was struck by the beauty of the moonlit night and the strangeness of the surrounding wilderness with its many night sounds. When the slave next wakened him, near dawn, the moon was still up, but the river was now covered with fog. A steamboat, appearing through the mists, announced herself as the *Talma* of Cincinnati. The crew quickly put out a plank, about a foot wide, and Lyell walked on board. As the plank moved slightly under his weight, he reflected that if he misstepped, he might easily find himself in the cool waters of the Mississippi.[34]

When Lyell registered as a passenger on the *Talma,* the clerk asked him whether he was the author of a work on geology. Lyell admitted that he was, and the clerk asked him whether he knew Thomas Macaulay. Lyell did. The clerk then showed him an article on Joseph Addison in a recent number of the *Edinburgh Review* and asked him whether Macaulay were the author. Lyell said that he was. The contrast between the wilderness on the riverbank, where he had just been, and the clerk's interest in London literary life struck him. A few minutes later Lyell was shown to his room and for the third time that night went to bed. As he crept under the mosquito net into the comfortable berth, he felt "the satisfaction you feel in

33. Lyell, Notebook 136, 76, Kinnordy MSS.

34. Ibid.

getting under shelter in heavy rain" and, despite the bells and other noises of the boat, was soon asleep.[35]

The next day, as Lyell watched the banks of the Mississippi from the deck, he pondered the upright cypress stumps, visible in many places at various levels in the banks. Since cypresses grew for centuries in the lowlands of the Mississippi Valley, their stumps would gradually be buried by the accumulation of sand and silt around them during the annual flooding of the river. When a cypress tree died and fell, or rotted down to the surface, its stump would be preserved in the sand and silt in which it was buried. In the meantime, young cypresses would grow on the new surface, several feet above the level at which the previous generation began its growth. When in time the Mississippi shifted its course to cut through such a cypress swamp, the old stumps, buried at various levels, would be revealed.

On the boat Lyell talked with the pilot about the efforts of engineers to shorten the Mississippi by making a cut across the base of the longer bends. Such attempts were often futile. When they did succeed, the river immediately began to cut away the bank to make its course as sinuous as before. At the mouth of the Red River, he observed the great bar, covered with willows and poplars, and the succession of bluffs along the eastern bank of the Mississippi. Late that night he landed at Natchez and joined Mary at the hotel.

Mary had left New Orleans Tuesday evening on the *Magnolia,* one of the largest and finest steamboats on the Mississippi. After escorting her to the boat with her luggage, Richard Wilde introduced her to the captain. He also found several acquaintances on board and asked one of them, Judge Arthur Hopkins[36] from Mobile, to look after Mary particularly. Before they parted, Wilde gave Mary a manuscript copy of one of Michelangelo's sonnets that he had translated from the Italian. He promised also to send her his comments on Lyell's father's essay on Dante, of which the Lyells had given him a copy. Mary wished that Wilde would come to London, because he would be "so exactly suited to the literary society there."[37] Since the death of his wife, he was lonely and at New Orleans few persons shared his interest in Italian literature.

35. Ibid., 74.

36. Arthur Francis Hopkins (1794–1865) was born in Pittsylvania County, Virginia, and was a distant relative of Thomas Jefferson. In 1816 he began the practice of law at Huntsville, Alabama. A very successful lawyer, he became a large landowner in Alabama and moved to Mobile.

37. Mary Lyell to Sophia Lyell, 20 March 1846, Kinnordy MSS.

The *Magnolia* was the most splendid steamboat Mary had yet seen. The staterooms were spacious and the main cabin richly furnished with crimson drapes, velvet-covered chairs and sofas, and a piano. At dinner there was "every luxury, turtle soup, ice cream, claret, bananas, pineapples & everything that could be thought of." Next day the boat steamed past one plantation after another, each with its neatly built houses for the Negroes. Mary decided that the slaves in Louisiana lived under conditions similar to those they had seen in Georgia, "where their lot is certainly not an unhappy one as compared to other poor people."[38] Late Wednesday night, the *Magnolia* reached Natchez. Judge Hopkins, her pleasant companion all day, escorted her to the hotel before taking his leave.

Next morning, Mary ventured out and found Natchez in the midst of its spring beauty with a luxuriance of flowers such as she had never seen before. Everywhere in woods and gardens the fragrant yellow jessamine bloomed with its bell-shaped flowers. Hedges of Cherokee rose were covered with cascades of large white flowers.[39] She brought with her letters of introduction to several persons at Natchez, one of them Joseph Davis, a rich planter and the elder brother of Jefferson Davis.[40] Joseph Davis immediately invited Mary to go driving in his carriage. They visited several houses with beautiful gardens, some laid out in a formal French style and adorned with statues. The houses were elegantly furnished; Mary noted that "their masters showed a refinement and intelligence which was unexpected to me in the State of Mississippi."[41]

On 13 March, his first day at Natchez, Lyell went with Dr. Montroville Dickeson to examine the river bluffs.[42] The previous year, Dickeson had sent to the meeting of the Association of American Geologists and Naturalists at New Haven, Connecticut, a description of the Natchez bluffs in which he distinguished twenty-two beds.[43] These bluffs were 230 feet high at low water on the Missis-

38. Ibid.

39. Ibid. The Cherokee rose is a smooth-stemmed white rose, *Rosa laevigata,* of Chinese origin. It has been cultivated extensively in the southern United States and is the state flower of Georgia.

40. Joseph Emory Davis (1784–1870) was born in Georgia but in 1811 moved with his father to Mississippi. Admitted to the bar in 1812, he was in active practice until 1827 when he became a planter. In 1820 he settled at Natchez, but he also had a plantation near Vicksburg. Lynch, *Bench and Bar of Mississippi,* 73–78; Davis, *Papers,* 1:19.

41. Mary Lyell to Sophia Lyell, 20 March 1846, Kinnordy MSS.

42. Montroville Wilson Dickeson (1810–?) was born at Philadelphia. About 1828 he began to study medicine. He wrote several papers on fossils from the Cretaceous beds of New Jersey. From 1837 to 1844 he traveled through the Ohio and Mississippi Valleys, excavating Indian mounds and collecting artifacts. He was at Natchez as early as 1842. See Culin, "Dickeson Collection," 113–14.

43. Dickerson [*sic*], "Geology of the Natchez Bluffs," 77–79.

sippi and 180 feet at high water. The beds of gravel and sand at the bottom of the bluffs lacked fossils, except for a few pieces of wood, corals, and other fossils derived from older rocks from which the sand and gravel had come. The upper sixty feet of the bluffs consisted of an unstratified yellow clay loam. In the loam Lyell found fossil shells of about twenty species of land snails, abundant and well preserved and all identical to species still living in the Mississippi Valley. The clay loam of the Natchez bluffs resembled closely the loess of the Rhine Valley that Lyell had studied near Heidelberg in 1832.[44] In both deposits, the genera of fossil shells were the same. In the Rhine Valley the loess containing land shells sometimes passed into a stratified deposit containing freshwater shells—clearly a sediment laid down in a freshwater lake. The Natchez loam formed a plateau about two hundred feet above the level of the Mississippi flood plain, twelve miles wide from east to west but extending a much greater distance north and south along the east bank of the Mississippi. Every stream crossing the plateau had carved out its own deep, narrow valley or ravine, providing cross sections through the loam deposit.

Lyell rode with Dickeson to the village of Washington, six miles east of Natchez, to see Colonel Benjamin Wailes, a planter interested in geology and natural history, whom Lyell had met previously at New Orleans. The same age as Lyell, Wailes had come with his parents to the Mississippi Territory in 1807 as a boy of ten. As a trustee of Jefferson College at Washington, Wailes promoted the teaching of geology and had established a museum with a collection of fossil shells and bones.[45] Like Dickeson, Wailes had sent a paper to the meeting of the Association of American Geologists and Naturalists in April 1845.[46] He showed Lyell his collection of fossil bones from the Mammoth Ravine near Washington, including those of a tapir found by Dr. Carpenter, and bones of horse, deer, bison, *Megatherium,* and mastodon. Lyell also examined Wailes's collection of fossil shells from the bluffs at Vicksburg. He was interested especially in the shells from a local marl containing such freshwater species as *Lymnea, Planorbis,* and *Cyclas,* in place of the land shells found in the Natchez bluffs. Thus the Natchez loam, like the loess

44. Lyell, "'Loess' of the Rhine."

45. B. W. Brown, "Geology in Mississippi." See Sydnor, *Wailes.*

46. Wailes, "Geology of Mississippi." Sydnor, on the basis of a report of the New Haven meeting in the *New York Daily Tribune,* 8 May 1846, wrote that Wailes had reported the occurrence of a portion of a human skeleton in the same bed as large unknown fossil bones described by Dickeson. There is no mention of human bones in the published version of Wailes's paper.

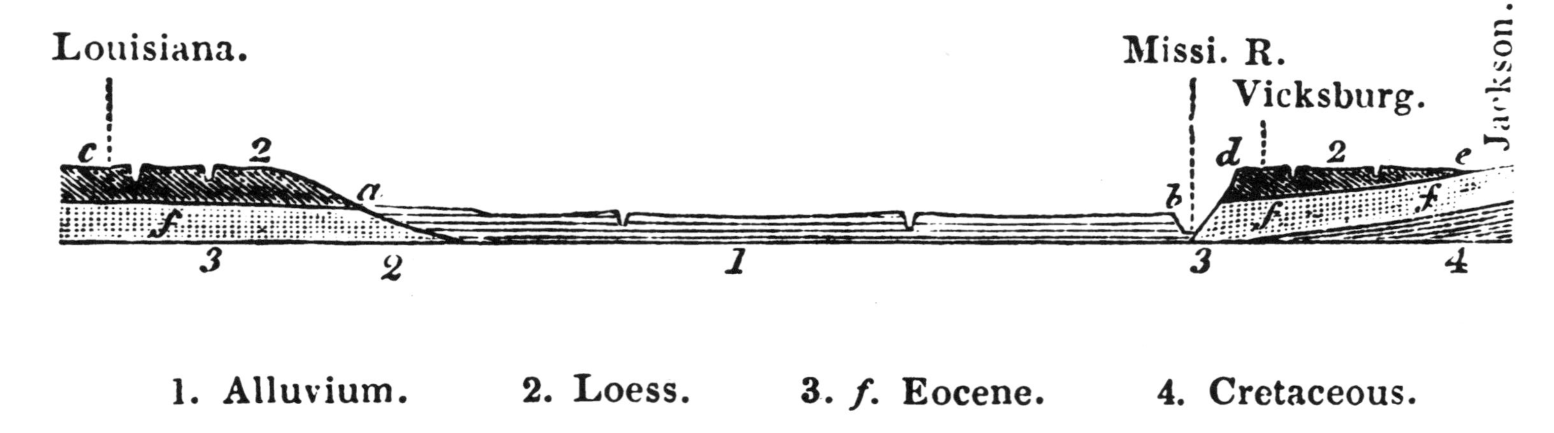

Valley of the Mississippi. Lyell's *Manual* (1851), p. 116.

of the Rhine Valley, changed in places from an unstratified deposit containing land shells to a layered sediment containing freshwater shells. The Natchez formation was so like the loess of the Rhine that, after seeing the marl at Washington, Lyell began to refer to the Natchez loam in his notebook for the first time as "loess," convinced that it was a loess deposit.[47]

The three men went to inspect the Mammoth Ravine, five miles from Natchez, where many bones had been found. As they rode through the countryside in the warm sunshine, the trees, draped with Spanish moss, were still leafless, but dogwood and wild plum were in bloom. Lyell noted various birds—a hawk, mockingbird, bluebird, and hummingbirds. While collecting fossil land shells in the Mammoth Ravine, Lyell was struck again by the resemblance of the yellow loess to that of the Rhine.[48] The loess also contained mastodon bones, while a bed of clay beneath it contained bones of the *Megalonyx,* horse, bear, stag, and other extinct species.

On their return to Natchez, Lyell learned from Colonel Wiley, a planter who had known the country for many years, that the Mammoth Ravine had been formed entirely since the great earthquake of 1812. Just before Lyell's visit, Dickeson had discovered in the ravine part of a fossil human pelvis. The discovery was sensational because the human bones appeared to be from the same deposit that contained the bones of the mastodon and other extinct animals. When Lyell asked Wailes and Dickeson exactly where the human bone had been found, they said that they had not seen it dug out of a particular bed. Lyell concluded that the fragment had simply been picked up in the stream at the bottom of the ravine. The bone was stained a dark color as if it had been buried close to the topsoil. Lyell thought that it had probably been washed out of an Indian burial ground at the top of the ravine. Wailes had come independently to the same opinion.[49] Dickeson, however, was too much excited by the possibility of the discovery of human bones contemporaneous with those of extinct fossil animals to accept such a mundane explanation.

On 16 March, the Lyells, Dr. Dickeson, and Joseph Davis crossed the Mississippi by ferry to Vidalia, a hamlet near where Davis owned three large plantations.[50] At one of his plantations, Davis showed

47. Lyell, Notebook 137, 9, 13, Kinnordy MSS.

48. Ibid., 28.

49. Lyell, "Alleged Coexistence of Man and the Megatherium," 269. This article was a reprinting of Lyell's letter to the editor of the *Times* of London, published in the *Times,* 8 December 1846.

50. Lyell misdated his expedition to Vidalia. Lyell, *Second Visit,* 2:202.

them the "neatly built and well whitewashed" houses in which the slaves lived. He told Lyell that in Louisiana and Mississippi, plantations were "just as well and kindly managed" as was James Hamilton Couper's plantation in Georgia. Richard Wilde had told Lyell the same thing. On one of Davis's plantations, Lyell was impressed by the large number of Negro children in the nursery, an indication to him that the slaves were well treated. At Vidalia the party was joined by Caleb G. Forshey, the Louisiana engineer who for several years had taught engineering and mathematics at Jefferson College. Forshey took the Lyells and Davis on an excursion to Lake Concordia, an oxbow lake fifteen miles long formed from a detached loop of the Mississippi River. Like Lake Solitude, Lake Concordia was surrounded by a dense cypress forest abounding in wildlife. On the water, they saw many ducks and a kingfisher perched on a pole driven into the mud near the shore. Two cormorants wheeled in the air, one with a fish in its mouth that the other was trying to snatch.[51] Forshey showed them a tree where, a few days before, a female alligator had come out of hibernation from beneath its roots with numerous young. Lake Concordia demonstrated the repeated shifting of the bed of the great river. Once the lake was cut off from the river, it rapidly became shallower. It was now only about 40 feet deep, whereas the Mississippi was as much as 150 feet deep. Every time the river flooded, the flood water deposited a layer of silt over the lake bottom; with no river current to remove it, the silt remained.

At Natchez on Monday morning, 17 March, the Lyells went down to the wharf boat to wait for the next steamboat going upriver. As they sat for hours watching the slowly moving water and talking casually to a man also waiting there, Lyell reflected that the Mississippi was

> a death-like stream—a true Lethe—no one is saved who falls in s[ai]d my companion. In the ocean you may be seen, may rise again; be taken up, but [on the Mississippi] sucked down by the eddies of the dark stream of mud, the true River of Oblivion, you are lost—snaggs—fogs, collisions, fires of cotton, explosions.[52]

As dusk descended, the attendant on the wharf boat brought two chairs so that they might have a place to sit in the bare wooden cab-

51. Lyell, Notebook 137, 48–50, 70, 86, Kinnordy MSS; cf. Lyell, *Second Visit,* 2:204.
52. Lyell, Notebook 137, 72, Kinnordy MSS.

in. A little later he provided the additional luxury of a small table and a candle. For a time during their long wait, the Lyells were amused by some boys who built a small raft of wood logs, heaped shavings on it, and set fire to them. The boys then pushed their floating bonfire out into the current to drift downriver through the gathering dusk. Finally, after dark, the steamer *Peytona* picked them up. As they entered the gaslit grandeur of the *Peytona*'s great saloon, the Lyells felt the sharp contrast with the bareness of the wharf boat. They appreciated especially the comfortable beds in their stateroom. In two days the river had risen twenty-five feet, and during the night the *Peytona* bumped against many logs floating downstream on the flood waters.

Early next morning the *Peytona* landed them at Grand Gulf, Mississippi, a small place but with a good inn. For the next two days, they rambled along the bluffs at Grand Gulf in the warm spring sunshine, collecting fossils and botanizing while butterflies fluttered in the air. Although similar to the strata of the Natchez bluffs, the beds at Grand Gulf were often hardened into solid rock to a degree unexpected in such recent strata. On the evening of 19 March they were picked up at Grand Gulf by the *Magnolia,* on which Mary had come to Natchez from New Orleans, and continued upriver to Vicksburg. The Vicksburg bluffs were similar to those at Natchez and Grand Gulf, with a thick cap of loess overlying beds of sandstone, but at the bottom of the bluff appeared strata of Eocene marls and limestones containing many shells and corals. The loess at Vicksburg, Grand Gulf, and Natchez was uniform throughout, a uniformity as marked, Lyell noted, as that of the loess of the Rhine.[53]

Lyell spent the next morning collecting fossils on the Vicksburg bluffs, many of which he had already seen in Colonel Wailes's collection, while Mary remained at the inn to write letters. Sitting in a room overlooking the Mississippi, she could look up now and then and see a steamboat being loaded with bales of cotton at the landing. Their last letters from England and Scotland had reached them at Mobile; she longed for news of home. She wrote to her sister Joanna of their travel-worn condition:

> I have got rather tanned from the hot air lately, in spite of bonnet, veil & parasol, but I am so much in the open air I cannot avoid it. My hair is turning so grey & makes a visible difference in the

53. Ibid., 104.

> colour, but it is time it should. [She was thirty-six.] I never felt better or stronger, & I think Charles looks exceedingly well. Our clothes are so worn out that we shall need a refit at Philadelphia, which is the first place we expect to stay long enough to have any thing made.[54]

That afternoon, leaving Mary at Vicksburg, Lyell took the train to Jackson, the capital of Mississippi, fifty-two miles inland. The railway line passed through woods, soft green with the budding of new leaves. Arriving at Jackson in late afternoon and knowing no one there, he went to a pharmacy, as he had been accustomed to do in villages in France, to ask whether anyone were interested in geology. The druggist said that a physician who lived upstairs might be able to help. The physician, a Dr. Gist, was at home. He had read Lyell's *Principles* and had made a collection of fossils from the countryside around Jackson. Within ten minutes of the time Lyell stepped from the train, he and Dr. Gist were on their way to explore the bed of a small stream. Lyell found an abundance of fossils, many familiar to him from Claiborne. The fossils at Jackson resembled the Claiborne fossils more than they did those of the Eocene beds at Vicksburg. The beds at Jackson were evidently older and lower than those at the base of the Vicksburg bluffs, indicating that in Mississippi the Eocene beds dipped to the west.

Lyell and Gist returned to Jackson after dark. When Lyell entered the hotel, the landlord, a former general in the Tennessee militia, was announcing dinner by proclaiming in a loud voice, "Gentlemen, we are a great people."[55] He then announced the various items on the menu and invited his guests, mostly lawyers attending the legislature, to partake. The landlord introduced Lyell to several of the lawyers, with whom he was soon engaged in conversation. Lyell rallied the Mississippians about the location of their capital in an inaccessible and inconvenient place in the midst of the wilderness.[56] They took his sallies in good humor and explained that because of its railway connection to Vicksburg and the very fertile country surrounding it, Jackson was flourishing, despite its initial disadvantages. That evening, Lyell realized how much he enjoyed the social ease and freedom of America. During the day he had

54. Mary Lyell to Joanna Horner, 20 March 1846, Kinnordy MSS.

55. Lyell, Notebook 138, 5, Kinnordy MSS. In *Second Visit,* 2:210, Lyell described the landlord as "General A____."

56. Lyell, *Second Visit,* 2:211.

Mississippi R.
Vicksburg.
Jackson.
Pearl R.

Length of section 50 miles.

1. Mud of alluvial plain of Mississippi.
2. Superficial drift.
3. Freshwater loam with land shells, &c.
4. Eocene strata.
5. Cretaceous strata.

Geological Section, Vicksburg to Jackson, Mississippi. Lyell, "On the Relative Age . . . Nummulite Limestone of Alabama" (1847).

talked with various men about Mississippi's repudiation of its debts in 1841. After dinner he reflected:

> No state exhibits democracy in a less inviting aspect than Miss. but when you meet with a physic[ia]n, a school master, a lawyer, a watch-maker, with several men, a democratic planter, with whom you talk & find their ideas in harmony with your own, freely condemning the faults of the [State] Constit[utio]n, you cannot despair of the State righting itself by degrees.[57]

The next morning, Lyell went with Gist to examine geological sites near the state penitentiary and along the Pearl River. Five miles south of Jackson on the west bank of the Pearl River, they examined a place where the skeleton of a *Zeuglodon* had been found. Afterward, they went to see the new statehouse and governor's mansion. In 1837 the state of Mississippi had spent more than a half million dollars on new buildings, for which it incurred debts repudiated in 1841. By referendum, the people had refused to approve the taxes needed to pay the state debt. Lyell was puzzled that repudiation could occur when everyone to whom he talked disapproved of it. "A lawyer told me," Lyell observed, "that the only way he could explain the universal condemnation of repudiation, or reconcile it with the large majorities in favor of it, was that selfishness took advantage of the ballot box to oppose their professions."[58]

On 22 March, Lyell returned to Vicksburg and with Mary boarded the steamboat *Andrew Jackson* to continue upriver. On the crowded boat, the passengers included more than a dozen children who made a great deal of noise. Many traveling actors were on board. Mary talked with a young American actress, a Mrs. Forster, who often acted second to Ellen Kean.[59] On the steamboat the Lyells saw more vividly than they had ever seen the rigid discrimination practiced against colored persons, even when their inheritance of color was so slight that they were indistinguishable from white persons. Traveling with Mrs. Forster was a young maidservant, a child of fourteen, who looked like a white girl but possessed some trace of color. When Mrs. Forster was away for a short time, the steamboat stewardess called the child to the second table in the dining room, at which the officers of the boat and the children ate. On her return Mrs. Forster, much disturbed, called the child away from the table.

57. Lyell, Notebook 138, 3, Kinnordy MSS.

58. Ibid., 9, 13.

59. Mary Lyell to Sophia Lyell, 20 March 1846, Kinnordy MSS.

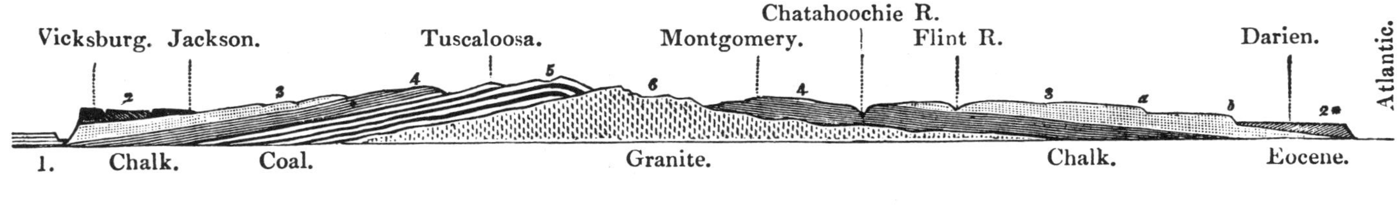

1. Modern alluvium of the Mississippi.
2. Ancient fluviatile deposit with recent shells and bones of extinct mammalia; loess.
2*. Marine and freshwater deposit with recent sea shells and bones of extinct land animals.
3. Eocene, or lower tertiary with Zeuglodon. *a. b.* Terraces. Vol. i. p. 345.
4. Cretaceous formation; gravel, sand, and argillaceous limestone.
5. Coal-measures of Alabama (Palæozoic). See vol. ii. p. 80.
6. Granite.

Section from the Valley of the Mississippi to the Atlantic, Crossing the States of Mississippi, Alabama, and Georgia. Lyell's *Second Visit* (1849), 2:262.

She explained to Mary that had she not, she would have been blamed, when the child's racial identity became known, for allowing a colored child to eat with white persons on equal terms. The child, who appeared to feel only that she had made a social mistake, was allowed to finish her meal in the pantry.

The sharp racial distinction was in strong contrast to the complete lack of social distinction among white people, even those differing greatly in wealth, education, manners, or culture. Mrs. Forster told Lyell that once at a boardinghouse in Cincinnati where she and her husband were staying, a young woman, hardly distinguishable from white persons, came in company with an Englishman. When the Forsters and other boarders perceived that the woman was colored, they threatened to leave the boardinghouse en masse. The proprietor, who said she suspected the girl's color, immediately told the couple to leave, and they departed for New Orleans. The Forsters and other boarders believed that the girl was the Englishman's mistress. Lyell pondered whether prostitution in New Orleans was as great as in London, where it was fostered by great differences in wealth. The contrast of race in America, Lyell reflected, exerted the same effect as the contrast of wealth in England. If freeing the slaves sunk the black race further into poverty, it would, Lyell thought, tend to increase the amount of prostitution among black women.[60]

At leisure on the steamboat, the Lyells contemplated the diverse population of westerners among whom they were traveling. Amid the variety of passengers, the obvious crudeness of many of them and the general freedom of manners, Lyell was conscious of a strictly maintained social order. Courtesy was given without fail to women passengers. The captain was responsible for order and decorum. Any passenger who broke the rules of good manners or defied the captain's authority would be set ashore at the next landing, even in the midst of a wilderness, and left to complete his journey as best he could. Much annoyed by the children on the boat, the Lyells were astonished at the parents' almost complete lack of control over them.

On 24 March, after two nights and a day on the steamboat, the Lyells landed at Memphis, Tennessee; Lyell spent a few hours examining the loess in the Memphis bluffs and found it similar to that at Natchez and Vicksburg. That night they boarded a smaller steam-

60. Lyell, Notebook 138, 90, Kinnordy MSS.

boat, the *Diadem,* to continue upriver another 170 miles to New Madrid, Missouri. An earthquake there in 1811–12 altered the course of the Mississippi River. Lyell had discussed the New Madrid earthquake in the *Principles* and now wished to see its effects first-hand.[61]

At ten o'clock the next night, they landed at New Madrid. There was no hotel. Lyell hoped they might spend the night in the wharf boat but when they arrived, the woman on the wharf boat was not prepared to provide beds for them. Her slave had not come to make them up and the woman sat in a rocking chair waiting for the slave. The slave did not appear and her absence was an insuperable obstacle to making up the beds. Although the Lyells would have been willing to make the beds themselves, the woman only rocked and wished that her slave would appear. Lyell then left Mary in the wharf boat to keep warm by the stove and went to hunt up a place to stay. After finding nothing at the tavern, a storekeeper suggested he try a German baker named Linton, who sometimes took in lodgers. Lyell woke up the Lintons only to learn that their spare room was already occupied, but the lodger, a retired sailor, cheerfully gave up his simple room with a mattress on the floor. During the night, they occasionally heard the puffing of steamboats on the river, but otherwise they slept well. In the morning they appeared for breakfast at half past eight only to learn that the Lintons had been waiting breakfast for them since six o'clock.

Linton, who had come to America from Strasbourg a few years before, began to wait on them at the table. Mrs. Linton, a native American, would not have him be unequal to his guests and insisted that he sit down. The Lintons had no slave, and the lack of a slave in a southern state implied that they were very poor. Their furnishings were few and simple, and their house, in Mary's private opinion, was rather dirty. Despite some inconvenience, the Lyells enjoyed their stay. Like the Lintons and their two children, they called the other lodger, an old sailor with only one eye, "Uncle John." They found him dignified and polite and, moreover, greatly entertaining. A native of Louisiana, Uncle John had visited ports in the United States and in Europe, including many around the Mediterranean. At breakfast he told Lyell that the Mississippi was now flowing where the town of New Madrid had stood before the earthquake in 1811.

After breakfast Lyell went out to examine the riverbank, which

61. Lyell, *Principles* (1830), 407–8.

was about twenty-five feet above the water. The river was running high from the melting of snow to the north. The bank was wearing away rapidly; just the week before, three houses had fallen into the river. The beds in the bank were modern deposits laid down by the Mississippi, and Lyell found them similar to the river deposits of earlier geological periods. With a resident of New Madrid, he went a short distance west of the town to see some of the sinkholes formed during the 1811 earthquake. They were circular depressions of varying depths, the deeper ones containing water, and were surrounded by mounds of earth, thrown out at the time of the earthquake. Lyell also went to see the bed of former Lake Eulalie, which had been drained and left dry by the earthquake. The old lake bed, about twelve to fifteen feet below the level of the surrounding plain, was now covered with a forest of poplars. Many of the trees had trunks two and a half feet thick.

That evening the Lintons told the Lyells of instances of remarkable cruelty toward Negro slaves in the country around New Madrid. "Is it not strange," Mrs. Linton remarked, "that people will give 600 [dollars] for a Negro & lose them & let them be frozen to death for neglect of warm clothing." The Lintons' stories suggested that one or two slaves owned by small farmers might suffer much more than those on the large plantations of Mississippi and Georgia. In his notebook, Lyell reflected:

> The out-of-the-way places in wh[ich] cruelty can be practised, the wreckless new settler who w[ould] stab or shoot any one who remonstrated or who represented the case to a magistrate. The newly come & speedily about to flit; if his character rendered here odious, he can decamp & carry his victims with him. If s[ai]d the old woman my master sh[oul]d go, sh[oul]d move & get me into the country, he will murder me. . . .
>
> The first impulse is to give way to unbounded feelings of horror at slavery when you hear of a case of cruelty, but we must reflect on the difficulty of guarding ag[ains]t the excesses of individual character in any country, apprentices, etc.[62]

The Lintons had come to New Madrid from Indiana only the previous autumn and had the attitude of northerners toward manual work. They felt it no disgrace to work since they were poor but were looked down on in Missouri because they did everything for themselves.

62. Lyell, Notebook 138, 88, 90, Kinnordy MSS.

The next day, Lyell borrowed a horse to ride westward along the Bayou St. John to see more of the effects of the earthquake. For the first three miles, he passed land being cleared. Many trees were girdled to kill them, but he saw no signs of the earthquake. Flights of blackbirds chattered, and he heard the hammering of a woodpecker. As Lyell rode along beside the bayou in solitude, the wildness of the flooded swamp struck him. The Bayou St. John was an isolated loop of the Mississippi, and Lyell wondered when the river might flow through there again. Presently he came to a farm. The owner told him that the first shock of the earthquake had occurred about midnight and people had fled from their houses to avoid falling chimneys. Some people, caught within houses, had to be dug out later; some were injured and some killed.[63] The people had felled trees across the line along which fissures tended to open in the earth, and then clung to the trunks to avoid being buried when fissures opened beneath them. He showed Lyell the remains of several such fissures, each a half mile long and two to three feet deep, running in a northwesterly-southeasterly direction. Originally the fissures were much deeper, but gradually they had become filled in.[64]

At another farm, Lyell saw fissures five feet deep and fifteen feet wide left by the earthquake, one by the farmhouse and another a half mile to the west. The farmer took Lyell to see "the sunk country," an area extending four or five miles along the Bayou St. John and fifty or sixty yards wide. It had sunk eight or ten feet at the time of the earthquake, converting it from dry land to swamp or flooded ground. The sunk country was part of a much larger area of subsidence, covering several hundred square miles in eastern Missouri and Arkansas. The sunken land was covered with young cypresses and cottonwoods. Among them were big, old dead trees—oaks, hickories, gums, and cypresses—killed at the time of the earthquake. The vast area of the sunk country had become lake, swamp, or marsh, abounding in fur-bearing animals such as the raccoon, muskrat, mink, bear, and otter. Each year thousands of their pelts were taken from the area to be sold to the fur trade. Lyell rode back to New Madrid by a different path through an oak forest. In the cloudy afternoon, the muddy road, the dark trunks of the bare trees, and the ground covered with brown oak leaves presented a

63. Ibid., 98; cf. Lyell, *Second Visit,* 2:234–35.

64. Lyell, Notebook 138, 98, Kinnordy MSS. In *Second Visit,* 2:235, Lyell estimated the depths of the fissures as "five or six feet."

somber scene. As he drew near the river, he saw a brightly colored woodpecker and heard the cheerful song of the mockingbird.

To provide for the comfort of the Lyells, the Lintons bought various household articles, including a tablecloth. During the day, Mary had offered to hem the new tablecloth for Mrs. Linton and, when Lyell returned, was at work with her needle. On her part, Mrs. Linton asked to copy a collar, embroidered in the latest Paris fashion, that Mary had bought in New Orleans. The Lintons' home was, Mary wrote, "the poorest house I think I ever staid in."[65] They ate bacon and eggs and coffee three times a day, living in close intimacy with the family. They were not very comfortable, but when Mrs. Dawson, who lent Lyell a horse, invited them to stay at her much larger and better furnished house, they declined. The Lintons had taken them in the first night and had gone to some trouble to provide them lodging, so they would remain there. The Lyells were again bewildered by the Lintons' failure to discipline their children, despite terrible threats to do so. Lyell noted:

> U.S. Mother at New Madrid
>
> "I'll beat you to death" to a child 5 & 7 years old—"I'll thump, I'll whip you if you soil the tablecloth today." All p[ai]d no attention to tho' an occasional thump etc. was applied.[66]

On 27 March, the Lyells went to the wharf boat to wait for the next upriver steamboat bound for the Ohio River. All night they waited. Various steamboats came by, but they were bound either for St. Louis or for Nashville on the Cumberland River, or were already full. During the night, they lay down without undressing so as to be ready to board a steamboat at a moment's notice. At one point, Lyell awoke to hear the puffing and splashing of a steamer going by. When he went on deck, he found the slave who was supposed to wake him fast asleep and the steamboat too far past to hail. He sat up the rest of the night to keep watch himself. The next morning, the wharfman brought Mary a French novel, which she read through "at a galloping pace" in the course of the long morning.[67] Finally, in the afternoon they were picked up by the steamer *Nimrod,* a large and luxurious boat with music, dancing, current newspapers and magazines, and, best of all, comfortable berths.

65. Mary Lyell to Sophia Lyell, 20–31 March 1846, Kinnordy MSS.
66. Lyell, Notebook 138, 112, Kinnordy MSS.
67. Mary Lyell to Sophia Lyell, 20–31 March 1846, Kinnordy MSS.

The Ohio River

During the night the *Nimrod* entered the Ohio River. When the Lyells went on deck Sunday morning, they were at Smithland, opposite the mouth of the Cumberland River. On either bank the land rose in forested hills to limestone bluffs, making the Ohio worthy of its name, "beautiful river." Late that afternoon the *Nimrod* reached the small town of Mount Vernon, Indiana, where the Lyells landed and stayed overnight at an inn.

At Mount Vernon, Lyell found a terrace of yellow loam like the loess of Memphis, Vicksburg, and Natchez, containing both freshwater and land shells. On 30 March, they sent their trunks by stagecoach and hired a carriage to drive about fifteen miles to New Harmony, Indiana, where the geologist David Dale Owen had invited them to visit. As they drove through the woods, they saw wildflowers in bloom beneath the trees. Each time Mary spotted a new wildflower, Lyell stopped the carriage and got down to pick some for her. At New Harmony, a village in a lovely valley beside the Wabash River, Owen and his wife welcomed them. Owen's house, a substantial brick mansion built by the founder of New Harmony, George Rapp, contrasted strikingly with the small log and frame houses of Indiana villages and farms.

New Harmony was a remarkable center of educational and scientific life that travelers might hardly have expected to find among the wooded hills of southern Indiana, hundreds of miles from any city. In 1815 a group of pious German immigrants from Württemberg, led by George Rapp, purchased twenty thousand acres of public land, together with several improved farms along the lower Wabash River. They named their new settlement Harmony. During their early years at Harmony, the Rappites suffered severely from malarial fevers with many deaths. Nevertheless, by 1819 they had between two and three thousand acres under cultivation. Gradually they replaced log houses with frame houses, and for himself Rapp built a large brick house. He also built a granary of stone and brick and, in 1822, a large brick church in the form of a Greek cross. The Rappites set up a sawmill, a distillery, and a mill for the manufacture of woolen cloth. They planted orchards and a vineyard.

Despite the prosperity of their settlement, the Rappites became dissatisfied and decided to move to Pennsylvania. In April 1825 they sold the whole property, including nearly thirty thousand acres of land, to the visionary social reformer Robert Owen of New Lanark,

Scotland. At the age of fifty-four, Owen was withdrawing from the management of the spinning mills at New Lanark, which he had made famous. He wanted to create a new kind of social community to embody his "New Moral World." Owen widely publicized his plan for the community that he named New Harmony. In the early summer of 1825, nearly a thousand people flocked to join it. After organizing a "Preliminary Society" at New Harmony, Owen returned to Scotland, but in September again sailed for America. At Philadelphia he joined forces with William Maclure, the philanthropist and scientist.[68] During European travels, Maclure had been impressed by the school for children at Yverdon, Switzerland, established by Johann Heinrich Pestalozzi. In his enthusiasm for the teaching at Yverdon, Maclure brought three teachers from Pestalozzi's school to Philadelphia. He set up a school for boys and another for girls on Pestalozzian principles.

Maclure purchased half of the town of New Harmony from Robert Owen and transferred his schools there. As an experiment in communal living, New Harmony failed immediately. In 1828 Robert Owen admitted the failure and returned to England, giving his property to his three children: Robert Dale, David Dale, and Jane.

David Dale Owen had studied chemistry and geology in London. At New Harmony he set up a chemical laboratory and began to give popular lectures in chemistry. He also became seriously interested in geology. In 1837 he was appointed state geologist and began a geological survey of Indiana. Owen described the Indiana coal basin and made geological maps of Indiana and adjoining states. Two years later the federal government appointed him to survey the geology of a vast area of the upper Mississippi Valley, comprising the mineral-bearing lands of what is now Illinois, Wisconsin, Iowa, and southern Minnesota. With the aid of a large corps of assistants, he completed the survey in just two months.

After Maclure's death in Mexico in 1840, his brother and sister gave David Dale Owen charge of Maclure's collection of mineralogical and geological specimens, gathered from all over the United States. Maclure's collection together with his own collections of minerals and fossils provided Owen with the materials for a valuable mineralogical and geological museum, which he housed in the large stone granary built by the Rappites. Lyell had met David Dale

68. Maclure, "Geology of the United States."

Owen at Boston in April 1842. He had arranged for his paper on the geology of the western states to be read at the Geological Society of London.[69] At New Harmony, Lyell now examined the specimens in Owen's museum and discussed with him the correlation of North American geological formations with those of Europe. They also went to see coal-bearing strata on a tributary of the Wabash River and examined a loess formation that Lyell thought must formerly have filled the Wabash Valley.

David Dale Owen had married Caroline Neef, a daughter of one of the Swiss teachers whom Maclure had brought to New Harmony. The atmosphere of their home was that of a cultivated European family. After the rough-and-ready life the Lyells had been leading, they greatly appreciated, wrote Mary, "the refinement and quiet of a gentleman's house." They delighted in the music and dancing at various parties, the German atmosphere reminding them pleasantly of Bonn. The Lyells particularly appreciated the good manners of the Owen children, whom, Mary wrote, "it is a pleasure to caress, for we have seen so many naughty children that Charles and I have become child-haters."[70]

Owen suggested that western children's precociousness and lack of respect for their parents were the result of their being around their parents so much. The parents were often themselves very young, the husband being perhaps twenty and the wife only fourteen or fifteen when they were married. They were poorly prepared either to provide for their children or to teach them. They might begin by cutting wood for a wooding station along the river to earn income and living in a log cabin. "They live in a very hard way," Lyell noted, "their children at first an expense, then a great wealth if they can keep their sons on till 20." After ten or fifteen years, they might buy a farm and build a frame house, without ever having learned to read or write. Yet Owen assured Lyell of the underlying stability of American democracy. Compared with England, where the laws were designed to favor the rich, in America, he believed, they could "more easily remedy abuses, that illiterate men respect learning & knowledge, can when they hear two sides discussed choose the best,—so that tho' less enlightened they will elect good men."[71]

On 3 April, the Lyells, accompanied by Owen and his assistant, drove in a carriage to Evansville on the Ohio River. About twelve

69. D. D. Owen, "Geology of the Western States."

70. Mary Lyell to Joanna Horner, 20–31 March 1846, Kinnordy MSS.

71. Lyell, Notebook 139, 60, 64, Kinnordy MSS.

miles from New Harmony, they stopped at Kimball's Mill on Big Creek, a tributary of the Wabash. During the excavation for a saw and grist mill, about twenty-five fossil tree stumps had been uncovered, embedded in upper Carboniferous strata. In 1843, Owen described the fossil tree stumps as those of palm trees. Lyell recognized the fossil stumps to be those of a *Sigillaria* species, although they lacked the vertical fluting characteristic of other *Sigillaria* fossils.[72] Owen had already excavated three of the stumps from the soft stratum of clay shale.[73] Owen, his assistant, and Lyell now borrowed shovels from the mill to dig the soft shale away from another buried tree. The exposed stump was more than four feet high and nearly two feet in diameter, with roots spreading out at the base in their natural position. The bark, converted to coal, showed the scars where leaves had been attached. The relationship between the fossil tree and the surrounding strata was similar to that found in the coalfields of Pennsylvania, Nova Scotia, and Europe. The fossil trees had grown in flat, low-lying land, under conditions similar to those of the Mississippi Delta, and later were buried in the same way that stumps of dead cypresses at New Madrid were being buried by successive floods of the Mississippi. From Kimball's Mill, the road to Evansville lay for the most part through a thick forest, in contrast to the northern part of Indiana and Illinois, which, Owen told Lyell, tended to be a great prairie. At Evansville, the steamboat *Sultana* was just coming into sight on the river, so the Lyells bade a hasty farewell to Owen and his assistant and again boarded a large and luxurious riverboat.

Despite their haste, Lyell observed that the upper eighteen feet of the bluff at Evansville consisted of yellow loess, containing a few *Helices* (fossil land snails). He made a note to send Owen a copy of his paper on the loess.[74] The next morning, opposite Leavenworth, Indiana, Lyell saw what appeared to be the same loess formation in the cliff. As the steamboat continued up the Ohio, the terrace of loess appeared sometimes on the Indiana side, sometimes on the Kentucky side.

On the *Sultana* Lyell again had an opportunity to study people and social customs. Western men, he was startled to learn, regularly carried knives and pistols for personal defense. If a quarrel arose

72. Ibid., 46.

73. D. D. Owen, "Palm Trees"; cf. Lyell, *Second Visit,* 2:272–73; Lyell, Notebook 139, 86–92, Kinnordy MSS.

74. Lyell, Notebook 139, 96–98, Kinnordy MSS; cf. Lyell, "'Loess' of the Rhine."

between men, particularly when they were drinking, such weapons would be brought out instantly and might be used with deadly effect. The rudeness of some passengers on the *Sultana* also struck him forcibly. One man came up to Lyell in the cabin, seized his eyeglass, and "asked how it looked through it." After staring through it, he returned it without a word of thanks. On the boat there was much talk of war with Great Britain. Lyell heard the opinion expressed that "We should suffer by a war, but we should gain by taking our place as one of the Great Powers." In his notebook he commented, "So much for the simplicity of democracy, its freedom from ambition, its civilizing peace-extending effect."[75]

On 4 April, the Lyells landed at New Albany, Indiana, just below the falls of the Ohio. As they went upriver, the strata along the river gradually became older, the rocks at the falls representing an ancient coral reef of Devonian age. At New Albany, Lyell called on Dr. Asahel Clapp, a physician who had settled there in 1817.[76] Clapp had made a large collection of fossil corals from the falls. He took Lyell to visit the falls at a place where the water was low enough to expose the rock ledge. The river tended to wash away the softer parts of the reef to reveal the hard limestone corals, as if they were living corals. Later, in Clapp's library, Lyell saw in one window a group of the fossil corals and in another a group of living corals from the West Indies, the fossil and the living corals looking so much alike that only a skilled zoologist would have been able to determine which group was ancient and which modern.[77] They demonstrated strikingly the uniformity of conditions in ancient and modern seas.

In the evening, the Lyells crossed the Ohio River to Louisville, Kentucky, where they found a very comfortable hotel, the Galt House. Lyell possessed a letter of introduction to Dr. Lunsford P. Yandell.[78] Sunday evening, 5 April, Dr. Yandell took them to the service at a Negro Methodist church, a large building brightly lighted by gas, crowded with a congregation of about four hundred people. The black preacher read the scriptures clearly and effectively. Near

75. Lyell, Notebook 139, 105, Kinnordy MSS.

76. Asahel Clapp (1792–1863) grew up in a backwoods settlement at Montgomery, Vermont, and studied medicine as an apprentice. "Obituary Notice of Asahel Clapp."

77. Lyell, *Second Visit,* 2:278.

78. Lundsford Pitts Yandell (1805–78), physician and paleontologist, was born in Tennessee and after studying medicine at Transylvania University, Lexington, Kentucky, and the University of Maryland, Baltimore, in 1831 became professor of chemistry and pharmacy at Transylvania University. In 1837 he helped to found the Louisiana Medical Institute. He collected fossils from the coral reefs of the falls of the Ohio and other Kentucky formations.

the end of his sermon, the minister said, "Sirs and Madams I observe many of you nodding. I advise every one to wake up your neighbour as the Sexton poor man has too much to do."[79] The sexton had been tapping many of the children on the head with his cane—rather hard, Lyell thought. There was a general stir in the congregation and the two Negro ladies between whom Mary was sitting turned to see whether she was awake. The minister then discussed the kidnaping from Ohio of Jerry Phinney, a Negro who had lived in Ohio as a free man for several years but who had been reclaimed by the heirs of his former owners and brought back to Kentucky. The minister advised the members of the congregation that they must not in their excitement "get into a scrape." Evidently, he feared an attempt to set Jerry Phinney free by force; instead, the black community soon raised five hundred dollars to purchase his freedom.

After many days of travel on steamboats and of life under frontier conditions, Lyell felt a sense of release at Louisville. In his notebook he pondered what it meant. It was not so much the greater physical comfort of life in a hotel, he thought, as the escape from the constant mingling on equal terms with uneducated and uncultivated people. In New England there was a respect for education and cultivation lacking in the West, except at special places such as New Harmony. At New Harmony, in the absence of any distinctions based on birth, rank, or family, there was "simply educ[atio]n & all the sympathies it brings with it even when opinions on religion & politics are different." In contrast, on western steamboats a formal etiquette was imposed by strict discipline on an otherwise rude, ignorant, and uncouth crowd of passengers. Lyell compared the strict etiquette and enforced politeness on river steamboats to the elaborate etiquette practiced at the court of Queen Victoria to protect the Queen from undue liberties. "Court & Society in a democracy resemble each other," reflected Lyell. "The divinity which hedges round a queen or on the many headed sovereign [a democratic people] requires equally barbarous restraints & unnatural formalities. Extreme civilization would tend to be without law or etiquette."[80]

On 7 April, the Lyells boarded the steamboat *Ben Franklin* to continue up the Ohio River to Cincinnati. Lyell's eyeglass again fascinated his fellow travelers. A man picked up the eyeglass and, looking through it, asked what it was for. The question had been asked

79. Lyell, Notebook 140, lc, Kinnordy MSS; cf. Lyell, *Second Visit,* 2:284.
80. Lyell, Notebook 140, 2–3, Kinnordy MSS.

too many times to surprise Lyell. He explained that he needed the glass because of his nearsightedness and was about to ask in return where the other man was from and what he did for a living. There was no need. "How many [thousand] hogs do you think I butchered last year?" asked the stranger. "Three?" said Lyell. "Eighteen—and all in thirty-five days." "You must have employed a great many hands?" queried Lyell. "160. One of my men can gut 800 hogs a day and before many neighbours who could not believe it, he gutted five in one minute."[81] The stranger then demonstrated for Lyell, with the aid of attitudes and gestures, how a hog was butchered.

That evening the *Ben Franklin* reached Cincinnati, having come steadily the 130 miles from Louisville at 12 miles an hour. The Lyells found that Cincinnati had grown and changed much since their visit in 1842. As the hog farmer on the steamboat had evinced, Cincinnati slaughterhouses were butchering great numbers of hogs. The German population, mainly Catholic, had increased greatly and in elections their political weight was felt. Several of their friends of 1842 again welcomed them. John G. Anthony possessed a collection of freshwater and land shells from the Ohio Valley loess with which Lyell compared his specimens from Natchez, Vicksburg, and elsewhere. At Nicholas Longworth's house, they admired a new marble bust that the sculptor Hiram Powers had sent Longworth from Rome. Near Cincinnati, Longworth had planted vineyards and was making wine that found a ready market among the German immigrants. On Friday they visited the new astronomical observatory, containing an equatorial telescope twelve inches in diameter and seventeen feet long. Completed in 1845 with funds raised by public subscription, it was the inspiration of Professor Ormsby MacKnight Mitchel of Cincinnati College.[82] Mitchel planned to explore part of the heavens too far to the south to be studied by the European observatories. He had already discovered many new double stars.[83] The Cincinnati businessman Mr. Buchanan took them on an excursion into the country to Mill Creek. In its banks was a layer of loam, containing freshwater shells, resembling the loess of the lower Ohio and Mississippi Valleys. Beneath the loam was a bed con-

81. Ibid, 22; cf. Lyell, *Second Visit,* 2:284.

82. Ormsby MacKnight Mitchel (1809–62) entered the United States Military Academy at West Point in 1825 and, after graduating in 1829, was appointed to teach mathematics. In 1832 he resigned his commission and moved to Cincinnati, where in 1836 he was appointed a professor in Cincinnati College.

83. Lyell, Notebook 140, 78, Kinnordy MSS.

taining fossilized leaves and tree trunks of the same age as the famous deposit at Big Bone Lick in Kentucky.

After nearly a week at Cincinnati, the Lyells boarded the steamboat *Clipper* on 13 April to continue up the Ohio River to Pittsburgh. As they left Cincinnati, they counted twenty steamboats tied up along the waterfront. The *Clipper* possessed a two-cylinder compound steam engine, which produced much less noise and vibration than the usual high-pressure engines found on river steamboats, and instead of a bell it had a steam whistle, which Lyell thought made a wild cry, "fearful by night & most harsh & dissonant." The boat traveled upriver at a rate of twelve miles an hour against a current of two miles an hour.[84] The next day, they passed the mouth of the Kanawha River, its blue waters contrasting with the yellow, muddy water of the Ohio. The narrow Ohio Valley, with its surrounding hills, was particularly pretty. At Moundsville, Virginia, Lyell observed the great Indian mound. Near Wheeling, the hills of coal-bearing limestone strata on either side grew higher and closer as the valley narrowed. The weather also grew cooler. The *Clipper* drew in to the wharf at Pittsburgh on the morning of 16 April, its upper deck covered with ice crystals from a hard frost during the night.

84. Ibid., 92, 102.

CHAPTER 8

Across Pennsylvania and Departure

PITTSBURGH STILL bore the scars from the 1845 fire that had consumed sixty acres of the central city, including the covered wooden bridge across the Monongahela River. From the railing of the *Clipper,* Lyell and Mary could see many new buildings and others under construction. A new light suspension bridge now replaced the old wooden bridge. The air was clouded by smoke from the chimneys of the many ironworks. Situated at the junction of two great rivers and surrounded by unlimited supplies of coal, Pittsburgh was about to become one of the great industrial cities of the United States.

Shortly after landing, Lyell hired a carriage and driver to take them thirty miles to Greensburg, Pennsylvania, where Lyell wished to examine fossil footprints in the local Carboniferous strata. They were soon rolling through the Pennsylvania countryside over a smooth macadam turnpike. In the spring sunshine, the countryside looked pretty, the hills and valleys reminding them a little of England.[1] The broad fields, enclosed by rail fences, were free of tree stumps, unlike the stump-filled clearings in Indiana. Each farmstead had its comfortable-looking house and apple orchard, while in the pale green fields of young winter wheat, flocks of sheep grazed. Everything had a settled and civilized air. At Natchez in February the swallows had already come, but in Pennsylvania in mid-April, they had just arrived. The spring was late and the trees still bare; only the occasional willow had begun to put forth buds. Near Tuttle Creek, two farmers took Lyell to see a seam of coal.[2] About fifteen miles out of Pittsburgh, Lyell and Mary saw across the horizon ahead of them a long, blue, unbroken line—the westernmost ridge of the Alleghenies. That afternoon, they reached Greensburg, a pleasant small town of brick houses.

They stayed at a comfortable, old-fashioned inn, run by the landlord and his family. The landlady was especially kind to Mary, who the next day remained at the inn while Lyell went to examine footmarks of a large fossil reptile. He was accompanied by Dr. Alfred

1. Mary Lyell to Frances Lyell, 18 April 1846, Kinnordy MSS.

2. Lyell, Notebook 141, 28, Kinnordy MSS.

King,[3] the young Greensburg physician who had reported the footmarks two years earlier, and a young Presbyterian clergyman, the Reverend Mr. Hackey. When they reached the abandoned quarry where the footprints were located, about six miles southeast of Greensburg, Lyell saw at once that the tracks were genuine. The layers of sandstone in the quarry were separated by thin layers of fine clay. Most of the footmarks occurred as pairs of casts, standing out in relief from the undersides of the sandstone beds. Each pair, consisting of the marks of a hindfoot and forefoot, was spaced nearly equidistant from the next pair.

Near Savannah, Georgia, in 1842, Lyell had observed that the tracks made by opossums and raccoons on the soft mud flats at low tide became filled with fine sand blown by the wind. Tracks thus filled would remain as impressions in the layer of mud and as casts on the undersurface of any overlying layer of sand. Similarly, in Nova Scotia, Lyell had seen at low tide in the Minas Basin how sandpipers, as they flitted along, left their tracks imprinted on the surface of the mud flats. When the mud dried and hardened in the summer sun, the tracks were preserved, to be filled in by a fresh layer of mud brought by the next incoming tide. The mud, when dried by the sun, also shrank and cracked, and the cracks, like the sandpiper tracks, were filled with mud by the next tide. At low tide the dried mud, when taken up, might be separated into successive layers, each layer having on its upper surface the impressions made by both bird tracks and sun cracks and on its lower surface the casts impressed into the layer beneath.

In the quarry near Greensburg, the strata containing the fossil footprints were in the middle of the Pennsylvania Coal formation, about a hundred feet below the main Pittsburgh seam. Such fossil plants as *Lepidodendron, Sigillaria, Stigmaria,* and *Calamites* occurred abundantly in beds both above and below the bed containing the footprints.[4] Thus the large reptile of Greensburg had lived in Carboniferous forests. It could not have been alone, Lyell suggested. Other species, genera, and even classes of land animals must also have been present, a varied fauna yet to be discovered in fossil form.[5] Dr. King realized that the animal that made the tracks must

3. King, "Fossil Foot Marks . . . Westmoreland County, Pennsylvania."

4. On 26 April 1846, Lyell wrote a paper on the fossil reptile footprints of Greensburg and sent it to Benjamin Silliman for publication. See Lyell, "Fossil Footprints of a Quadruped Allied to the Cheirotherium." The next day, he also sent a full account of the fossil reptile footprints to Leonard Horner, intending it to be read at the Geological Society of London. Lyell to Leonard Horner, 17 April 1842, Kinnordy MSS.

5. Lyell, *Second Visit,* 2:308–9; cf. Lyell, *Travels,* 1:167 and 2:168.

have been able to walk on land and was therefore air breathing. If the earth's atmosphere were able to support air-breathing animals, it must not have contained much more carbon dioxide than the modern atmosphere. The reptile footprints disproved a suggestion by Adolphe Brongniart and others that during the Carboniferous period the earth's atmosphere was much richer in carbon dioxide than at present in order to support the luxuriant growth of the coal plants.[6]

Dr. King and Mr. Hackney also took Lyell to see other footprints on the summit of a ridge near Derry, about fourteen miles north of Greensburg. The latter footmarks, if true fossil footprints, would be even more remarkable, because they included the tracks of birds, deer, and dogs. When Lyell examined the tracks, he saw that they were made by men. The tracks were on the underside of overhanging rock strata in a ledge of white sandstone. They were usually large, deep, and sharp, showing few effects of weathering. Some of them appeared on the edges of the exposed strata where they could not possibly have been made by ancient birds or other animals. Lyell decided that the tracks had been carved by Indians. They may have represented the animals also depicted in paintings on rock faces at various places in America. Near the sandstone ledge at Derry, there were many Indian graves and one of the main Indian trails across the Allegheny Mountains.[7]

When Dr. King announced his discovery of the fossil footmarks, a Catholic clergyman who was president of St. Mary's College at Greensburg had attacked him violently because the footmarks and other Carboniferous fossils provided evidence for the great antiquity of the earth. Both Catholics and Protestants in Pennsylvania believed that the account of Creation in the Bible was literally true. Several German-educated Lutheran clergymen, who knew that geology had clearly established a great age for the earth, were afraid to explain the significance of geology to their congregations. Even the Reverend Mr. Hackney, who had helped Dr. King in his research and understood the geological evidence thoroughly, did not support him openly. Lyell thought that such silence made the clergy in some measure accomplices in the persecution of science. It did nothing to help the younger generation understand the meaning of geology.

On 19 April, with seven other passengers, the Lyells boarded

6. King, "Fossil Foot Marks"; cf. *American Journal of Science* 48 (1845): 343–52.

7. King, "Fossil Foot Marks," 299–300.

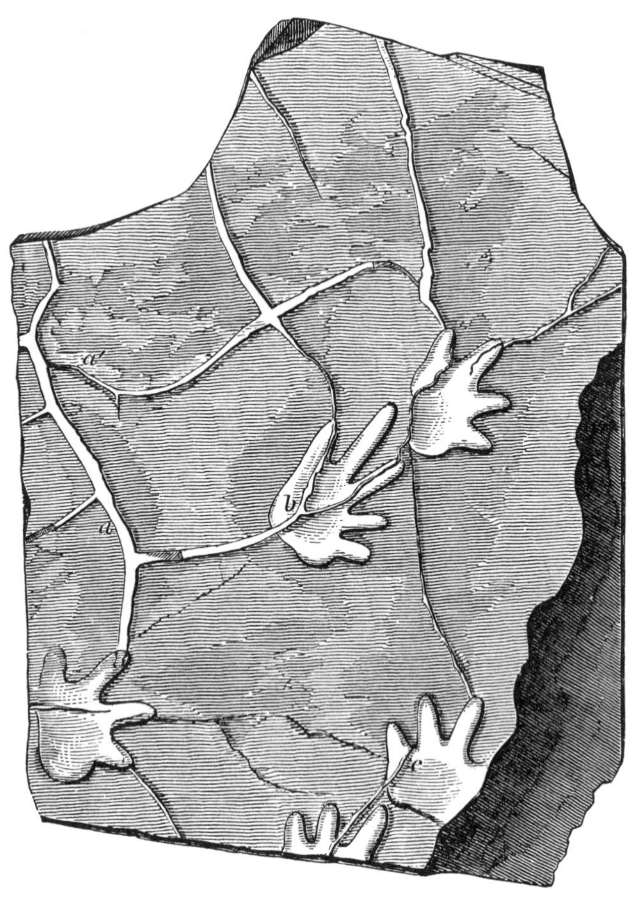

Fossil Reptile Footprints, Greensburg, Pennsylvania. Lyell's *Second Visit* (1849), Fig. 12.

a stagecoach for Philadelphia. Through the night they traveled eastward over the forested ridges of the Alleghenies through Youngstown, Stonystown, and Shellsburg to Bedford. Twice they had to change coaches. When a wheel broke on the third coach, the driver cut a thick branch from a tree and tied it to the axle with ropes to hold up the vehicle. As they walked behind the coach to the next town, some of the passengers sang psalms, reminding Lyell of those sung by James Hamilton Couper's Negro boatmen in Georgia. Dawn was just breaking when they were on a beautiful pine-covered ridge. Beneath the pines grew laurels, azaleas, kalmias, and rhododendrons that Lyell had not seen anywhere in the Mississippi or Ohio Valleys nor even at Greensburg.[8] In the valleys of the Alleghenies, the green meadows beside the rivers contrasted strongly with the dark forests of the ridges. Many of the valleys appeared to contain lakes, but they were really sheets of fog that the rising sun would soon evaporate. To Lyell the remarkable feature of the Appalachian valleys was that they contained no lakes. From the highest ridge of the Alleghenies, the coach descended by a road along the banks of the Juniata River to Chambersburg. There they boarded a train for Harrisburg. At Harrisburg the railway bridge across the Susquehanna River had burned. They crossed the river by ferry to board another train to take them to Philadelphia. They arrived late at night on 22 April.

After six months of travel, the Lyells appreciated Philadelphia. In the April sunshine, its regular streets of brick houses with double rows of shade trees, well-dressed pedestrians, and well-provided shops had a comfortably civilized air. They saw a good deal of their friends William McIlvaine and Clement Biddle. Letters from Britain informing them that Lyell's younger brother Henry had been severely wounded in India at the battle of Sobraon gave them a sleepless night and for days afterward much anxiety. The Philadelphia newspapers were full of talk of a possible war with Great Britain over Oregon. Lyell, acutely aware of the personal cost of war, marveled at the enthusiasm. Clement Biddle told him of a Philadelphian, a member of the Democratic party, who when asked, "What makes you for war; you want to be taxed?" had responded, "No, Jim, but I should get a large contract."[9] Some of Lyell's Philadelphia ac-

8. Lyell, Notebook 141, 76, Kinnordy MSS. In *Second Visit,* 2:320, Lyell wrote, "I had seen none of these evergreens since I left Indiana," but in his notebook he observed more correctly of the laurels, azaleas, kalmias, and rhododendrons that there were "none in Indiana, Ohio or Greensburg." Lyell thus recognized the distinctive character of the Appalachian flora.

9. Lyell, Notebook 141, 95, Kinnordy MSS.

quaintances were pessimistic about the prospects of the United States as a result of what seemed to them extreme Democratic influence. William McIlvaine assured Lyell that "there were always Federalists who croaked as much about the progress of democracy as they can now" and that things were really no worse than they had always been.[10] Lyell spent a morning discussing geological questions with Dr. Samuel George Morton. On Sunday they attended a black Episcopal church, where the clergyman conducted the service in a dignified and proper way.

On 27 April, Lyell struck out on his own to visit sites in Virginia and Maryland, especially to study further the Oolitic coal bed at Blackheath near Richmond, Virginia, and the Virginia Tertiary formations. At Baltimore, he again stopped to see Julius Ducatel's fossil collections and to discuss Maryland's geology with him.[11] Ducatel told Lyell that the marls, sands, and lignite along the railroad lines on either side of Baltimore were Cretaceous. When the Delaware and Chesapeake Canal was dug ten years before, they dug up Cretaceous fossils such as ammonites in beds just below the lignite.

At Washington, Lyell visited an exhibition of articles manufactured in the United States, intended to demonstrate the cheapness after seven years of a protective tariff of American manufactured goods, as compared with similar British products. Before leaving England, Lyell had seen a similar exhibition in London intended to demonstrate the advantages of free trade over protective tariffs. From Washington, he took a steamboat down the Potomac River to Acquia Creek. On the railroad to Fredericksburg, Virginia, Lyell was so impressed by the dazzling whiteness of the dogwood in bloom that he obtained seedlings to take back to England. After spending two days with Augustus Gifford studying the coal beds north and south of Blackheath, Lyell returned to Richmond. Jeffries Wyman took him to see the state capitol and the Medical College, and together Lyell and Wyman studied the beds along Shockoe Creek.

On 4 May, Wyman accompanied Lyell to Fredericksburg, where they obtained a handcar from the railroad company. With two Negroes to pump the handles, they went along the line of the newly built railroad from Fredericksburg to Acquia Creek, studying the geology of the country. Lyell wished to examine the cliffs along the

10. Lyell, Notebook 142, 10, Kinnordy MSS.

11. Ibid., 8.

Potomac River. When they had gone as far as possible by the railway, they left the handcar to walk along a road through the woods. White oaks were just coming into leaf. In the brilliant sunshine, butterflies fluttered through the air and dragonflies zoomed about. A dog, which had followed them from the inn, came upon a snake four feet long, resembling a swamp moccasin. The snake coiled and struck repeatedly, but the dog avoided each of its attacks, barking all the time, as if it enjoyed the sport. At one place in the sandy road, Lyell paused to watch beetles roll along and bury balls of dung in which they had deposited their eggs. About a mile below Acquia Creek landing, they collected fossils from the cliffs. On the shore, Lyell saw a heap of gar pike that fishermen had taken in their nets along with herring. Considered inedible, the gar fish were used to fertilize the fields. Near the inn at Acquia Creek, muskrat houses dotted the marshes. Lyell and Wyman waded through the water to one muskrat house and took it apart to see what the living quarters of a muskrat were like. They found three chambers connected together by passages and to a vertical passage that opened to the exterior underwater.

That evening, Lyell took the steamboat to Washington and the next day returned to Philadelphia. Mary had gone to New York to join the Ticknors who were there for a holiday. At Philadelphia, Lyell found a letter from Mary with the news from India that Henry Lyell was recovering from his wounds. With no one to talk to, Lyell went to see Samuel George Morton, to have him explain the nature of Henry's wounds. After paying a farewell visit to Lardner Vanuxem at Burlington, New Jersey, Lyell went on to New York on 7 May to join Mary.

At New York the Lyells had many friends and acquaintances. One evening, Lyell went with Joseph Cogswell to a party at the house of Albert Gallatin, then eighty-four years old. As secretary of the treasury under Presidents Jefferson and Madison, Gallatin had reorganized the finances of the United States. In 1784, as a young Swiss immigrant, Gallatin had cleared a farm on the Monongahela River in the wilderness of western Pennsylvania. He became interested in the culture of the Indians, his principal neighbors, and in 1842 founded the American Ethnological Society. Gallatin told Lyell that the development of a high degree of civilization among the Indians of Mexico and Peru was based mainly on the cultivation of Indian corn, or maize. One of the advantages of maize was that it could be stored for years so as to provide a reserve of food in case of famine.

Similarly, Gallatin thought that Indian corn had permitted the rapid settlement of Ohio, Indiana, Illinois, and Michigan. He could remember as a young man when there were no English-speaking white people in the vast territory of those four states, the only white settlement being that of the French at Detroit. During his early days in the West, he had seen a settler felling trees in March. By October the settler had harvested a crop of Indian corn from the newly cleared ground. With the aid of Indian corn, the new settler was able to get himself established quickly; he could not have done so if he had had to wait twelve to thirteen months to harvest a crop of wheat.[12]

In the cause of peace, Gallatin, despite his advanced age, had written and spoken against the threatened war with Great Britain over the Oregon question. When ambassador to Great Britain, he had negotiated the original Oregon treaty, so he could say with authority that the terms offered by the British government were fair and might be accepted with honor. As a former secretary of the treasury, Gallatin could also estimate what a war with Britain might cost. He explained the cost to the public and said what taxes would be needed to pay it. When war broke out with Mexico the next day, Gallatin opposed it with equal firmness. With calm courage, he spoke at a public meeting called by the mayor of New York in support of the war, maintaining his unflinching opposition in the face of a hostile crowd.

Lyell made an excursion to visit John James Audubon at his country estate on the Hudson and to see the new Croton aqueduct that supplied New York City with water. On 11 May he left for Albany, where he had been invited by James Hall and Ebenezer Emmons to examine the Taconic strata. Mary remained with the Ticknors at New York. The next day Lyell and Emmons went to Troy to examine some of the geological sections in question.[13] In 1838 Emmons had reported that the oldest sedimentary rock in the New York system was the Potsdam Sandstone and that no other sedimentary formation existed between the Potsdam Sandstone and the Primary rocks. When Emmons examined the New York strata east of the Hudson River, he found a different series of stratified formations, the lowest member resting, like the Potsdam Sandstone, directly on the Primary rocks. The stratified formations east of the Hudson, he

12. Lyell, Notebook 143, 39, Kinnordy MSS; cf. Lyell, *Second Visit,* 2:342–43.

13. For a somewhat extravagant estimate of the historical significance of Emmons, see Schneer, "Emmons."

decided, must represent a series lower and older than the Potsdam Sandstone. He named them the Taconic system, after the Taconic Hills where they occurred.[14] Other geologists, such as James Hall and Edward Hitchcock, thought that the rocks of the Taconic Hills were simply an eastward extension of the Silurian rocks of the New York system that had undergone much metamorphosis and disturbance. Emmons took Lyell first to see limestone strata at Troy, characterized by a particular fossil, *Isotelus.* The *Isotelus* indicated that the strata were of the same age as the Trenton Limestone. If they were the same age, Emmons concluded that the green slate underlying the Troy Limestone would be a rock below the Trenton and might therefore correspond to nothing in the New York system. In a quarry a half mile east of Troy, Emmons showed Lyell a slate formation that he considered part of the Taconic system. The slate was interspersed with igneous intrusions and metamorphic limestone, reminding Lyell of certain metamorphic limestones in the Pyrenees. James Hall told Lyell that mixed limestones, similar to those at Troy, occurred also in the Utica and Hudson River slate formations of the New York system. On the basis of fossils, Hall felt obliged to unite the Trenton and Hudson River groups of strata into one formation.[15] The Trenton corresponded to the Caradoc Sandstone in European Silurian formations, but in New York the Chazy and Potsdam formations lay beneath the Trenton. Lyell spent two days studying Emmons's Taconic sections in the field and discussing with Emmons and Hall the geological questions involved. He wrote to Mary: "I feel already that his Taconic system cannot hold, but I wish to do him justice & hear him out & it is very improving to me for Hall is anxious to explain the other side of the question."[16]

On 14 May, Hall and Emmons took Lyell to Saratoga Springs, "a pleasant looking town of white painted wooden houses," with many flowers in bloom. North of Galesville, they examined the Bald Mountains, where Emmons considered the strata part of the Taconic system. The Bald Mountain strata were steeply inclined but formed a continuous series. If the Taconic strata were perfectly conformable to the Silurian strata of the New York system, Lyell thought that they must be considered simply a lower portion of the New York system that had undergone metamorphosis.[17] In 1843 in his report

14. Emmons, *Second Geological District,* 135–64.

15. Lyell, Notebook 143, 64, 74, 76, Kinnordy MSS; cf. Hall, *Fourth Geological District,* 30.

16. Lyell to Mary Lyell, 13 May 1846, Kinnordy MSS.

17. Lyell, Notebook 144, 11–12, Kinnordy MSS.

of the geology of the First District, William Mather had adopted this view. Later, Mather compared the dispute over the Taconic system with the difficulty in defining the boundary between Roderick Murchison's Silurian system and Adam Sedgwick's Cambrian system in Great Britain.[18] The difficulty lay partly in a confusion over the meaning of the term *system*. When Murchison had described the Silurian system in 1838, he assumed that the series of Silurian strata represented a chronological unit with definite boundaries marked by geological unconformities. Nevertheless, as he studied the sequence of Silurian strata downward, he found no distinct boundary to separate strata that he considered Lower Silurian from strata that Sedgwick had termed Cambrian.[19] Similarly, in New York State, the strata formed a continuous series from the Potsdam Sandstone upward. Although today the New York strata are considered to belong to four geological periods—the Cambrian, Ordovician, Silurian, and Devonian—the strata of the successive periods still form a continuous series. Each period is defined in terms of a characteristic fossil fauna, but considerable faunal continuity exists between periods. In many respects, Paleozoic time, as represented by the New York strata, was one long unbroken period.

The Taconic strata were ultimately found to contain fossils ranging from the Lower Cambrian to the Middle Ordovician in the modern geological sequence. In 1846 Joachim Barrande published the first of his papers on fossil trilobites found in Bohemia.[20] When their counterparts were found also in Taconic strata, Barrande's trilobites demonstrated that the Taconic series was indeed lower and older than the Silurian formations of New York State. The relationships and significance of the Taconic rocks remained a question of prolonged and often bitter dispute throughout the nineteenth century and even into recent times.[21] On 15 May, Lyell returned to Albany with Hall and Emmons. At Albany he examined Hall's collection of fossils from the New York strata. In the evening, he boarded the night steamboat for New York City, where next morning he rejoined Mary.

At New York, the Lyells went over the various boxes of specimens they had collected during their travels and arranged for their ship-

18. Mather, *First Geological District*, 438.

19. For a full discussion of the controversy between Roderick Murchison and Adam Sedgwick over the boundary between Cambrian and Silurian, see Secord, *Controversy in Victorian Geology*.

20. Barrande, *Le système silurien et les trilobites de Bohème*.

21. Schneer, "Great Taconic Controversy."

ment to London. Whenever possible, their New York friends entertained them. Theodore Sedgwick, a widely traveled lawyer and man of letters, invited them to his house, where Lyell found the library "quite a relief after the gorgeous houses and palaces of the merchants." Lyell visited Harper's publishing and printing house, publisher of many American editions of British books. Harper's printed editions in large numbers and sold them cheaply, the great scale of their publishing business demonstrating the enormous size of the reading public in the United States.

One evening, Lyell attended an open-air meeting in front of the city hall called by the mayor of New York to rally support for the federal government in the war with Mexico. Four persons addressed the crowd at once, three of them from separate wooden platforms with a flag flying from each. With the fountains playing and two large churches in the background, Lyell thought "the crowd on the steps of the hall & in the Park had a good effect on a fine evening at 7 o'clock, but no enthusiasm & it was scarce possible to raise a cheer except by bringing in Oregon or the British gold as the cause of the hostility of the Mexicans."[22]

Early on the morning of 21 May, the Lyells left New York for Boston, traveling by rail the length of Long Island at the rapid rate of thirty miles an hour. After crossing the Sound to New London, Connecticut, they continued by rail to Boston. During the next ten days at Boston, they saw many friends, particularly William Prescott. One day they went to Ipswich to see William Oakes and visit Wenham Pond, famous for the ice cut from its surface in winter. Lyell went to see two reconstructed skeletons of unusually large mastodons brought to Boston, one from New Jersey and the other from New York. From a sample of the stomach contents of one of the mastodons, he determined later in London that it had fed on the young shoots of the eastern white cedar (*Thuja occidentalis*).

On the first of June, Lyell and Mary sailed from Boston in the Cunard steamer *Britannia*. After a smooth voyage they landed at Liverpool on 13 June, nine months and nine days from the time they left. "I am so pleased to be in Old England again," Mary wrote to her mother, "& yet have left a great piece of my heart in Boston."[23]

22. Lyell, Notebook 144, 28, 33, Kinnordy MSS.

23. Mary Lyell to Mrs. Leonard Horner, 14 June 1846, Kinnordy MSS.

CHAPTER 9

Fame, Authority, and Influence

During Lyell's visit of nearly a year in America, his *Travels in America* had spread his reputation among many people unfamiliar with his geological books. Although ordinary British readers knew little of America—and that little was often wrong—they were fascinated by accounts of its vast mountains, forests, and plains, its lakes and rivers, and its restlessly active people. Of such things Lyell gave a detailed, credible account. British readers might be startled to find also among Lyell's American observations a trenchant criticism of English universities.

While Lyell's *Travels* increased his reputation among the general public, the work with its geological map and its accounts of the state surveys, Niagara Falls, and American Tertiary geology, similarly strengthened his scientific reputation. Lyell went to America in 1841 as a well-known author of geological books that contained radical and slightly dubious doctrines. In 1846 he returned from his second American tour not only as an acknowledged authority on America, but with his scientific position reinforced by wide-ranging studies of American geology. Over the next few years, he would be consulted by cabinet ministers, become a confidante of Prince Albert, receive a knighthood, and for a second time assume the presidency of the Geological Society of London.

New Directions, 1846–1847

During the first days at London, Lyell was occupied with nine months' accumulation of correspondence and other business, with unpacking, labeling, and repacking the boxes of fossils sent from America, and with plans for the immediate future. The *Principles* was out of print and needed a new edition; the second edition of the *Elements,* published in 1841, was almost completely sold. Even more urgent than new editions of his works was the Lyells' need for a larger house. Sixteen Hart Street, where they had lived since just after their return from their wedding trip fourteen years before, had grown more and more crowded. Successive tours on the Continent, in various parts of Britain, and two long periods of travel in America had each resulted in boxes of fossil and living shells, bones, rocks, and other specimens, all brought back to 16 Hart Street. Lyell also accumulated the scientific books and journals necessary for his work as a scientist, and especially as a scientific author. The boxes of fossils sent from America and brought with them on the *Britannia* decided Lyell that a new and larger house must be his first concern. Before the end of June he began house hunting.

Among houses in London reasonably close to their friends and to the scientific societies, they found several available at moderate rents around Cavendish Square, Harley Street, Wimpole Street, and Welbeck Street. Rents in the area had fallen 25 to 40 percent in the 1840s as a result of a general move to the more fashionable neighborhoods of Belgravia and Hyde Park Square. On 4 July they went "from garret to kitchen" over a house at 11 Harley Street that strongly tempted them, although Lyell wondered whether they could afford it. The house was listed at a rental of 180 pounds a year, but they learned they could have it for 150 pounds a year and might, if they wished, purchase the seven years remaining on the existing lease for 700 pounds. The coach house and stable for three horses might be sublet for eleven pounds ten shillings a year to reduce the annual rental of the house. They would also have to pay twelve pounds a year in ground rent, taxes, and various other charges. Uncertain as to the total annual cost of such a house, and with his father-in-law out of town, Lyell wrote to Charles Babbage to ask for advice and help.[1] In his notebook he compared the projected costs of the house on Harley Street with those of their house at 16 Hart Street:

	16 Hart Street	11 Harley Street
Rent	£80–0–0	£123–10–0 (with rent of 11–10–0 for stable subtracted)
Sewer		2–18–0
Water	2–10–0	15–15–3
Insurance	1–1–6	3–10–8
Assessed taxes	15–0–0	15–15–3
Poor rate	8–3–4	13–10–8
Paving and lighting	2–12–6	5–4–0
Pew Rent and Easter offering	1–11–6	1–12–3
Property tax	19–7–0	40–15–3
Total	£130–5–8	£212–9–5

1. Lyell to Babbage, 4 July 1846, Babbage MSS.

The annual cost of 11 Harley Street would thus be about eighty-two pounds greater than 16 Hart Street.[2] When he contemplated moving, Lyell had also to include paying for the actual move, the cost of fixtures and additional furniture for the new house, fees to a lawyer for the lease and an agent for inspection, and the cost of renovation of the old house. When he had taken out the lease on 16 Hart Street in 1832, the house was freshly painted throughout. At the expiration of a lease, the leaseholder was expected to return the property to the owner in a condition as good as when he received it. To pay the cost of the move, Lyell would have to use part of his savings in the Funds (British government securities). Yet, like other people in search of a new home, the Lyells seemed to have made, perhaps unconsciously, the decision to take 11 Harley Street when they first looked at it. After that it was a matter of checking details, calculating costs, and finding ways and means.

When they first looked at houses, Lyell had intended to postpone taking a lease on another house until the autumn. More houses would then be on the market and his lease on 16 Hart Street would expire. However, once launched on negotiations over 11 Harley Street, he did not stop. On 9 July he reckoned as closely as possible his expenses if they moved.

Moving £50, lease £25, agent £5	£80
Purchase of six years and nine months lease, including fixtures	700
Expenses to the end of the year 1846	300
	£1,080

He had 500 pounds in his account at Coutts Bank and would have to take the remaining money needed from the Funds.[3] It would require 50 pounds to renovate 16 Hart Street, and they would have to buy some new furniture for the larger house.

Lyell estimated his income for 1847 at 549 pounds, including 25 pounds from Wiley and Putnam in New York for the American edition of his *Travels in America.* In that year he did not expect to receive any income from the sale of the *Principles* or the *Elements.* By his calculation, he thought expenses would exceed income for the year by sixty-eight pounds. That would mean drawing further on savings in the Funds, savings built up from the sale of successive edi-

2. Lyell, Notebook 147, 20, Kinnordy MSS.

3. Ibid., 29.

tions of the *Principles* and the *Elements.* He was concerned enough about money to make an estimate of their expenses in 1844, taking it to be an average year, and then estimated their probable cost of living for 1847, 1848, and 1849. The gloomy conclusion was that they would run a deficit each year, so that by the end of 1849 his savings would be reduced from 800 pounds to 170 pounds.[4] Clearly, it would be important for Lyell to earn additional income if he were to continue to travel and do scientific work. Despite such discouraging calculations, which filled pages of his notebook with figures, on 10 July Lyell entered into serious negotiations for 11 Harley Street, which, he wrote, "I hope I may get, tho' it is doubtful."[5]

On 14 July, he signed an agreement for the house.[6] Two days later, he and Mary sailed by steamboat for Dundee to spend a long vacation with Lyell's family at Kinnordy. Lyell took with him all the notebooks from his American tour, plus many other books and journals. Before he left London, he arranged with John Murray to bring out a seventh edition of the *Principles of Geology* in one volume, keeping changes to a minimum, merely revising and adding notes where new developments in geology and natural history made them necessary. In thanking Joseph Hooker for his comments on botanical points in the *Principles,* Lyell wrote: "I believe I told you that I was not recasting the book so that I shall not enter into such great questions as I am glad to discuss with our friend Darwin as to whether species have had many birthplaces, or each separate stock etc."[7]

As he got into the revision of the *Principles,* which he had not touched since 1840, Lyell became enthusiastic. The corrections and suggestions sent by Hooker, Charles Darwin, Charles Bunbury, and others all served to stimulate him.[8] On 4 August, he wrote to Murray to determine whether he would be ready to go to press with the new edition when Lyell returned to London in September, adding:

> I think as yet the occasional omissions nearly compensate the additions, but not quite. The "Falls of Niagara," "Delta of the Mississippi" & other American subjects require a recast, both my U.S. tours being since 1840 (the time of publishing the 6th. Edn. & the new discoveries in fossil Botany, & mammalia [fossils], the

4. Ibid., 32.
5. Lyell to Bunbury, 10 July 1846, Kinnordy MSS.
6. Lyell to Babbage, 14 July 1846, Babbage MSS.
7. Lyell to Hooker, 23 July 1846, Kinnordy MSS.
8. Ibid.; cf. Darwin to Lyell, [8 August 1846] Darwin-Lyell MSS; printed in Darwin, *Correspondence,* 3:330–32.

Antarctic voyages in 1841–3 & a variety of different topics) give me in the first volume alone a great deal to amend, on points which it was impossible to know in 1840.[9]

On 25 August, Lyell and Mary left Kinnordy to spend a few days in Edinburgh, where Lyell had his photograph taken by the photographers Hill and Adamson.[10] They visited Francis Jeffrey, former editor of the *Edinburgh Review*, at Craigcrook, his country house outside Edinburgh, a lively house. Lyell enjoyed discussions of education, religion, and other questions with Jeffrey and Lord Cockburn. At the beginning of September they returned to London.

At the annual meeting of the British Association at Southampton on 14 September, Lyell delivered a lecture on the delta and valley of the Mississippi River.[11] On their return to London, they were caught up in the process of moving to 11 Harley Street. On 23 September, with the help of Mary's sister Leonora, Lyell was arranging his books on their new shelves.[12] A week later they were still in the midst of upholsterers, carpenters, and other workmen.[13] They installed a new carpet in the drawing room and oil cloth on the floors of Lyell's museum and the halls. With new furniture, curtains, lamps, pictures, and bookcases, they had much to do to get settled.[14] When Lyell had his library and museum straightened away, he returned to work on the *Principles*. He also read a paper sent to him by Andrew Crombie Ramsay, a young assistant on the Geological Survey of Great Britain, that had just appeared in the first volume of the survey's *Memoirs*.[15] Established in 1835, the Geological Survey had been at work for more than ten years.

When Henry De La Beche completed the geological survey of Cornwall, Devon, and Somerset in 1839, he began a similar survey of South Wales. In 1841, he added Ramsay to the survey staff. At the British Association meeting in Glasgow in 1840, Ramsay exhibited a geological map, sections, and model of the isle of Arran that impressed the geologists present. In the survey of South Wales, Ramsay had found coal-bearing strata arranged in a series of anticlinal and synclinal folds of which only remnants survived, the tops hav-

9. Lyell to Murray, 4 August 1846, Murray MSS.
10. Lyell, Notebook 147, 70, Kinnordy MSS.
11. Lyell, "Delta and Alluvial Deposits of the Mississippi."
12. Lyell to Bunbury, 23 September 1846, Kinnordy MSS.
13. Lyell to Mantell, 29 September 1846, Mantell MSS.
14. Lyell, Notebook 147, 126, Kinnordy MSS.
15. Great Britain, *Geological Survey: Memoirs.*

Charles Lyell, August 1846, at Edinburgh. Photograph by David Octavius Hill. Courtesy of the Department of Geology, King's College, London.

ing been shaved away by denudation. Working carefully to scale, Ramsay reconstructed the original height of the anticlinal folds to demonstrate that thousands of feet of rock strata had been removed. He thought that the strata had been worn down and removed slowly and gradually over a long period of time. By contrast he thought that the elevation and folding of the strata had occurred suddenly at the end of the Carboniferous period.

Ramsay's ideas reflected those of his chief, De La Beche, who had long insisted that the elevation and folding of strata could have been brought about only by forces much greater than those of modern earthquakes.[16] De La Beche frequently ridiculed Lyell's view of gradual elevation. Nevertheless, he did not bring forward geological evidence or reasoning to counter Lyell's evidence and arguments. De La Beche reacted to criticism by drawing cartoons caricaturing his opponents, instead of responding to their criticisms by reasoned arguments. An able, but isolated and unhappy, man he felt toward Lyell a mixture of bitter envy and grudging respect. To his subordinates on the Geological Survey, his bitterness toward Lyell was evident.[17]

Lyell read Ramsay's paper with admiration but decided he must protest Ramsay's opinion that the Carboniferous strata had been elevated and folded suddenly and catastrophically. After consulting Darwin, Lyell wrote to Ramsay complimenting him on his paper but also pointing out that he did not allow for the possibility that the strata were being worn away at the same time as they were being elevated and folded. In some places Ramsay described the Carboniferous beds as first being elevated and folded and then gradually denuded of several thousand feet of sediments. In other places he described denudation as having occurred in parallel with elevation and folding. If denudation occurred in parallel with elevation and folding, Lyell thought that there was no reason why the elevation and folding must have occurred suddenly and violently. It might have occurred gradually over a long period of time, just as it was doing at present in regions of volcanic activity. Lyell concluded by thanking Ramsay "for one of the very best papers I ever read."[18]

Ramsay replied from Bala, where he was doing fieldwork for the survey. He agreed that upheaval might occur slowly and gradually to any height. What had impressed him about the remarkable folds

16. De La Beche, *Researches in Theoretical Geology,* 109.

17. Secord, "Geological Survey," 240, 245.

18. Lyell to Ramsay, 8 October 1846, printed in Geikie, *Ramsay,* 85–88; copy in Kinnordy MSS.

and curvatures of the coal-bearing strata of South Wales was that they were not accompanied by signs of igneous action. The curvatures of the strata had not been produced, therefore, "by any efforts of melted matter to escape."[19] Ramsay attributed the folding to a contraction of the earth's crust brought about by cooling of the interior, a theory expounded by De La Beche in 1834.[20] Ramsay's views were thus linked to the theory of a cooling and contracting earth. The difficulty with that theory, as Lyell had shown repeatedly, was that there was no evidence either that the earth's climate had cooled over geological time or that the rate or intensity of volcanic action had declined. His geological experience limited to Scotland and Wales, Ramsay had never seen a country of active volcanoes and earthquakes, nor one that had undergone recent elevation and disturbance. Darwin thought that if Ramsay could see South America, his views would be different.[21]

Lyell wrote again to Ramsay, noting that in Somersetshire Ramsay said that the Mendip Hills had been elevated after some Carboniferous strata had been deposited but before younger horizontal strata of New Red Sandstone had been laid down over the edges of the upturned Carboniferous beds. Lyell pointed out that, between the elevation and tilting of the Carboniferous beds and the deposition of the horizontal sandstone, more than four thousand feet of strata must have been removed by denudation, a process requiring a long interval of time. If that were the case, Lyell could not understand how Ramsay could prove that the elevation and folding of the Mendips had occurred suddenly.[22] In reply, Ramsay acknowledged that the time required for the elevation and folding of the Mendip strata "may have been very great as we measure time."[23]

Lyell also wrote to Edward Forbes about Forbes's paper in the survey *Memoirs*, connecting the geographical distribution of species in the British Isles to the recent occurrence of glaciation. Lyell was fond of Forbes, a generous and open-hearted man. He complimented him most highly on the paper but regretted that Forbes had not mentioned his early discussion in the *Principles* of the relation between the present geographical distribution of species and geo-

19. Geikie, *Ramsay*, 91.

20. De La Beche, *Researches in Theoretical Geology*, 109; cf. Fourier, "Le refroidissement seculaire du globe terrestre."

21. Darwin to Ramsay, 10 October [1846], printed in Geikie, *Ramsay*, 85, and in Darwin, *Correspondence*, 3:352–53.

22. Lyell to Ramsay, 20 October 1846 [copy], Kinnordy MSS.

23. Ramsay to Lyell, 25 October 1846, Lyell MSS. Emphasis Ramsay's.

logical changes.[24] Lyell acknowledged that no one was more generous than Forbes in recognizing his contributions to geology. Lyell also mentioned his papers on the Crag and on Touraine. In them he had argued for the existence of a much colder climate during the recent geological past and a former land bridge between France and England across the straits of Dover. He thought Forbes might have cited these references, but he could understand how difficult it was to cite everything bearing on a particular problem.

In reply, Forbes explained that he had written the essay in March 1846, after he had been ill for two months and when he was also preparing two courses of lectures. Too weak from illness to go to libraries and with his own books packed up in various places, Forbes had to write out his argument as best he could and "back it by such references as were at hand, or as I had already taken notes of, when backing seemed to me absolutely necessary." He had not mentioned the *Principles* because he was writing for geologists "who are, or ought to be, as familiar with them as myself."[25] He acknowledged that he wished he had been able to cite Lyell's Touraine and Crag papers, because they would have provided strong support for his argument.

Through the autumn of 1846, Lyell worked on the seventh edition of the *Principles*. By late November he thought himself halfway finished. Meanwhile, their social life became more active. In November the Lyells attended a dinner party at Canon Milman's. The first of December they themselves gave a small dinner party for eight, followed by a large evening party with thirty-one additional guests, including the American minister and his wife, a number of scientists and scholars, and a large group of relations and connections—Bunburys, Napiers, and Horners. The same day, Lyell received from his father a substantial gift that "put them at ease" in their new house.[26] On 8 December, the *Times* published a letter from Lyell on the report of a fossil human bone discovered by Dr. Montroville Dickeson and Colonel Benjamin Wailes in the loess at Natchez, Mississippi. He denied that the human bone was from the same beds as those containing bones of *Megatherium* and other extinct animals, suggesting that it was washed out of an Indian burial ground above the ravine in which it was found.

In mid-December, the Lyells went on the Great Western Railway

24. Lyell to Forbes, 14 October 1846, printed in K. M. Lyell, *Life, Letters, and Journals,* 2:110–13.
25. Forbes to Lyell, 18 October 1846, printed in K. M. Lyell, *Life, Letters, and Journals,* 2:110–13.
26. Lyell to his father, 2 December 1846, Kinnordy MSS.

to Calne in Wiltshire to join a weekend house party at Bowood, Lord Lansdowne's country house.[27] The train carriages were extremely cold, causing Lyell to recall wistfully the stoves in American railway cars that kept everyone warm. The invitation to Bowood was a result of Lord Lansdowne's interest in education. When Lord John Russell's government returned to power in July 1846, Lord Lansdowne resumed the office of lord president of the Council. He sought to develop a national system of education for Great Britain. Lord Lansdowne was also interested in science and had read Lyell's *Travels in America,* especially the chapters concerned with education. At Bowood the house party included several leaders of the government and W. P. A. Delane, editor of the *Times.* One day Lyell took three of his fellow guests on a geological field trip to the top of the Chalk downs, five or six miles away. Lyell was flattered to meet government ministers socially and to find them interested in his science. He spent some time talking to Lord Clarendon about the United States and the Americans.

Early in February 1847, Lyell completed the seventh edition of the *Principles of Geology,* for which Murray paid him 250 pounds.[28] He then turned to his boxes of American fossils and began to write a paper on the coalfield at Richmond, Virginia. The fossil plants of the Virginia Coal seemed to indicate that it was intermediate in age between the Liassic and Triassic. In March he received from Gideon Mantell a copy of his book on the Isle of Wight, in which Mantell referred to Lyell's work in 1822 on the Isle of Wight beds.[29] Lyell was touched and, in writing to thank him, said that the reference to their joint work on the Wealden recalled "some very happy early days in the field."[30] On 11 March, as relaxation after finishing the *Principles,* the Lyells attended a literary breakfast at the house of the banker and poet Samuel Rogers. Rogers, then in his eighties, was noted for his taste, his kindness, and his sarcastic wit. Such breakfasts, lasting from about ten o'clock till past noon, were lively occasions with a great flow of conversation and anecdote.

On 30 April, Lyell delivered a Friday evening lecture at the Royal Institution on a new estimate of the age of some of the most recent

27. Henry Petty-Fitzmaurice (1780–1863), third marquis of Lansdowne, was educated at the University of Edinburgh and at Trinity College, Cambridge. Elected to Parliament in 1802, Lansdowne spent all of his life in politics, becoming Lord President of the Council in 1830 under Lord Grey and later serving in the same office under Lord Melbourne.

28. Lyell to Murray, 22 February 1847, Kinnordy MSS.

29. Mantell, *Isle of Wight.*

30. Lyell to Mantell, 9 March 1847, Mantell MSS.

volcanoes of the Auvergne, based on fossil bones associated with them.[31] In 1843, when Lyell revisited Nechers in the Auvergne, the Abbé Croizet showed him where fossil bones had been found in a bed of sand near the lower end of the great lava flow from the extinct volcano of Tartaret. Later the Auvergne geologist Auguste Bravard showed that the deposit extended beneath the lava. He had collected from it fossil bones of extinct species belonging to some fourteen mammalian genera. Bravard sent some of the Tartaret fossil bones to Lyell, who asked Richard Owen to examine them. Owen identified two species, *Equus fossilis* and *Tarendus priscus,* that occurred also among English cave fossils. Thus, the Tartaret deposit was about the same age as the English cave deposits. The Tartaret sand also contained a number of fossil land shells, all of living species, suggesting that it was relatively recent. Nevertheless, in North America Lyell had found deposits containing only living species of land and freshwater shells that were very old. Thousands of years had been required for the recession of Niagara Falls and the formation of the Mississippi Delta, yet both were younger than nearby deposits containing only living species of fossils. The Auvergne fossil animals were drowned in a flood resulting from an eruption of Tartaret. Tartaret belonged to the youngest group of extinct volcanoes in the Auvergne. Determination of its age would establish the period of the most recent volcanic activity in the region. Volcanoes such as Tartaret possessed such perfect craters and cones, with loose volcanic scoriae on their slopes, that many persons thought they might have been in eruption within historic times. Lyell showed that the youngest volcanoes of the Auvergne were as old as the cave deposits of England.

On 10 May 1847, Lyell began to write the travel account of his second visit to the United States. Frequently he was interrupted.[32] Robert Winthrop, visiting in London from Massachusetts, called occasionally at 11 Harley Street. The mail from America brought many letters to be answered. Their guest rooms were almost always occupied by a succession of relatives. The famine in Ireland brought about a general increase in the cost of living. The price of bread rose to a shilling a loaf, and Mary strove to make economies in the kitchen.[33] On 28 May, Lyell's brother Henry returned from India, recovered from his wounds and looking well, but thin. He arrived

31. Lyell, "Age of the Volcanoes of Auvergne."

32. Lyell, Notebook 152, 22, Kinnordy MSS.

33. Mary Lyell to Katherine Horner, 26–29 May 1847, Kinnordy MSS.

after dinner, just in time to accompany Lyell and Mary to Bedford Place to say farewell to the Leonard Horners, about to leave for an extended holiday in Germany.

At the Geological Society on 26 May, Sir Roderick Murchison read a letter from Louis Agassiz at Boston, Massachusetts. Agassiz's assistant Edward Desor had shown that the rotten limestone of Alabama was neither Cretaceous, as the Philadelphia naturalists Morton and Conrad had thought earlier, nor Eocene, as Lyell had identified it in 1846.[34] Instead, Desor said that the rotten limestone corresponded to a formation at Biarritz in France, characterized by multitudes of minute disclike fossils resembling small coins and therefore called *nummulites.* Several French geologists considered the Nummulitic formation of Biarritz intermediate between Cretaceous and Eocene (an opinion Lyell considered mistaken). Since the rotten limestone of Alabama also contained nummulites, Desor postulated that it, too, must be a formation intermediate between Cretaceous and Eocene.

Lyell was certain that Desor was wrong. When in Alabama, Lyell ascertained not only that the rotten limestone was Eocene, but also that it lay above the Eocene limestone of the Claiborne bluff. It was, therefore, among the younger of American Eocene formations. Samuel George Morton of Philadelphia had described the nummulitic fossil from the Alabama limestone as *Nummulites mantelli.* Lyell now showed specimens of the rotten limestone from Alabama to Forbes, who recognized immediately that the small disc-shaped fossils were not nummulites. They appeared similar to some minute living species of zoophytes that Joseph Jukes had recently brought back from the northeast coast of Australia. William Lonsdale agreed with Forbes that the Alabama fossils were not nummulites. They were related to *Orbitolites,* a fossil genus from the Paris Basin first described by Lamarck and later identified as a minute species of coral. When Forbes compared the Alabama fossils with the living Australian species, he found that both belonged to the genus *Orbitolites.* At the Geological Society on 9 June, Lyell read a paper on the so-called Nummulitic Limestone of Alabama, presenting Forbes's and Lonsdale's identifications. Lyell also sent specimens of the Alabama fossil to Alcide d'Orbigny at Paris. Orbigny replied promptly that he was very familiar with the fossil, which was certainly not a nummulite, though often confused with it. He had

34. Lyell, "Newer Deposits of the Southern States" (1848).

named it *Orbitoides*. Lyell decided that the fossil should be known as *Orbitoides mantelli*, retaining the specific name given originally by Morton. He sent his paper, with Orbigny's letter inserted, to Benjamin Silliman to be published in the *American Journal of Science;* early in 1848 the Geological Society also published it in the *Quarterly Journal.*[35]

On 23 June, Lyell went to Oxford to attend the annual meeting of the British Association, traveling on the newly completed railway, "most of the time at the rate of a mile a minute." A journey at such unprecedented speed seemed like a magical transfer from place to place.[36] At the inn in which he was staying at Oxford, Lyell encountered Professor Sven Nilsson of Lund, whom he had met in 1834 during his tour of Sweden. Nilsson told Lyell that he had found conclusive evidence that man had lived at the same time as the cave bear and other extinct mammals in southern Sweden. At Lund they had a skeleton of the extinct wild ox (*Bos primigenius*) with an ancient arrow embedded in it. In a peat bog they had discovered a skeleton of a cave bear together with ancient human hunting weapons.[37] Nilsson had fossil teeth, bones, arrowheads, and so forth with him. Such evidence was contrary to Lyell's long held belief in the recent origin of man. Not fully convinced by Nilsson's evidence, nevertheless Lyell thought it the best he had yet seen.

At Oxford, Lyell went with Charles Darwin to call on the Reverend William Jacobson, vice principal of Magdalen Hall, with whom Darwin was staying during the meeting.[38] Jacobson, an able and extraordinarily conscientious man, had worked for many years to improve the teaching in the college and the standards in university examinations. After Lyell left, Darwin noticed a copy of Lyell's *Travels in America* on Jacobson's shelf and asked him how Lyell's chapter on the universities had been received at Oxford. Jacobson said it had been received very well. Many people considered Lyell's discussion temperate and thoughtful. Even many Oxford dons who disagreed completely with Lyell's conclusions praised the chapter.

Lyell dined with the Reverend Baden Powell, Savilian Professor

35. Lyell, "Nummulite Limestone of Alabama."

36. Lyell to his father, 25 June 1847, Kinnordy MSS; printed in K. M. Lyell, *Life, Letters, and Journals*, 2:129–31.

37. Nilsson, "Primitive Inhabitants of Scandinavia."

38. William Jacobson (1803–84), a native of Great Yarmouth in Norfolk, graduated with a bachelor's degree from Oxford in 1827 and in 1830 was elected a fellow of Exeter College. He was appointed principal of Magdalen Hall in 1832, Regius Professor of Divinity at Oxford in 1848, and bishop of Chester in 1865.

of Geometry.[39] Baden Powell was one of the professors who had lost most of the students from his lectures. He was sympathetic to the cause of university reform and welcomed Lyell's treatment of the question in his *Travels*. After dinner, Lyell attended a large party for members of the British Association in the Botanic Garden. As the evening drew on and the moon came up, the party in the garden became livelier. Frank Buckland, the Reverend William Buckland's son, led around a bear dressed in cap and gown as a student of Christ Church. He introduced the bear formally to members of the association, including Lyell, and the bear sucked their hands affectionately. "Amid our shouts of laughter in the garden by moonlight," Lyell wrote, "it was diverting to see two or three of the dons, who were very shy, not knowing how far their dignity was compromised."[40]

The next morning, Lyell breakfasted with Philip Duncan, keeper of the Ashmolean Museum, who was working for the establishment of the natural sciences and mathematics as part of Oxford studies.[41] At eleven o'clock the Geological Section opened with a paper by Robert Chambers on raised beaches. Various geologists criticized Chambers severely for drawing extravagant conclusions from his evidence. They knew that he was the author of the anonymous work *Vestiges of the Natural History of Creation*. Lyell attacked Chambers deliberately to show him that the kinds of arguments he used in the *Vestiges* would not be tolerated among scientists.[42] That evening Lyell returned to London exhilarated by the meeting, perhaps especially by his learning that the chapter on university education in the *Travels* was known and appreciated at Oxford.

Reform of the Royal Society

In 1846 Lyell's father-in-law, Leonard Horner, led a successful movement to reform the Royal Society of London by limiting the election of new fellows each year to fifteen candidates, to be chosen on the basis of their contributions to science.[43] The reform was the

39. Baden Powell (1796–1860) was born at Stamford Hill, Kent, and in 1814 entered Oriel College, Oxford, where he graduated with a bachelor's degree with first class honors in mathematics. He carried out research on optics and radiation in collaboration with Sir John Herschel, Charles Babbage, and Sir George Airy. In 1824 he was elected a fellow of the Royal Society and appointed Savilian Professor of Geometry at Oxford.

40. Lyell to his father, 25 June 1847, Kinnordy MSS.

41. Philip Bury Duncan (1772–1863) became keeper of the Ashmolean Museum in 1826. He urged the inclusion of science and mathematics at Oxford.

42. Geikie, *Ramsay*, 103. Chambers evidently withdrew his paper; it was not published in the proceedings of the meeting.

43. M. B. Hall, *All Scientists Now*, 80–82.

culmination of many years of effort to make the society a genuinely scientific body. Previously, it had been dominated by fellows with only a mild interest in science, elected for their wealth or rank. Achieved against formidable opposition, the reform remained of uncertain future. To ensure its permanency, in April 1847 twenty-seven of the active scientists in the society, including Horner and Lyell, met to form the Philosophical Club. During May and June, Lyell was involved in selecting men to be invited to join the club.[44]

Elected to the council of the Royal Society the following autumn, Lyell worked to continue the process of reform. Early in 1848 the president of the Royal Society, Lord Northampton, announced his intention to resign. Lyell spoke in favor of selecting a distinguished scientist as the next president.[45] He and Colonel Edward Sabine were asked to invite Sir John Herschel to serve as president, but Herschel declined the presidency immediately and firmly.[46] Sabine then wrote to the astronomer Lord Rosse to invite him to become president.[47] In November 1847 the two secretaries of the Royal Society decided to resign, and during 1848 Lyell was involved in a contest over the election of new secretaries. He worked for the election of W. R. Grove, the physicist, as one of the secretaries, but Grove's opponent was elected. Although they had lost a battle, Lyell felt that they would ultimately win the war because each year the new fellows being elected were serious scientists.[48] So it worked out. By 1860, a majority of the Royal Society and almost all the members of the council were active scientists.[49]

University Reform: Lyell and Prince Albert

On 12 March 1847, at a dinner party given by Sir James Clark, personal physician to Queen Victoria, Lyell conversed a good deal with Baron Stockmar.[50] Stockmar had originally studied medicine at Jena, but for many years he had been an adviser to Leopold I, King of the Belgians. In 1836 he was sent by Leopold to England to act as an adviser to the young Princess Victoria. After Victoria's accession

44. Bonney, *Philosophical Club,* 1.

45. Lyell to Grove, 14 February 1848 [copy]; Lyell to Leonard Horner, 11 March 1848, Kinnordy MSS.

46. M. B. Hall, *All Scientists Now,* 89; Herschel to Lyell, 10 March 1848 [copy], Herschel MSS; Lyell to De La Beche, 14 March 1848, De La Beche MSS.

47. Lyell to Grove, 27 March 1848 [copy], Kinnordy MSS.

48. Lyell to Grove, 30 November 1848 [copy], Kinnordy MSS.

49. Lyons, *Royal Society,* 270.

50. Christian Friedrich, Baron Stockmar (1787–1863).

to the throne the following year, he continued to exert a good deal of influence at court, and he was in frequent contact with the Queen and Prince Albert. Stockmar had been reading Lyell's *Travels in America* with great interest, the book having been recommended to him by Chevalier Bunsen, secretary to the Prussian embassy at London.[51] He was particularly interested in Lyell's account of the English universities and surprised to learn that various features of Oxford and Cambridge had not come into being until after the Reformation.[52] Stockmar told Lyell that Prince Albert had been reading the *Travels* and advised Lyell to present the Prince with a copy of the new seventh edition of the *Principles of Geology*. Accordingly, Lyell took a copy of his book to Karl Meyer, Prince Albert's librarian.[53] Two days later, Dr. Meyer called at 11 Harley Street with a note saying that Prince Albert was delighted with the gift and wished to make the author's acquaintance. The Prince invited Lyell to come to Buckingham Palace to meet with him on Sunday, 28 March.

Prince Albert had joined the Queen in her audiences with her ministers for almost eight years and participated informally, but effectively, in the conduct of state affairs. In September 1841 he was appointed chairman of a royal commission on the encouragement of the fine arts in the United Kingdom. This position brought him into contact with other members of the commission, including leading members of the government, the poet Samuel Rogers, and historians Thomas Macaulay and Arthur Hallam. Prince Albert went to Cambridge University in the autumn of 1843 to receive an honorary degree. While he was there, Adam Sedgwick, the professor of geology, showed him around the geological museum. During a visit to Germany in August 1845, the Prince arranged for the German chemist August Wilhelm von Hofmann to direct the newly founded Royal College of Chemistry in London. In February 1847, just a month before issuing his invitation to Lyell, Prince Albert was elected chancellor of Cambridge University.[54] With Prince Albert's interest in science and his new responsibilities as chancellor of Cambridge, he was interested both in Lyell's geology and in his opinions about university education.

51. Christian Karl Josias Bunsen (1791–1860) was educated at Göttingen. He began his diplomatic career in 1818 as secretary to the Prussian embassy at Rome and in 1842 was transferred to the Prussian embassy at London. Bunsen was a scholar who studied the history of the ancient world, particularly Egypt.

52. Mary Lyell to Marianne Lyell, 13 March 1847, Kinnordy MSS.

53. Friedrich Karl Meyer (1805–84), historian.

54. Pound, *Albert*.

When Lyell went to the palace, a dark and rainy day, he met Prince Albert and Dr. Meyer in the library. After an exchange of pleasantries, the Prince thanked Lyell for the copy of the *Principles,* and Dr. Meyer commented on Lyell's chapters on the history of geology. The Prince then turned to Lyell's discussion of Oxford and Cambridge in the *Travels.* He had found it very useful. As chancellor of Cambridge, Prince Albert was directly concerned with university education. He queried Lyell whether the giving of prizes in the physical sciences, history, and other subjects, would be helpful in raising the level of university education. William Whewell, master of Trinity College, Cambridge, had suggested the establishment of such prizes. Lyell replied that prizes would not be sufficient. Entrance examinations and examinations for honors degrees in the sciences were essential for effective reform. Lyell suggested that public opinion had already moved toward acceptance of some type of reform. If the Prince, as chancellor, pushed for essential reforms, there was a good chance that they would be accomplished. Lyell commented that it "was not a question between Whig and Tory or churchman & dissenter," but simply a question of whether professors would be allowed to teach effectively. The Prince replied that in Whewell's judgment, if the professors gave examinations on their lectures, students would stop attending them. Lyell agreed, explaining "that had Buckland examined me, I might have desisted from geology—as I wanted to gain honours." The Prince mentioned that in German universities lectures were open to anyone who wished to attend. He thought this policy would be useful in England, too. The Prince confided that he would like to attend lectures at Cambridge, but with the official demands on his life, he did not want to go through the ceremony of a royal reception. He asked Lyell about official records of the reforms attempted by the heads of houses at Oxford in 1839, mentioned in the *Travels,* and whether any heads of houses at Cambridge, other than Whewell, favored reform. Whewell was so unpopular at Cambridge that many members of the university had voted against Prince Albert for chancellor simply because Whewell was for him. The conversation then turned to the content of entrance examinations. Lyell advocated the inclusion of physical science, and the prince believed modern history should be included as well. Prince Albert noted that frequently students who gained the highest honors at the university later did nothing in public life; he hoped that a better understanding of modern history would remedy this situation. The Prince appreciated how difficult it was to bring about change. The few reforms made so far were insufficient.[55] After about an hour and a quarter, Lyell took his leave. He was delighted with the interview and much pleased by Prince Albert's simple and unassuming manners. "When talking of things & people interesting to him," the Prince was much more animated and enthusiastic than on public occasions.[56] His command of English had also improved greatly since Lyell had first heard him talk several years before.

Lyell had been concerned with the subject of university reform for more than twenty years. In 1827 in an article for the *Quarterly Review,* he drew attention to the narrow curricula of the English universities. At Oxford it was restricted to the study of Latin and Greek and at Cambridge to Latin, Greek, and mathematics. Education for law, theology, and medicine was excluded almost completely. Such subjects as mathematics, natural history, and natural philosophy, taught regularly in universities on the Continent and in Scotland, were neglected in the English universities. The exception was mathematics at Cambridge, but Cambridge mathematics did not include the more advanced branches developed on the Continent during the eighteenth century. In 1827 Lyell argued for the study of the physical sciences on the grounds that they enlarged knowledge of the works of the Creator and served to develop the mind. In England the public schools, rather than the universities, played the essential role in the education of the higher classes. They produced men who sought distinction based on their ability and talent and who combined "with elegance of manners a high feeling of honor and a manly spirit of independence."[57]

Oxford and Cambridge had once performed the functions now fulfilled by the public schools; boys had entered the universities at the age of eight or nine, when they now went to school. To illustrate how young university students once had been, Lyell cited the Cambridge statute forbidding undergraduates to play marbles on the steps of the Senate House. With the passage of time, students had tended to enter the universities later and later with more and better preliminary schooling. In the early nineteenth century, students entered Oxford and Cambridge at eighteen or nineteen, an age at which in earlier centuries they would have been about to graduate. Yet little change had been made in the curriculum to allow for the

55. Lyell, Notebook 151, 59, 63, Kinnordy MSS.

56. Mary Lyell and Lyell to Marianne Lyell, 29 March 1847, Kinnordy MSS.

57. [Lyell], "State of the Universities," 225.

students' greater maturity and better preliminary education. Thus students spent three years at university continuing to study the same subjects they had studied at school and at nearly the same level. Young men who wished to study law or medicine could not begin their professional studies until after they had graduated from the university. The result was either to make the course for professional training very long and expensive or to eliminate university education from the training of lawyers and doctors. In 1827, only about a hundred physicians practicing in England had been educated at Oxford or Cambridge.[58] Most of the practice of law in England was in the hands of some eight thousand attorneys, almost none of whom had studied at a university.

A consequence of the limited and essentially superfluous nature of English university education was to restrict Oxford and Cambridge largely to young men of the upper classes. It increased greatly the cost of university education, not the fixed costs for tuition, room, and board, but the more pervasive, but inescapable, expenses incurred by living among wealthy men—the need to maintain certain standards of dress, entertainment, and hospitality. Lyell thought that the expense of attending Oxford and Cambridge could be reduced only if the average income of the students were lowered by including more students of modest means.

If education at Oxford and Cambridge were to include professional training, the number and range of subjects taught would have to be greatly increased. To keep abreast of the modern development of his subject, each teacher would have to give his whole time and attention to it. The tutors at Oxford and Cambridge colleges were unable to specialize in one subject. All had to teach the subjects in which the students were examined for degrees, and the examinations covered the works of only a limited number of ancient authors. Since the tutors had studied the same authors for their own degrees, their teaching did not expand the range of their learning. In contrast, teachers in German universities specialized in a particular branch of knowledge. The German universities taught a broad range of subjects, and each professor contributed to his subject by original scholarship or scientific research.

In the 1820s the English universities made modest moves to broaden the range of studies they offered: Oxford included mathematics by creating a separate board of mathematical examiners,

58. Ibid., 235.

and Cambridge established a new tripos examination in mathematics. Examinations for degrees became more rigorous, but an effect of the higher standards of examination was to concentrate the attention of students on the subjects in which they would be examined. Consequently, at Oxford attendance at lectures in chemistry, natural philosophy, astronomy, anatomy, and botany fell; sometimes there was no audience at all. At Cambridge, interest in science was greater. Professor John Henslow's lectures on botany, Professor William Farish's lectures on the applications of chemistry, and Professor Adam Sedgwick's lectures on geology had each attracted enthusiastic classes, although the students came more for relaxation than for serious study.

In 1827 Lyell called for what he judged a moderate reform, the creation at each university of examining boards in the faculties of law, theology, and medicine. This step would enable Oxford and Cambridge to become centers of professional training and attract larger numbers of students. In a series of articles in the *Edinburgh Review* between 1831 and 1835, Sir William Hamilton, the eminent mathematician, attacked the English universities even more severely than Lyell. His attacks had called forth from Oxford a lengthy reply, which Hamilton saw as a "tissue of disingenuous concealments, false assertions, forged quotations and infuriate railings."[59] Hamilton's criticisms also influenced Lord Radnor in 1835 to move in the House of Lords for a royal commission to inquire into the universities, a move that failed immediately in the face of opposition from Bishop Copleston and the Duke of Wellington, then chancellor of Oxford.[60]

Hamilton's articles stimulated William Whewell to write a book to defend the existing system of university education, particularly at Cambridge. Whewell argued that university education fitted men for practical life. Classics and mathematics were practical, because they required students to participate actively by translating Latin or Greek authors or in mathematics by solving problems. In contrast, at professorial lectures in history, political economy, natural philosophy, or other modern subjects, students were merely passive listeners. Classics and mathematics were fixed and permanent in their subject matter. Other subjects were constantly progressing and changing. Whewell thought it would be difficult for students to re-

59. Quoted in Tillyard, *University Reform,* 51.
60. Edward Copleston (1776–1849), bishop of Llandaff.

spect the teachers of subjects so constantly in flux. Yet, Whewell accepted a need for some change in the system of teaching. The examinations should be connected with the lectures, and the only way to do so was to give lecturers considerable influence over the examinations. He intended to give such control only to the college lecturers, that is, to the college tutors. Whewell thought that English students had little inclination to attend lectures. Furthermore, the examinations tended to absorb all the students' capacity for study. The undergraduate course of study should not be changed in any drastic way. To an astonishing degree, Whewell was content to omit subjects in an active state of intellectual development. The change and progress in the social and natural sciences seemed to alarm him. "If, for instance, we adopt Political Economy as one of our subjects," Whewell asked, "who can tell us what kind of science political economy will be fifty years hence?"[61]

In his *Travels in America,* Lyell wrote that when he had asked questions in the United States about the teaching at American colleges, he had often been asked in turn to explain university education at Oxford and Cambridge.[62] The more he attempted to explain English university education, the stranger it appeared to him. It was not only Lyell's difficulty in explaining the English system but also the complete failure of a recent attempted reform at Oxford that had led him to discuss university education. In 1839 the professors of natural science at Oxford had sent a memorial to the heads of houses saying that they were unable to perform their duties.[63] Students were kept away from their classes by the need to study for the university examinations. The heads of houses proposed to require all candidates for a bachelor's degree to attend at least two courses of lectures given by a university professor. This moderate attempt at reform was rejected by the Oxford Convocation, an assembly of graduates in which the college tutors exerted a dominant influence. The convocation did enact a statute that the various university professors should be required to give their lectures. Thus, while the convocation insisted that lectures be given, their other measures prevented students from attending them. The effect of the 1839 statute was the virtual elimination of science teaching from Oxford. In the face of the expulsion of the natural sciences togeth-

61. Whewell, *English University Education,* 129.

62. Lyell, *Travels,* 1:215.

63. Ward, *Victorian Oxford,* 108–9; cf. Lyell, *Travels,* 1:232.

er with all higher learning from Oxford, Lyell felt he must protest. The means at hand was the pages of his *Travels in America*.

Lyell's treatment of university education in the *Travels* was relatively brief, occupying only thirty-seven pages, but it possessed the clarity, conciseness, and remorseless logic of a legal brief. His indictment of the English universities was widely read and was the chief reason why he had been invited to meet Prince Albert. Among the authors whom Lyell cited was William Whewell. Lyell revealed the inconsistencies in Whewell's defense of the Cambridge system. Lyell suggested that Whewell's attitude was "not of that uncompromising kind which should make us despair of his co-operation in all future academical reforms."[64]

Lyell's references to Whewell infuriated the latter. In 1846 Whewell published a second book on the Cambridge system in which he charged Lyell with ignoring his statements that the progressive sciences are "highly valuable acquisitions to the student," even though they could not serve to cultivate the mind. He thought that Lyell should not have accused him of "defending the exclusive monopoly enjoyed by Classics and Mathematics in the education of young men at Oxford and Cambridge."[65] It is difficult to follow Whewell's argument. He seemed to think that his statements that the progressive sciences were valuable to the student might be taken to imply that they should be included in university education, even though he himself excluded them as subjects for a degree. On various points, he charged Lyell with confusion. What Whewell objected to especially was Lyell's proposal for a royal commission on the universities. He thought a commission might destroy the existing system without creating a new and better one.

When Whewell and Lyell met socially at the home of Reverend Henry Milman in March 1846, Whewell brought up the subject of university education, but their conversation was necessarily brief. Whewell followed up the encounter with an indignant letter, complaining to Lyell that he had treated him unjustly in the *Travels*.[66] On 17 April, Lyell replied to Whewell in a long letter of eighteen pages. He was sorry that Whewell had been offended by certain passages, from which he thought Whewell had construed a stronger

64. Lyell, *Travels*, 1:245.

65. Whewell, *Liberal Education*. I have used the second edition, of which the first edition forms Part I. Whewell devoted chapter 2, section 2 (117–32), to Lyell.

66. Whewell to Lyell, 31 March 1847, Lyell MSS.

meaning than the words implied and certainly more than Lyell had intended. Although Lyell wrote in a conciliatory tone, he remained firm in his own position. "I should not have alluded to your book," he told Whewell, "if I had not regarded it as one of the most authoritative and able defenses of 'things as they are,' for so in the main, whether I was mistaken or not, I understand the spirit of your treatise."[67] He congratulated Whewell on his suggestion for a new tripos examination in the inductive sciences at Cambridge, but he criticized Whewell's opinion that ideas in science should not be taught in universities until they were established beyond all possibility of controversy. Lyell continued:

> I now come to that part of your reply to my 13th Chapter (Travels, v. 1) in which you distinctly join issue with me as to my representation of facts, on the correctness of which I willingly admit that the whole force of my reasoning depends. . . . You then add, "The College system which I defend agrees with the professorial system which he [Mr. Lyell] recommends, in the feature which he thinks most important, the division of departments. This division has long existed at Cambridge. The private tutors &, for the more advanced students, the public tutors also, have their separate subjects. The tutors each confine their lectures to their separate provinces in which their favourite studies have lain, and to these they give their labours as especially as if they were Professors of their selected subjects." p. 119 Camb. Stud. In continuation you observe, "I do not conceive that any such advantage would be gained by calling our College Lecturers Professor instead of Tutors."[68]
>
> I confess I was never more astonished than on first reading these passages. They contain a distinct denial of the statements made by me respecting the present tutorial organization of Oxford and Cambridge on which my principal objections to the system were founded. I have throughout maintained that the college-tutor machinery differed from a Professorial staff belonging to the University at large not only in name, but so essentially in substance as to be wholly incompatible with the subdivision of the sciences & different branches of human learning which I desire to see cultivated. Some of my friends both Ameri-

67. Lyell to Whewell, 17 April 1847, Whewell MSS.

68. Whewell, *Liberal Education,* 120. There are minor inaccuracies in Lyell's quotation.

> can and English have naturally asked me whether I who left college a quarter of a century ago, could possibly be as well informed as the Master of Trinity writing in 1846. I reply that my representations are strictly true, not only as to Oxford & all the smaller colleges at Cambridge, but even as to Dr. Whewell's own college Trinity itself.

Lyell went on to recapitulate what he had said in the *Travels* and to demonstrate from the *Cambridge Calendar* for 1847 that apart from St. John's and Trinity, the Cambridge colleges each had only one or two tutors and one or two assistant tutors. Lyell continued:

> I am informed that in some of these colleges somewhat more than a binary division is accomplished by making one of the classical teachers devote his time chiefly to "Moral Philosophy" yet on the other hand the tutorial staff is really weaker than it would appear in the Table [from the *Cambridge Calendar*], because one of the tutors, & commonly the senior, has to attend to the keeping of the accounts of the young men & the discipline of the college.

Lyell argued that it would be feasible to maintain the discipline of students under a professorial system if students were required to attend the professors' lectures and if the professors possessed an effective voice in the examinations for honors. In reply to Whewell's argument that students were not interested in "the progressive sciences" and that even in London professors could not attract a regular audience for scientific lectures, Lyell said:

> This is not the case—witness Professor Lindley on Botany, Graham on Chemistry & many others where a steady nucleus is secured to the class by the medical students who are obliged to attend & where others are attracted by the fame & ability of the lecturers. Put the larger part of your Cambridge Tutors in a similar position in London & I should not be surprised if the nucleus alone remained while the splendors of the tail vanished.

Although Lyell controverted those points on which Whewell had publicly contradicted him, he welcomed what he thought were positive suggestions in Whewell's book on Cambridge studies: the proposals for university entrance examinations and a tripos in the inductive sciences. Lyell thought such reforms unlikely to be achieved because of opposition from the clergy, who dominated both Oxford and Cambridge and worked to exclude "all the progressive branches of knowledge."

> I know there are many individual ecclesiastics who can rise above all professional prejudices & prepossessions, but their training & the study of dogmatic theology imparts to their minds in general as history proves & as Oxford now exemplifies, a Medievalist rather than a Baconian cast, & loving extreme deference to authority rather than free investigation, they will oppose themselves to every attempt you may make to render the study of the Inductive Sciences an effectual part of their system.[69]

Similarly the colleges, considered simply as richly endowed corporations, would resist university reform, because it would involve the wider sharing of the income from endowments.

Before Lyell sent this long letter to Whewell, he had several copies made. One copy he sent to Chevalier Bunsen and another to Sir James Clark, clearly with the idea that it would be read by Prince Albert. He also sent copies to Leonard Horner and to his father at Kinnordy.[70] Later he sent a copy to George Ticknor with the request that he show it to Edward Everett, president of Harvard. The letter was much more than a private letter to Whewell; it was a detailed answer to Whewell's public strictures on the *Travels,* to be read by persons closely concerned with university questions.

A few days after he received Lyell's letter, Whewell replied in a friendly manner. He thought that the distinction between classical and mathematical teachers in most Cambridge colleges was true specialization. On the whole, Whewell left most of Lyell's points unanswered and thought it not worthwhile to explain his views further, telling Lyell, "you have already taken up opinions, not only with regard to the condition of the university, but also with regard to the motives, 'instincts,' and the like, of those who conduct or defend the universities, which would make it impossible for us to go on together."[71] They did not discuss the subject further. Lyell gave his opinion of Whewell's reply when he wrote to George Ticknor. "I have an answer from Whewell which is quite a knock-under as to every disputed fact and extremely pacific."[72]

69. Lyell to Whewell, 17 April 1847, Whewell MSS.

70. Lyell, Notebook 152, 5, Kinnordy MSS.

71. Whewell to Lyell, 26 April 1847, Lyell MSS.

72. Lyell to Ticknor, 3 May 1847, printed in K. M. Lyell, *Life, Letters, and Journals,* 2:128–29. The copy of Lyell's letter to Whewell of 17 April 1847, sent to Ticknor, is in the Ticknor Papers.

When Lyell sent copies of his correspondence with Whewell to Chevalier Bunsen, he said that Bunsen might show them to anyone interested in the university question, but he did not wish to compromise Whewell, who was in his own way working for reform.[73] Lyell also sent a copy of the letter to Charles Darwin, who returned it with the comment: "I like your letter very much & admire your cool boldness. The Master of Trinity must have been surprised at being argued with on terms of such perfect equality; & if he thought you meant to publish it, I am not the least surprised at his calls & civilities to avert anything so unpleasant."[74]

Lyell became active in drawing up a petition for a royal commission to inquire into the universities. On 24 April 1848 he sent a copy of the petition to Dr. Meyer, Prince Albert's librarian, saying that they had already obtained "many good signatures" on it.[75] Lyell felt keenly the need for university reform as part of a move for intellectual liberation in England. In the wake of the recent revolution in France, he wrote to Ticknor:

> I tell the Tories here that they may now thank those who compelled them to adopt reform here in 1830. But we are in a state of great intellectual thraldom here with no immediate prospect of being aided by the present European movement. I mean we are so church-ridden. The publications suppressed, articles for reviews maimed & cut down, works compelled to steal out anonymously or posthumously & the stealthy reading of the same & fear of admitting the fact, the increasing enforcement by some Bishops of more severe examinations on articles of faith, the retrograde Puseyite movement, check progress & the love of truth & make our educational & academical reforms slow & scarce possible. We must wait till Germany & Massachusetts & I hope France shall shame us into more practical mental freedom.

Lyell found himself increasingly impatient with the pretensions of the Church of England. He continued:

> I am disgusted everyday at the ignorance in our pulpits. Last Sunday we had to listen to the 6th veil of wrath & the three frogs of the revelation, viz. infidelity, the spirit of lawlessness & Popery which these gloomy times of Revolution exemplify showing that

73. Lyell to Bunsen, 19 May 1847 [copy], Windsor MSS.
74. Darwin to Lyell, 17 June 1847, in Darwin, *Correspondence,* 4:43–44.
75. Lyell to Meyer, 24 April 1848, Windsor MSS.

> the world is approaching its end, which means that old mother church is rather nervous, seeing that the world over the water will not remain stationary like our liturgy.[76]

On 28 June 1848, Lyell went a second time to Buckingham Palace to talk with Prince Albert. He was ushered into the library. At exactly half past two, the time appointed, Prince Albert came in with Dr. Meyer. They first discussed the theories put forward to account for the parallel roads of Glen Roy. The Prince sent Dr. Meyer to get a map of Switzerland so that he might show Lyell where a Swiss lake near Brientz, the Lungern See, had burst its barrier. He thought the lowering of the Lungern See might be an event comparable to what had occurred in Glen Roy.

As they went on to talk of other subjects, Dr. Meyer mentioned the recent political changes on the Continent, and Prince Albert drew a geological analogy to political upheavals. He recalled Lyell's comparison of the fossil reptile footprints of Pennsylvania to the footprints of birds and the cracks in the sun-dried mud of tidal flats in Nova Scotia and said of European politics, "Yes, but where the mud has settled, there will be those cracks—they take time to fill up & before another even stratum is laid over that which has been rent." Lyell mentioned that Leopold von Buch had said that in England individuals had less freedom to publish their opinions on science and philosophy than did professors in German universities, who lived under a despotic monarchy. The reason was "that a representative government, by giving power to the less educated & more prejudiced classes, fettered the higher ones." When Dr. Meyer protested against such a paradox, Lyell said that he did not wish to defend it, but Prince Albert took up the point seriously:

> I wish it may not be too true. You see that the jury-men who condemned Mitchell in Dublin are ruined, for no one will deal at their shops, such is the tyranny of the people. Now if in a country where there is a political censorship or a church inquisition, you express yourself freely, you can carry with you public sympathy for the people support you against the authority which frets and restrains them, but when they have the power themselves you have no appeal & it requires more daring to oppose the many.

The remedy for the problem, Lyell thought, was to educate the people. He described how impressed he was with the system of general

76. Lyell to Ticknor, May 1848 [copy], Kinnordy MSS.

public education in Massachusetts. Prince Albert agreed that education was "the true way of removing the evil & many others, but unfortunately many who think so will not, for their own reasons, promote education."[77] Lyell mentioned that in the southern United States the slave owners opposed education for the Negroes because they thought it would make them discontented with their lot. In England many educated people said exactly the same thing about education for the lower classes.[78] Concerning the reform of the universities, Lyell observed that "all changes in England were very slow, but if required & there was perseverance, they were carried at last." Prince Albert asked which English statesman it was who said that a truth must be repeated again and again to the English public before they would take it in, and that they needed "a great deal of soaking." Lyell remembered that it was Charles James Fox.[79]

Lyell found Prince Albert "very animated, seemed stronger than last year & his ideas flow more rapidly in English."[80] Both had amusing stories to tell and the hour passed pleasantly and quickly, but their conversation had a serious purpose. Both men were engaged in the movement for university reform. As chancellor, Prince Albert was using his authority and influence to bring about reforms at Cambridge. He had invited Lyell to the palace in part to discuss measures then pending at Cambridge. After the Prince's election as chancellor the year before, William Whewell had sent him suggestions for the extension of the range of studies at Cambridge. The Prince, knowing that Whewell was unpopular, wished to form his own opinions.[81] He had already read the discussion on universities in Lyell's *Travels* and talked with Lyell. When he and Queen Victoria visited Cambridge in July 1847, he discussed university affairs with the vice-chancellor of the university, Henry Philpott.[82] The following October, Philpott sent him a table describing the teaching offered at Cambridge. In November 1847 the Prince invited Philpott to stay at Windsor Castle. Prince Albert was gathering data on

77. Lyell to Eleanor Lyell, 29 June 1848, Kinnordy MSS. Lyell was quoting Prince Albert from memory and may not have recalled his exact words.

78. Lyell, Notebook 156, 12, Kinnordy MSS.

79. Lyell to Eleanor Lyell, 29 June 1848, Kinnordy MSS.

80. Ibid.

81. Winstanley, *Early Victorian Cambridge.*

82. Henry Philpott (1807–92) matriculated at St. Catherine's Hall, Cambridge, in 1825 and graduated with a bachelor's degree in 1829. Elected a fellow of St. Catherine's in 1829, Philpott served successively as assistant tutor and tutor before his election as master in 1845. In 1846 he was elected vice-chancellor of Cambridge University.

the details of education at Cambridge and submitting the information to informed persons for their comments.

Prince Albert was also in correspondence with the prime minister, Lord John Russell, about the results of his Cambridge inquiries. On 12 November, the prime minister informed the Prince that he planned to establish a commission of inquiry into schools and colleges, but the Prince persuaded him to defer a commission until Philpott might see what he could do to initiate reforms. Meanwhile, the Prince hinted to Philpott that if Cambridge did not move to reform itself, the government might act.

On 26 February 1848, the new vice-chancellor of Cambridge, Dr. Phelps, presented a plan for various changes to the syndicate, a board composed of the heads of the colleges. Phelps's plan was incorporated into a report of the syndicate published in April. The report recommended that all students be required to attend a course of lectures given by a professor and to pass an examination set by him. It would establish new honors examinations in natural sciences and moral sciences to be administered by the professors of the subjects concerned. The moral sciences were to include moral philosophy, political economy, modern history, general jurisprudence, and English law.[83] Although limited in extent, the recommendations were in the spirit of Lyell's suggestions in his *Travels in America.*

Fearing rejection, the Cambridge syndicate decided to withhold the report from the university senate until the following October. Early in April 1848, Phelps learned from Prince Albert's secretary that a group of individuals was preparing a memorial to the government requesting a royal commission on the universities. Informing the syndicate of the memorial, Phelps used it as a reason for the immediate publication of the syndicate's report. On 8 April, the *Athenaeum* noted the circulation of the memorial among graduates of both Oxford and Cambridge.[84] It was signed by 224 university graduates. Lyell supported the memorial actively. On 10 July he was one of a committee of about a dozen designated to present the petition to the prime minister. After meeting at a club on St. James Street where they elected Lyell to act as spokesman, the committee proceeded to 10 Downing Street. There Lyell addressed the prime minister on behalf of the committee. Referring especially to Oxford, Lyell argued that only a royal commission could rescue the

83. Winstanley, *Early Victorian Cambridge,* 207–9.

84. *Athenaeum,* 1848, 367.

universities from the Anglican clergy. Edward Bunbury described the situation at Cambridge, explaining that the constitution of the university prevented the whole body of graduates from even discussing measures for reform. A single member of the syndicate could veto any measure without giving a reason, and once a measure was presented to the senate it could not be amended, but must either be passed unchanged or rejected. Such formidable obstacles to change meant that Cambridge would need the help of a royal commission to achieve reform. Lord John Russell promised that the government would give the memorial serious consideration.[85]

Although he did not act immediately, the prime minister's favorable reception of the memorial was well known at Cambridge. Furthermore, the number of Cambridge graduates who had signed the petition indicated that it had broad support. At Cambridge the changes proposed by the syndicate in April aroused great excitement, and on 31 October many nonresident graduates came from a distance to vote on them in the Senate House. To the surprise of almost everyone, the measures passed by substantial majorities.[86] Cambridge had managed to launch serious reform without government intervention; Oxford would ultimately need a royal commission to accomplish the same thing.

Family Matters; Literary and Scientific Work

At the end of June 1847, the Lyells left London for a summer vacation at Kinnordy. Lyell's mother had been unwell for some months, and they wished to be with her. Lyell took along his notebooks to continue writing his second American travels. Henry Lyell was also at Kinnordy, looking rested and cheerful. For the first ten days the weather was beautiful, and on 6 July Lyell found in Caddam Wood a patch of the rare and lovely twinflower, *Linnaea borealis,* which he had last seen in the woods at the foot of Mount Washington in America.

At Kinnordy Lyell discussed the financial affairs of the estate with his father. The repeal of the Corn Laws had caused a fall in grain prices, which in turn reduced the rents the estate took in from the farms. The estate was encumbered with debt. The rise of interest rates during the 1840s greatly increased the effective burden of the debt. With various other obligations, including Lyell's allowance guaranteed under his marriage contract and smaller allowances to

85. Lyell to his father, 10 July 1848, Kinnordy MSS.
86. Winstanley, *Early Victorian Cambridge,* 213.

Harry and Tom, Mr. Lyell's income was seriously reduced. Since his father was now seventy-eight and no longer well, Lyell talked to the estate factor and the family solicitor to see what might be done, but to no avail.[87] He had to persuade his father to act, and, impractical as ever, Mr. Lyell was unwilling to be guided by his son's advice.

On 1 August, Mrs. Lyell suffered a severe paralytic stroke. During the preceding year she had had several lesser strokes, but now she was critically ill. She remained conscious, although blind, but for an hour the family was gathered around her bed, expecting her end. She recovered partially but remained an invalid. Despite anxiety about his mother and about the finances of the Kinnordy estate, on which his income depended, Lyell worked steadily on the account of their second visit to the United States.

At the beginning of September, Lyell and Mary returned to London. A week later, Lyell had completed more than fifty pages of manuscript for his travels and was writing steadily.[88] By January 1848, he had written almost seven hundred pages. Such intense writing meant that they could accept few invitations. Apart from scientific meetings and Lyell's occasional visits to the Athenaeum, they lived quietly. For Christmas, they went to Leonard Horner's new house in the country, Rivermede, on the bank of the Thames at Hampton Wick. Lyell took his notebooks with him and continued to work over the holiday. Near the end of 1847, Sir Henry De La Beche, president of the Geological Society, told Lyell that he and the council of the society were counting on him to be the next president. Lyell was willing to serve for the honor of the post, but for two years it would require him to be in London from November to June and would prevent his making any long expeditions until the end of 1851.[89]

On 12 January 1848, the Lyells again went for several days to Bowood, where Lyell enjoyed the fellowship and good conversation. One day he went to Bath to see William Lonsdale and discuss the local geology with him. Two weeks later, they were at Rivermede for the wedding of Henry Lyell and Katherine Horner. The marriage was a joy to them, particularly to Mary, who, since childhood, had felt responsible for her younger sisters.[90] The newlyweds went to Hastings for a brief honeymoon before leaving for India. Lyell

87. Lyell, Notebook 153, 45–55.

88. Ibid., 101.

89. Lyell, Notebook 154, 63, Kinnordy MSS.

90. Anne Susan Horner, "Memories of My Life," 26, Malcolm Lyell MSS.

and Mary, accompanied by Leonard Horner, returned to London in bitterly cold weather. In the evening, while Lyell and Horner were at a meeting of the Philosophical Club, Mary felt the quietness of the house after so many busy days when Harry, as the family affectionately called Henry, was with them. She wrote to Katherine that "Harry must have been a great scribe while he was here, for really when I came into this room, I quite looked to see him on the sofa behind the writing table. He must have been very busy buying the tea & sugar & candles [for India]."[91]

On 4 February, Lyell gave a Friday evening lecture at the Royal Institution. The audience of about four hundred included all the geologists and many of the literary figures in London. For his subject, the fossil reptile footprints in Pennsylvania, he had prepared many maps and illustrations. The lecture was received so well that he did not regret the time required to prepare it and planned to incorporate a portion of the lecture into his American travels. He wrote to his sister Eleanor:

> The consciousness that no one else, either in Europe or the U.S., could from actual observation have given the same account of the nature of the evidence . . . respecting the proofs of the first quadruped or air-breathing reptile ever found in such ancient rocks as the Coal Strata gave me confidence and spirit as I knew it would be of interest to all the geologists present.[92]

Before the lecture, Michael Faraday, the true master of popular scientific lecturing in England, gave Lyell hints about his own lecture methods, particularly for timing the various parts of the lecture.

In mid-February Lyell paid a brief visit to Charles Darwin at Down. They discussed American geology, comparing the loess of the Mississippi Valley to the red mud of the South American pampas and the fossil bones found in the two deposits. They also talked over the circumstances required for the preservation of animal footprints on sea beaches.[93]

The anniversary meeting of the Geological Society was always a signal event on 28 February for British geologists. Following their established custom, the geologists dined together, and after dinner the president summarized the year's activities. At the end of Sir

91. Mary Lyell to Katherine Lyell, 27 January 1848, Kinnordy MSS.

92. Lyell to Eleanor Lyell, 7 February 1848, Kinnordy MSS; see K. M. Lyell, *Life, Letters, and Journals,* 2:138.

93. Lyell, Notebook 155, 22–28.

Henry De La Beche's presidential address, Lyell rose and proposed a toast. He described De La Beche's excellent work in organizing the Geological Survey, congratulating him on appointing such competent scientists as Edward Forbes to the staff. Aware of the influence of De La Beche's catastrophic views on some of his staff, Lyell also sounded a cautionary note. Although government support of scientific research might confer great benefits, independent scientific societies like the Geological Society were needed to prevent official position and age from exerting too much authority over theoretical questions. Lyell had seen at Paris the restrictive influence that powerful men like Cuvier could exert on the expression of scientific ideas. He thought that "we must have in the republic of Science the most unshackled expression of opinion for the truth to make way & prevail."[94] The fact that De La Beche was at that moment both president of the Geological Society and director general of the Geological Survey enabled Lyell to make his point courteously. His remarks were well received and he was complimented by some who spoke afterward. From the very beginning of the Geological Survey in 1836, Lyell had feared that a government scientific establishment might inhibit freedom of scientific discussion. It had exerted that effect at Paris. Young French geologists were well aware that, if they were to receive jobs and promotions, their theoretical opinions must support those of their superiors.

Through the winter and spring of 1848, Lyell continued to write the account of their second American tour. He was also involved in university reform. In late July he and Mary left for Scotland. At Kinnordy during the next few weeks, Lyell wrote steadily on their second visit to the United States, only occasionally putting down his pen to take a botanical or geological excursion into the surrounding hills. He and Mary returned to Caddam Wood to examine the patch of twinflower he had found the year before. Early in September, he received a letter from Lord Lansdowne saying that in consideration of Lyell's distinguished position and scientific reputation, he had recommended to Queen Victoria that Lyell be given a knighthood. The Queen, Lansdowne added, "understands that it is without any solicitation on your part."[95] Lyell was too far from London to receive the honor immediately, Lord Lansdowne thought, but the Queen arranged that there should be no delay. Earlier in

94. Lyell to Eleanor Lyell, 19 February 1848, Kinnordy MSS.

95. Quoted from Lord Lansdowne's letter in Lyell to Leonard Horner, 11 September 1848, Kinnordy MSS.

the summer, Prince Albert had purchased from Lord Aberdeen a long-term lease of Balmoral Castle, including an estate of ten thousand acres, in the hills of Aberdeenshire beside the river Dee. In September Queen Victoria and Prince Albert were to come to Balmoral for their first autumn holiday in the Scottish Highlands. Although a long way around by road, Balmoral was not far from Kinnordy in a straight line across the Grampian Hills. In June, when Lyell was at Buckingham Palace, Prince Albert, excited about his new acquisition of heather and forest, had asked him questions about Scotland. Lyell had shown him on a map where Kinnordy was located. Now, the Queen invited Lyell to come to Balmoral to receive his knighthood and to stay overnight.

On 18 September, Lyell packed his saddlebags with various necessities and, taking a notebook, a "tin box for plants, a compass and geological hammer," left Kinnordy on horseback. Riding up Glen Clova, with its high, steep hills on either side, Lyell noted the striking evidences of glaciation. He stayed overnight at the Kirkton of Clova and the next day, accompanied by a guide, rode across the hills. On the tops of the Grampians he observed that "here as on Mt. Washington & in the White Mountains the decomposing granite boulders & the bare surfaces of disintegrating granite are not scored with glacial furrows or polished."[96]

At Balmoral, Lyell found a congenial group, including Lord John Russell and Sir James Clark, as well as Queen Victoria, Prince Albert, and their personal party. He stayed at an inn about a mile from the castle, but the Queen sent a carriage to take him back and forth. In the relaxed atmosphere at Balmoral, among intelligent, well-informed people, Lyell found the conversation and laughter at dinner very pleasant. After dinner the first evening, in a brief but formal ceremony, Lord John Russell presented Lyell to the Queen, and she conferred knighthood on him. Sir James Clark told Lyell that when Prince Albert learned of Lord Lansdowne's recommendation, he said that he wanted Lyell to be the first person knighted at Balmoral. The Queen asked Lyell to remain an additional two days. The next day, Lyell and the Prince went on a geological excursion among the surrounding hills, Lyell explaining the local geological features.[97] Prince Albert never forgot the geology he learned from Lyell and when at Balmoral continued to study the local rocks.

On his return to Kinnordy, Lyell found waiting for him a letter

96. Lyell, Notebook 157, 12, Kinnordy MSS.
97. Martin, *Prince Consort*, 2:111.

from Gideon Mantell, who, sick and in constant pain, overcame his own unhappiness to congratulate Lyell on his knighthood.[98] Charles Darwin also wrote promptly on hearing the news to express his delight, saying that for some years he had mentally "most heartily abused the Government for not having long ago, by so small an outward token, marked your public estimation."[99] At Kinnordy the servants were already taking pains to address Mary as "My Lady." On 8 October, the day before her fortieth birthday, she wrote to her father, "Surely no one ever had a more prosperous life than mine has been. It almost frightens me, & I pray that when the dark days come, I may be able to say God's will be done."[100] Late in October they returned to London by the steamboat from Dundee.

Once settled again at 11 Harley Street, Lyell began the final stages of preparing the manuscript of his *Second Visit to the United States* for the press. He was immediately beset by the demands of London life—meetings of the Philosophical Club, the Royal Society, the Geological Society, and other societies, the contested election for secretary of the Royal Society, and invitations to breakfast and dinner. In addition, Joseph Cogswell, whom they had met at New York, was in London. Lyell nominated him for membership in the Athenaeum Club and sent him invitations. As he put the finishing touches on the *Second Visit,* he also plied Cogswell with questions about America.

Authority on America, 1849–1850

Lyell's *Second Visit to the United States* appeared in June 1849. For more than two years the book had been Lyell's chief preoccupation. He was relieved to have it completed. He could now turn to the new books and articles on geology that he had been unable to read while he was writing. The *Second Visit* was a more general work of travel, with less emphasis on geology than the *Travels in America.* The reviews were almost universally favorable. The reviewers liked Lyell's portrait of America—of a vigorous nation growing with astonishing rapidity in a vast and fertile continent, richly endowed with coal and iron ore. The reviewer for the *Athenaeum* thought that Lyell carried with him "the best habits of scientific observation into other strata

98. For Lyell's reply, see Lyell to Mantell, 25 September 1848, Mantell MSS; printed in K. M. Lyell, *Life, Letters, and Journals,* 2:148–49 [misdated 24 September].

99. Darwin to Lyell, [postmarked 25 September 1848], Kinnordy MSS.

100. Mary Lyell to Leonard Horner, 8 October 1848, Kinnordy MSS.

than those of clay," quoting at length passages from the book to illustrate the variety and interest of his observations.[101]

Even the *Quarterly Review,* to Lyell's slight surprise, contained a long and favorable article on the *Second Visit,* written by his friend Henry Milman. Milman was impressed "by the strong predominance of good-will, by the total absence of acrimony, though now and then there is a touch of sly, perhaps involuntary satire (in some of the quiet anecdotes there is singular force and poignancy)." He noted the large amounts of public money spent on education in Massachusetts and New York, and the mutual toleration among so many religious sects in America. Though Lyell attributed this toleration to a reaction against the Calvinism of the earlier Puritans, Milman was inclined to attribute it simply to practical necessity imposed by the sheer number of sects. He also cited Lyell's observation that a consequence of every man's possessing a vote in the United States was "the irresistible temptation it affords to a needy set of adventurers to make politics a trade, and to devote all their time to agitation, electioneering, and flattering the passions of the multitude."[102] He thought that the United States must make a special effort to bring its best men into public affairs.

Early in July, Sir Charles and Lady Lyell went to Scotland for another long vacation at Kinnordy. They found his father somewhat better, but his mother had failed considerably from the year before. In Caddam Wood the patch of *Linnaea borealis* was again in bloom. Lyell counted some thirty of the tiny twinflowers, "remarkably fragrant," he thought, "for so small a flower."[103] At Kinnordy he tried to catch up on his geological reading, especially because he would have to review the state of the science in his address as president of the Geological Society the following February. He was also reading in preparation for a new edition of the *Elements of Geology.* After suffering from an aching tooth, he finally had it pulled. In August he made several geological excursions to follow further westward along the border of the Highlands the dike of serpentine on the Carity Burn, on which he had published his first paper in 1825. Late in the month, Charles and Frances Bunbury came to Kinnordy for a visit. Lyell discussed fossil botany with Bunbury, and they went on plant-collecting expeditions together.

101. *Athenaeum,* 1849, 640–42.

102. Milman, review of *A Second Visit,* 184, 198, 204; quoted from Lyell, *Second Visit,* 1:101.

103. Lyell to Frances Bunbury, 9 July 1849, Kinnordy MSS.

At the end of August, Lyell again was invited to Balmoral. Queen Victoria and Prince Albert were enjoying a vacation after their tour of Ireland. This time Mary accompanied him. They went by a westerly route around the hills by Blairgowrie and up Glen Shee to join Sir James Clark and his family at Birkhall in Glen Muick.[104] On 31 August, Clark and Lyell left their wives at Birkhall and rode to Balmoral, where they were to remain for two nights. Balmoral Castle being then quite small, the Queen put them up in a small building apart from the main house. After their arrival, Clark and Lyell accompanied the Prince of Wales, a lively boy of seven, and his tutor on a long walk. Prince Edward described to them the feats of a magician who had been at Balmoral a day or two before. When he was alone with Lyell, the Prince asked him the names of plants and to see spiders through his magnifying glass. After two and a half hours of rambling, a messenger asked Lyell to join the Queen, Prince Albert, and Lord John Russell, who were fishing for trout and salmon by a deep pool on the river Dee. On their walk back to the castle, they stopped to see a quarry in which a trap dike had cut through limestone, producing garnets along its edges. After the Queen left them, Lyell and Prince Albert became so deep in conversation that they forgot about time, until the Queen sent a messenger to say that it was only a quarter of an hour until dinner. Prince Albert had just finished reading Macaulay's *History of England.* He was also widely read on serious subjects in German. From him Lyell learned much about German thought. Lyell was impressed by "the steady development of Prince Albert's mind in a great variety of directions."[105] The visit had an underlying seriousness. Prince Albert was developing a plan for bringing various colleges in Ireland together into one united Irish university. He was also planning a world industrial exhibition, to be held in London in 1851. The Prince had already held a meeting at Buckingham Palace to discuss with members of the Society of Arts his plan for an exhibition of the products of every country, whether of raw materials, machinery, mechanical inventions, or manufactured goods, to illustrate the application of art in industry.[106] Lyell quite enjoyed his second visit to Balmoral.

When he returned to Birkhall, Lyell went to the top of Quoil, one

104. Lyell, Notebook 159, 52–56, Kinnordy MSS; cf. Lyell to Susan Horner, 5 September 1849, Royal Society MSS.

105. Lyell to Leonard Horner, 3 September 1849, Kinnordy MSS; printed in K. M. Lyell, *Life, Letters, and Journals,* 2:156.

106. Martin, *Prince Consort,* 2:216–17.

of three large hills of serpentine rock in the neighborhood, to compare the plants growing on the serpentine hills of Aberdeenshire with those growing on the serpentine rock along the banks of the Carity near Kinnordy.[107] A few days after their return to Kinnordy, Sir Charles and Lady Lyell left to attend the British Association meeting at Birmingham, where Lyell presided over the Geological Section. On his return to London in late September, Lyell found when he called at Murray's that the seventh edition of the *Principles* was nearly sold out and that during the summer two-thirds of the copies of the *Second Visit* had also sold. He was particularly pleased by the latter news. Summer was "the dullest of all seasons" for the sale of books, with "a book being no more in season in the months which want an R in the name than an oyster."[108]

On 15 October, puzzled by some of Hugh Miller's assertions in his *Footprints of the Creator,* Lyell spent the morning with Gideon Mantell going over questions in vertebrate paleontology.[109] The next day, he and Mary went to Down. Lyell wanted to discuss South American geology with Darwin. He also took the opportunity to ask him about James Dwight Dana's recent work on coral reefs. Just returned from seven months at Malvern taking the water cure, Darwin looked better than he had for years. The weather was beautiful, and Lyell went for a long walk to the High Beeches with young William Darwin and others. Early on 17 October the Lyells returned to London for Lyell to attend a meeting of the council of the Royal Society.

At Kinnordy, Mr. Lyell had become seriously ill with influenza. At first his family was not alarmed. About the middle of October, although weak, he seemed to be recovering. Near the end of the month he suffered a relapse, and it became apparent that he was gravely ill. On 31 October, Lyell was summoned urgently to come. When he and Mary reached Kinnordy the next day, they found Lyell's father fully conscious. "He was much pleased to see us," Mary wrote, "& has expressed it frequently."[110] Growing weaker by the hour, he could not sleep and suffered constant pain. He knew he was dying and was content to do so; at the same time he realized, he told his eldest son, that it would be "a hard & somewhat long struggle that could kill him."[111] All of his eight children and his two

107. Lyell, Notebook 159, 96–97, Kinnordy MSS.
108. Lyell to Katherine Lyell, 21 September 1849, Kinnordy MSS.
109. Lyell, Notebook 160, 16, Kinnordy MSS.
110. Mary Lyell to Mantell, 7 November 1849, Mantell MSS.
111. Lyell to Leonard Horner, 10 November 1849, Kinnordy MSS.

daughters-in-law were now at Kinnordy. The Lyell daughters took turns caring for their father and, in watching him suffer, suffered themselves. "The poor girls look wretchedly," Mary wrote. "They will not spare themselves even as much as I think they ought to do for the sake of others."[112] Mrs. Lyell, senior, her mind impaired by earlier strokes, was unaware of her husband's illness.

During those dark November days at Kinnordy, the weather was extremely cold. On the sixth there was a driving snowstorm outside the windows as Mary wrote:

> It is a sort of waiting for death such as I cannot describe. Each time a door opens we start up, & this may go on several days. Poor Mrs. Lyell is happily unconscious & looks pleased if Katherine or I sit rubbing her feet & repeating little rhymes to her. Her maid has come out very much under this trial & grudges no trouble or fatigue in her service.[113]

Two days later, on 8 November, Mr. Lyell died.

Thus passed Charles Lyell, Esq. of Kinnordy—Dante scholar, bibliophile, amateur botanist, and Scottish laird. In the pursuit of his various interests he had tended to remain in each a dilettante. The persistent labor required for major achievement in any of them would have interfered too much with the leisure and comfort he maintained as a country gentleman. Good-natured, generous, easygoing, Mr. Lyell may have been puzzled and perhaps secretly jealous of his son's success and fame as a scientist. During the fifty-two years that he had been laird of Kinnordy, he had burdened the estate with an ever-increasing debt. In 1840 he had mortgaged almost the whole of his lands for 30,000 pounds. It is difficult now to judge in what measure Lyell's father had been extravagant and in what measure he had simply been unfortunate. The management of an estate like Kinnordy required continual capital investment in new farm buildings, draining and fencing of fields, planting of trees, and similar projects. Thus his expenditures may have been largely for the improvement of the estate. Yet such a large debt could hardly have been wise. In his death he not only left Kinnordy debt ridden, but he struck his eldest son a bitter blow, for which, from the grave, there could be neither explanation nor apology.

In the mode of inheritance customary among landed families in Scotland, Lyell, as the eldest son, would have been his father's prin-

112. Mary Lyell to Leonard Horner, 6 November 1849, Kinnordy MSS.

113. Ibid.

cipal heir and the new laird of Kinnordy. His father had always referred to Lyell as his "son and heir," and Lyell had fully expected to inherit Kinnordy, with due provision made for his brothers and sisters. But when his father's will was read, Lyell was shocked to learn that his father had not left the Kinnordy estate to him. Instead, the will established a trust from which the eight Lyell children were to receive equal shares of the income and were also to act equally as trustees of the estate. Lyell himself would receive the rent from the home farm of Kinnordy, which was not included in the trust, and his one-eighth share of the trust income. He would also continue to receive the annual 500 pounds that he had received since his marriage, a prior obligation of the estate under the marriage contract. His income was increased by his father's will, but less so than if he had received the estate intact, with annuities provided for each of his brothers and sisters. The loss of additional income meant less to him than the humiliation implied by disinheritance and the sense that whatever disagreements he may have had from time to time with his father, he had done nothing to deserve such treatment. The fact that he was disinherited could not be concealed. It would become known throughout Forfarshire and perhaps beyond. What hurt Lyell even more was that his father's will had been drawn in 1836, and his sisters had had some knowledge of its contents but had not told him. He felt that they had acquiesced in its injustice. His father's will required Lyell to take part in extended discussions with his brothers and sisters and with the family solicitor about the steps necessary to establish the trust. Such talks were sometimes painful and for Lyell extremely frustrating. They kept him and Mary at Kinnordy until late November.

When they returned to London, Lyell resumed preparation of his presidential address for the anniversary meeting of the Geological Society. On 19 December he read to the society a long paper on the formation of volcanoes, attacking the theory of craters of elevation. Leopold von Buch had put forward the theory in 1825 to explain the origin of large bowl-shaped depressions associated with volcanoes, such as the Caldera of Palma in the Canary Islands and the Gulf of Santorin in the Grecian archipelago.[114] Buch postulated that the Caldera of Palma and similar bowl-shaped valleys had been formed by a sudden, explosive upheaval within the volcano, elevating accumulated layers of lava and leaving a large circular crater at

114. Lyell, "Craters of Denudation."

the site of the explosion. The theory was intended to explain both the great circular valley at the center of a volcanic mountain and the fact that on all sides the sheets of lava sloped away from the central valley.

In 1830 in the first volume of the *Principles,* Lyell objected to the theory, noting that it was not founded on any comparable effect produced by a modern volcano or earthquake. All modern volcanic cones and craters were produced by volcanic eruptions. None resulted in a truncated cone with a great cavity in the center. Yet Buch's craters of elevation always occurred in the midst of extinct volcanoes. The theory required that volcanic rocks must first accumulate in horizontal beds to a depth of several thousand feet, an accumulation that could occur only in the vicinity of a volcanic vent. Then by a sudden explosion the beds were heaved up thousands of feet. Lyell objected that "instead of being shattered, contorted, and thrown into the utmost disorder, [the beds] have acquired that gentle inclination, and that regular and symmetrical arrangement, which characterize the flanks of a large cone of eruption, like Etna!" He argued that in every large hollow, such as the crater of a truncated volcanic cone, there would be a channel to drain off water and that, in the course of ages, such a channel would become a deep ravine like the gorge that opened on one side of the Caldera of Palma.[115] In 1832 in the second edition of the *Principles,* Lyell suggested that the gorge on Palma might have been formed by the action of the tide while Palma was emerging gradually from the sea.[116] Lyell thought that such great circular valleys as the Caldera of Palma and the Val del Bove on Mount Etna in Sicily had probably been formed by subsidence of the valley floor, caused perhaps by the removal of an underlying lake of lava.[117]

Buch's theory of craters of elevation was consistently supported by Léonce Élie de Beaumont, who related it to his own theory of the sudden elevation of parallel mountain chains. In 1834 Élie de Beaumont argued that the great extent of the thick, compact sheets of basalt occurring on the sides of volcanoes such as Etna and Vesuvius indicated that they must originally have been poured out as liquid lava on a surface inclined at a much lower angle than that of their present position.[118] During a visit to Mount Etna, Élie de Beaumont

115. Lyell, *Principles,* 1:389, 395.

116. Lyell, *Principles,* 2d ed., 1:452.

117. Lyell, *Principles,* 3:95–97.

118. Élie de Beaumont, "Cratères de soulèvement."

studied actual lava streams to learn the factors that determined their flow. He also tried to show that the various dikes in the precipices surrounding the Val del Bove constituted evidence that the mass of Etna had been elevated and distended by the later intrusion of lava from below.[119]

In 1847 in the seventh edition of the *Principles,* Lyell accepted Élie de Beaumont's argument that the beds of lava in Etna had originally flowed out and solidified in a less inclined position than they now held. Lyell thought that the beds had been upraised to become more steeply inclined as a result of successive small disturbances accompanying volcanic eruptions and not by a single sudden upthrow.[120] He continued to deny the reality of craters of elevation. Lyell held that volcanoes like Etna had grown by the accumulation of layers of lava, produced in a long succession of volcanic eruptions, such as were still going on. Thus in 1847, Lyell accepted part of what he took to be Élie de Beaumont's evidence for craters of elevation, while continuing to reject the theory itself. He reviewed the recent hydrographical survey of Santorin by Captain Graves for the British Admiralty. Graves's soundings showed that if Santorin were raised above the sea it would possess a great central crater containing three smaller volcanoes and a deep gorge passing through the rim of the crater. Lyell thought that the gorge had been formed by the action of the tides when Santorin stood at a higher level. To illustrate the possible role of the tide in maintaining an opening into the crater, he provided the example of the island of St. Paul in the Indian Ocean, surveyed by the British Navy in 1842. Similarly, he suggested that the truncated cone of Somma around the northern side of Vesuvius might have originated as a submarine volcano. His emphasis on the possible role of the tide in excavating the calderas associated with volcanoes appears to have been influenced by his observation of the action of the tides of the Bay of Fundy in eroding the coastline of the Minas Basin. He speculated that the great cavity of the Val del Bove in Etna might also have been produced by the action of the sea, when Etna was at a much lower level with respect to the Mediterranean than it now was. Although there were no traces of seashells within the Val del Bove, there were fossil seashells in beds on the outer slopes of Etna eight hundred feet above the sea.

After writing his paper on the structure of volcanoes in 1849,

119. Élie de Beaumont, "Mont Etna."

120. Lyell, *Principles,* 7th ed., 401.

Lyell received and read a copy of James Dwight Dana's *Geology of the United States Exploring Expedition.*[121] He then came to doubt even the evidence that Élie de Beaumont had adduced for the theory of craters of elevation. In the Hawaiian Islands, Dana described two huge volcanic cones, Mauna Loa and Mauna Kea, each fifteen thousand feet high and representing two and a half times the volume of volcanic rock found in Etna. From vents near the summit of Mount Loa streams of lava, sometimes more than two miles wide and twenty-six miles long, poured down in every direction. Dana's observations of the crater of Kilauea showed that currents of lava might cool so quickly that they could solidify even on slopes as steep as fifty or sixty degrees. Even layers of cinders might accumulate on slopes as steep as thirty-five or forty degrees. Lyell concluded that the laws governing the flow and solidification of sheets of basaltic lava had not yet been fully determined.[122] If Dana's observations were true, Élie de Beaumont's assumption that sheets of basaltic lava could solidify only on gentle slopes was unfounded. Dana himself rejected the theory of craters of elevation.[123]

The Anniversary Address, 1850

The paper on volcanoes that Lyell read to the Geological Society in December was only a prelude to the reconsideration and reaffirmation of the theoretical views that he would present in his anniversary address to the society on 15 February 1850. Nearly twenty years had passed since the first appearance of the *Principles of Geology,* in which he set forth the view "that the existing causes of change in the animate and inanimate world might be similar, not only in kind, but in degree, to those which have prevailed during many successive modifications of the earth's crust."[124] In 1850, after two decades of extraordinarily rapid progress in geology, the theory seemed far less visionary and extravagant than it had in 1830. Since many geologists still opposed Lyell's doctrine, he decided to examine in his address some of the more important recent discoveries bearing upon it.

The elevation of mountain ranges was usually attributed to paroxysmal forces. Geologists who thought that extraordinary forces must have been required to raise mountains disagreed whether

121. Dana, *Geology.*

122. Lyell, "Craters of Denudation," 233.

123. Dana, *Geology,* 369–70.

124. Lyell, "Anniversary Address" (1850), xxxii.

such forces had been most intensely violent when they had first occurred and had later declined, or whether they had occurred repeatedly in short bursts of violent convulsions, each followed by a long interval of quiescence. In 1848 Sir Roderick Murchison showed that the enormous movements by which the Alps had been elevated had occurred at least as recently as the Eocene. The Alps, therefore, were relatively modern.[125] In 1847 near the end of a prolonged controversy over whether the Nummulitic formation was Cretaceous or Eocene, the Austrian geologist Ami Boué identified it as Eocene.[126] The Nummulitic formation occurred in some of the highest peaks in the Alps, at altitudes of more than ten thousand feet; it showed, therefore, that the Alps had not been elevated until late in the Eocene, or later. The elevation of the Alps in a very modern period of geological history invalidated, said Lyell, "all theories which ascribe such magnificent displays of mechanical force to very remote epochs."[127] Furthermore, since the Nummulitic formation was also found high in the Pyrenees, Apennines, and Carpathians, and in the mountains of North Africa, Persia, and India, the elevation of mountain ranges in the Tertiary period had occurred extensively throughout the world.

Metamorphosis of Eocene formations accompanied the elevation of the Alps. Bands of granite and other igneous rocks were intruded among them. Murchison had described in the Alps the widespread, often dramatic, folding of great thicknesses of strata. In some instances, they were overturned completely, placing masses of Jurassic limestone unconformably over the edges of Eocene Nummulitic strata. Murchison regarded such inversions of stratified rocks as clear evidence of a sudden operation or catastrophe, indicating "that in those days the crust of the earth was affected by forces of infinitely greater intensity than those which now prevail."[128] In reply, Lyell pointed out that the Alps had originated entirely since the beginning of the Tertiary period. They showed that the causes producing disturbances in the earth's crust were just as active and just as powerful in the Tertiary as they had been in the Secondary or Primary periods. The discovery in the United States and Russia of Carboniferous, Devonian, and Silurian strata, horizontal and

125. Murchison, "Structure of the Alps."

126. Boué, "Nummuliten-Ablagerungen."

127. Lyell, "Anniversary Address" (1850), xxxviii.

128. Murchison, "Structure of the Alps," 258; quoted by Lyell, "Anniversary Address" (1850), xlii.

undisturbed over vast areas, demonstrated that even in the most ancient geological times, the disturbing forces in the earth's crust operated only in limited localities. The results of the New York and other American surveys and of Murchison's own work in Russia proved that there had been no directional decline in the extent or intensity of geological disturbing forces. The alternative for geologists who wished to retain paroxysmal hypotheses was to postulate that although geological forces had not declined through geological time, they acted in brief periods of extraordinarily violent disturbance in particular areas, such as the Alps. During such disturbances, geological changes occurred on a scale and with an intensity entirely different from anything seen in historical times, or from anything occurring in the long periods of rest between convulsions.

Many geologists believed in periodic episodes of convulsive activity. In Adolphe d'Archiac's history of the progress of geology from 1834 to 1845, Archiac asserted that the geological causes now active were insufficient to explain geological phenomena. Other more powerful and even paroxysmal forces were needed.[129] In contrast, Lyell argued that the inability to explain certain geological phenomena immediately was not a reason to invoke violent catastrophes to account for them. Geologists remained ignorant of many changes going on today in the depths of the earth or beneath the oceans, yet such changes must steadily be producing geological results. In view of our ignorance of geological processes in the interior of the earth, it would be remarkable if nothing were obscure or difficult about the geological monuments produced by them. Furthermore, the results of geological processes were cumulative. The overall effect of the geological changes that had occurred throughout the history of the earth, or even within one geological period, would be much greater than the effect of the geological changes that had occurred within historic times. The real question was the amount of geological change brought about in a given period of time. "It is not the magnitude of the effects, however gigantic their proportions," Lyell said, "which can inform us in the slightest degree whether the operation was sudden or gradual, insensible or paroxysmal. It must be shown that a slow process could never in any series of ages give rise to the same results."[130]

If geological changes recognized as active today—the recurrent eruptions of volcanoes, the accumulation of sediment in deltas, the

129. Archiac, *La géologie de 1834 à 1845.*

130. Lyell, "Anniversary Address" (1850), xlv.

wasting of sea cliffs, and the rise of the land in Scandinavia—were to continue at the same rates for thousands of years, they would produce large-scale geological changes. Slow and gradual as they were, geological changes tended to occur much more rapidly than changes in the animal and plant kingdoms. Yet, even in the living world, change was visible in modern times. Some species had become extinct within the past three centuries. In the Alps, movements of both elevation and subsidence had occurred, not at one time but through a succession of geological periods, extending from the Jurassic to the Pliocene. Similarly, in the Pyrenees, which Élie de Beaumont once supposed to have been elevated in a single upthrow (*un seul jet*), he now distinguished six, if not seven, distinct series of movements that had occurred at different times. Faults in which rocks had shifted several thousand feet upward or downward were often assumed to have occurred suddenly. In fact, the manner in which their walls were rubbed, polished, and striated suggested that the movements had occurred slowly and may have been repeated many times.

The erect fossil trees and tree roots in successive layers of Carboniferous strata, often thousands of feet thick, showed that as the Carboniferous beds accumulated, they must have subsided slowly, at about the same rate as they accumulated. The subsidence was slow enough for forests of large trees to grow, collect beds of peat around their roots, die, and fall down. The slow, long continued subsidence of the coal beds was similar to the subsidence of the ocean floor that Darwin and Dana had shown occurred in the formation of coral atolls. Such large-scale movements of subsidence might often result in great flexures of the strata. Although it might be difficult to imagine the length of time required to produce the great folds and overturns of the strata in the higher Alps, it would be equally difficult to imagine the time required to form the pebbles of a conglomerate formation eight thousand feet thick. "In this case, however," Lyell said, "there is no mode of evading the obvious conclusion, since every pebble tells its own tale."[131]

Henry Rogers of Philadelphia had noted a remarkable resemblance between the folds of the Alps and those of the Appalachians. The longer limb of each anticlinal fold dipped inward toward the central axis of the mountain range, while the steeper, shorter side of the fold faced outward. Arnold Guyot of Switzerland had sug-

131. Ibid., lv.

gested that the movements producing the folds might have been the result of a contraction of the earth's surface caused by gradual cooling. Lyell suggested that the movements might have occurred as insensibly as the "creeps" that occurred in coal mines as a result of the collapse of the overlying strata after the removal of the coal. The bends and curvatures of strata produced by creeps showed that strata could become bent and curved, slowly and imperceptibly.

Élie de Beaumont had suggested that the granites of the higher Alps were extruded at the surface and that in earlier geological periods, the granites contained more quartz than in later times.[132] Lyell, in contrast, explained the mineral composition of granite and other igneous rocks by supposing them to have originated deep within the earth.[133] He reaffirmed his belief that the granites produced within the earth were essentially similar at every geological period. The many active volcanoes throughout the world showed that within the earth there must be large masses of molten rock, cooling slowly to form solid crystalline rocks. The chemical composition of modern lavas extruded at the surface was similar to that of granite. Masses of crystalline rocks forming within the earth would be essentially identical with the masses of granite revealed in mountains. Granite was formed only deep within the earth. Therefore, masses and veins of granite would be revealed only after large amounts of overlying rock had been removed by denudation. The older stratified rocks, subjected more frequently to extensive denudation over long periods, thus appeared to contain granite more frequently than younger rocks.

In concluding his anniversary address, Lyell spoke of the glacial hypothesis, accepted ever more widely to account for the distribution of erratic boulders over northern Europe and North America. Nineteenth-century geologists had repeatedly cited erratic boulders as evidence of "violent earthquakes, waves, sudden deluges, rapid and overwhelming currents of mud, and other extraordinary agencies." Lyell said that the glacial theory showed that one natural agent—frost—had been much more intense at one period than at others. Future changes in the altitude and area of polar and equatorial lands might make the world climate colder or warmer than it was at present or than it had ever been. "But speculations of this kind," said Lyell, "belong equally to the future, the past and the pres-

132. Élie de Beaumont, "Les émanations volcaniques," 1299.
133. Lyell, *Elements,* 487–510.

ent, and imply no inconstancy in the general condition of our planet, such as is assumed in the hypothesis of its passage from a chaotic to a fixed, stable and perfect state."[134]

While preparing the anniversary address early in January 1850, Lyell was also assembling corrections for a second edition of the *Second Visit to the United States,* the first edition having sold out. He wrote to George Ticknor and others in the United States to send him any errors they might have detected, but he had little to change. In February he received a royalty cheque for eighty-eight pounds, ten shillings, from Harpers in New York, publisher of the American edition. They had already sold thirty-five hundred copies.[135]

On 4 March, Lyell's mother died at Kinnordy. Her death relieved her daughters of the need for her constant care, but the final loss of both parents was to Lyell and his brothers and sisters a somber event. Lyell received word on 6 March and left for Scotland the next day. While at Kinnordy he attempted to deal further with some of the problems created by the trust established under his father's will. It was an increasingly sad task for him. Much of the furniture and paintings had to be removed from Kinnordy, some to be sold, the rest to be stored. The house was to be let, and Lyell's sisters rented a smaller house, Drumkilbo, at Meigle some miles away. His brother Tom, who remained a bachelor, farmed at Shielhill, a farm on the estate. When Lyell left Kinnordy after his mother's funeral, he was leaving for the last time. He never stayed at Kinnordy again. His travels would take him to France, Germany, Italy, Switzerland, the Canary Islands, and America, but it was almost ten years before he returned to Scotland, even for a brief visit. His letters to his sisters became fewer, although they remained affectionate. His father's will left wounds that healed slowly.

When Lyell returned to London in mid-March, he was immediately busy with seeing the eighth edition of the *Principles* into publication, a task interrupted by his mother's death. In June, Murray offered him four hundred pounds to publish the eighth edition in two thousand copies, with an additional five hundred copies for the American market.[136]

On 28 April, the Lyells went to Down to spend a few days at Charles Darwin's house. Lyell wished to discuss with him the changes required in a new edition of the *Elements of Geology.* He

134. Lyell, "Anniversary Address" (1850), lxii, lxv.

135. Lyell to Murray, 6 February 1850, Murray MSS.

136. Lyell to Murray, 8 June 1850, Murray MSS.

planned to include as a frontispiece an illustration of Siccar Point near St. Abb's Head on the east coast of Scotland. At Siccar Point in 1788, James Hutton pointed out to his friends John Playfair and Sir James Hall of Dunglass a striking unconformity of the strata and explained to them what the unconformity meant in terms of the history of the earth. Lyell asked the artist James Hall, the youngest son of Sir James Hall of Dunglass, to allow his painting of Siccar Point to be used for the woodcut for the frontispiece.[137]

Early in 1850 Lyell was appointed to the Royal Commission for the Exhibition of 1851. Through the winter and spring of 1850, in addition to scientific work and family concerns, he attended meetings of that body. Corrections for the eighth edition of the *Principles,* plans for a new edition of the *Elements,* his duties as president of the Geological Society, the planning of the exhibition, and his family duties gave him more than enough to do.[138] In June, William Prescott and his son arrived in London from America. Prescott stayed at Mivart's Hotel but dined with the Lyells. They took him to Rivermede, Leonard Horner's house in the country, where Prescott delighted in the lively conversation.[139] On 26 July, Lyell broke off work on the new edition of the *Elements* at the beginning of the nineteenth chapter. The next day, he and Mary took the train to Dover for a tour on the Continent.

137. Lyell to James Hall, 28 April 1850 and 3 June 1850, University of Pennsylvania MSS. James Hall (1800–54) was an advocate and amateur painter, who painted a number of Scottish landscapes.

138. Lyell to Fleming, 2 May 1850 [copy], Kinnordy MSS.

139. Lyell, Notebook 161, 94, Kinnordy MSS.

CHAPTER 10

The *Manual* and Belgian Tertiary Geology, 1850–1851

THE LYELLS' TOUR of Germany in the summer of 1850 was in part a vacation and in part a geological tour. Lyell wanted to see collections of Tertiary fossils in Belgium, to study the granite of the Harz Mountains, and to examine the fossils of the Triassic formations, of which Germany possessed the type localities. In contrast to their last tour in Germany in 1835, this time Lyell and Mary traveled almost entirely in large, comfortable carriages on the new network of railways that had spread across Europe during the 1840s.

After landing at Calais the evening of 27 July, they rolled across the plains of northern France and Belgium, a country as flat as the Mississippi Delta. After a brief visit to Maestricht to examine the fossils of St. Peter's Mount, they crossed the Rhine to board their first German train. They had a whole carriage to themselves. They found the cushioned seats, table, and mirrors of the first-class compartment luxurious, and the Prussian railway conductors attentive and polite.

At Minden, Lyell studied the loess deposits above the Weser River. At Brunswick, a local geologist, August von Strombeck, took him to see an anticlinal fold in which the three formations of the Trias were exposed on either side of the axis. Strata of Lias and Chalk lay above them.[1] Lyell wished to study the granite of the Harz Mountains from which Leopold von Buch had derived his theory of the rise of granite en masse from below. From Clausthal, at the foot of the Harz Mountains, they made a carriage tour to the top of the Brocken. Lyell found no sign of glaciation, just as he had found none on the top of Mount Washington in New Hampshire. In beautiful weather they spent several days skirting the base of the Harz Mountains, flanked with Devonian slates. Lyell studied the slates and decided that the Devonian strata had formerly extended over the top of the Harz and that the granite was not laid bare until after

1. Lyell to Leonard Horner, 8 August 1850, Kinnordy MSS; printed in K. M. Lyell, *Life, Letters, and Journals,* 2:160–63. August von Strombeck (1809–1900), geologist and paleontologist, was a native of Brunswick.

the Triassic formations had been laid down. Consequently, the granite had formed beneath the Devonian strata, deep within the earth.

At Halberstadt on 6 August, they boarded a train to go to Berlin, obtaining refreshments when they changed trains at Magdeburg. German women, they noted, now dressed the same as in England, while more of the men wore beards than in 1835. The democratic tone of the newspapers reminded Lyell of the United States. They enjoyed the journey. To her sister Mary wrote:

> I was thinking yesterday what a refined picture I must have made to any fashionable English friend, with a tall glass tankard of beer in one hand and a bitter brod with sausage in the other standing on the platform [at Magdeburg] while the carriages were changing. The beer is so good here. We never drink anything but that or Seltzer water and are both feeling uncommonly well.[2]

At Berlin Lyell called on scientists and visited museums. Eilhardt Mitscherlich, the professor of chemistry, invited them to tea and to meet the Berlin geologists who were not away during August. On Sunday they went to Potsdam to see the elderly but still vigorous Alexander von Humboldt, who arranged for them to tour the royal palaces. During the following week they visited Dresden, Leipzig, Halle, and Weimar, Lyell finding at each place collections of fossils from the German formations and often local geologists with whom to talk. At Weimar he made a collection of loess fossil shells, and at a quarry workmen sold him some fossil bones from the loess. From Weimar they traveled via Erfurt to Gotha, finding that as they moved from one duchy to another, they did not need to show passports nor have their baggage examined.

At Eisenach on 23 August, they visited the castle on the Wartburg, a wooded hill, where in 1521 the elector of Saxony concealed Martin Luther for nearly a year. That afternoon they took the train to Kassel and on 25 August traveled on the first passenger train over the new railway line to Giessen. A large, cheerful crowd greeted the train at the station. After calling on August von Klipstein, the professor of geology, and Justus von Liebig, the famous chemist, the Lyells resorted to horse-drawn travel as far as Friedberg, where they took a train into Frankfurt. From Frankfurt they traveled by steamboat down the Rhine to Cologne, spending the whole day on deck,

2. Mary Lyell to Katherine Lyell, 8 August 1850, Kinnordy MSS.

watching the lights and shadows on the passing hills. They returned across Belgium by railway, stopping at Liège and Louvain to talk to Belgian paleontologists and examine collections of fossils from Belgian Tertiary formations. On 31 August, they arrived back at 11 Harley Street.

Lyell immediately took up work on the new third edition of the *Elements of Geology*, sending part of the revised text to the printer on 9 September. Besides revising the old text, he was rewriting considerable portions and making additions. On 25 September, he began work on the chapters dealing with the Eocene.[3] Through the autumn he continued to work steadily at the book, during the quiet time before the Geological Society and other societies began to meet in November and while many of his fellow scientists and friends were still in the country.

Henry and Katherine Lyell had taken a house on Duke Street, and Katherine was expecting a baby. On 21 October she gave birth to a son, whom they named Leonard after his grandfather Leonard Horner. The birth of their first nephew was a great event for Lyell and Mary, who had realized long before that they would have no children. In November, Mary wrote to William Prescott in America, "As I think you are not a child-fancier, it is well you are not here at present, for I can think of little besides my little nephew."[4]

On 14 November, the entire Horner family came to breakfast for Lyell's fifty-third birthday, "a merry party." Through November and December Lyell and Mary went out little, Lyell remaining intensely absorbed in the third edition of the *Elements*, which he strove to finish.[5] At Christmastime they went to Suffolk to spend the holiday with the Bunburys at Mildenhall amid a large family party. After their return to London on the first day of the New Year, Lyell sent to the printer the preface to *A Manual of Elementary Geology*, the new title he chose for the third edition of the *Elements*. The book was finished.

A Manual of Elementary Geology

From one small duodecimo volume in 543 pages in the first edition in 1838, the *Elements of Geology* had grown in the second edition in 1841 to two duodecimo volumes containing a total of 897 pages, illustrated by 439 woodcuts, and from twenty-five to thirty-six chap-

3. Lyell, Notebook 164, 125, 133, Kinnordy MSS.
4. Mary Lyell to Prescott, 13 November 1850, Prescott MSS.
5. Mary Lyell to Prescott, 13 December 1850, Prescott MSS.

ters. The third edition, *A Manual of Elementary Geology*, was a single octavo volume of 512 pages, set in a slightly smaller type. Despite its compact form, the book had grown still larger. Three chapters were new, and some chapters were transposed. Chapter 12 of the second edition, dealing with the classification of Tertiary formations, became Chapter 10 in the third edition. The former Chapter 13, "Newer and Older Pliocene Formations," became "Newer Pliocene Strata and Cavern Deposits," and Chapter 14, which had dealt solely with the Miocene, was now entitled "Older Pliocene and Miocene Formations." In the second edition Chapter 22 was entitled "New Red Sandstone Group"; in the third edition its title became "Trias or New Red Sandstone Group," reflecting work published on the Triassic formations of Germany during the intervening decade. Chapter 13 in the third edition, "Permian or Magnesian Limestone Group," was a new chapter based on Sir Roderick Murchison's discovery of the distinct nature of the formations in the Russian province of Perm during his travels there in 1840 and 1841. The final chapter, "Mineral Veins," was also entirely new.

The tendency for his readers to confuse the *Elements* with the *Principles* was a perennial problem for Lyell. In the preface to the *Manual,* he took pains to point out that they were distinct works. The *Principles* dealt with the modern changes now proceeding on the earth's surface; the *Elements* dealt with the geological record preserved in successive stratified formations and in the igneous rocks frequently associated with them.

One of the important advances in the decade since 1841 was in the understanding of the conditions under which coal deposits had accumulated. In 1841 Lyell had written: "Some of the coal-measures are of freshwater origin, and may have been formed in lakes; others seem to have been deposited in estuaries, or at the mouths of rivers, in spaces alternately occupied by fresh and salt water."[6] Together with other geologists at that time, he had thought of coal as having accumulated as a sediment underwater. From the work of William Logan, Lyell, and others in England and America, geologists now knew that each coal bed represented an ancient swamp. The coal had accumulated originally as a bed of peat that was converted later to coal by the pressure of superimposed strata. Sometimes it was modified further by heat and the pressure of earth movements. The *Stigmariae* in the underclay were simply the roots of trees that had

6. Lyell, *Elements,* 2d ed., 2:106–7.

grown in the ancient swamps. In the Coal of Cape Breton Island in Nova Scotia, Richard Brown had discovered *Stigmariae* attached to the trunks of fossil *Sigillaria* trees.[7]

As geologists became more familiar with phenomena that had long been part of the experience of coal miners, they learned of the "coal pipes." Each coal pipe was the cylindrical stone cast of an upright fossil tree trunk extending upward from a coal bed, often through many layers of the overlying strata. As a bed of coal was removed in mining, such coal pipes, deprived of their foundation and weighing several tons, frequently slipped downward into the mine. Miners dreaded their sudden descents, which often resulted in deaths or severe injuries. In July 1842 Lyell saw upright fossil trees in the cliffs at Joggins on the shore of the Chignecto Basin in Nova Scotia. Though a few of the fossil trees were rooted in beds of clay and shale, most were rooted in seams of coal. The fact that the fossil trees were based in beds of coal demonstrated that the coal had accumulated in ancient forests of such trees.[8] Thus between 1841 and 1851, Lyell's and other geologists' view of the origin of the Coal had undergone a complete change. It was further enriched by a greatly increased knowledge of the many species of coal plants and of the conditions under which they grew.

During the 1840s, a number of fossil reptiles were discovered in Carboniferous strata. Although many large fossil reptiles were known from the Jurassic, reptiles had not been thought to exist as early as the Coal period. The discovery of air-breathing land animals brought about a major change in the interpretation of the Coal period. In 1844 Hermann von Meyer described a fossil reptile skeleton from the Coal of Münster-Appel in Rhenish Bavaria,[9] and two years later Lyell examined reptile footprints at Greensburg, Pennsylvania. The following year Ernst Heinrich Carl von Dechen discovered in the Coal of Saarbruck three new species of fossil air-breathing reptiles, named *Archegosaurus* by Georg August Goldfuss.[10] Von Meyer considered *Archegosaurus* the most important fossil reptile discovery since the discovery of *Pterodactyl* in the

7. R. Brown, "Upright Lepidodendron with Stigmaria Roots"; R. Brown, "Sigillariae with Conical Tap Roots."

8. Lyell, *Manual*, 321–22.

9. Meyer, "Reptiles of the Coal," 52. The article is a review by Meyer of a book by Goldfuss, *Beitrage zur vorweltlichen Fauna des Steinkohlengebirges;* cf. Goldfuss, "Apateon pedestris, aus der Steinkohlen formation."

10. Goldfuss, "Einen Krokodilier und einige neue fossile Fische."

1820s.[11] He thought that it was more closely related to labyrinthodonts, of which several fossil species were known from the Triassic, than to crocodiles. The discovery of fossil reptiles in the Carboniferous—a period much earlier than any in which reptiles previously were thought to exist—strengthened Lyell's conviction that knowledge of the fossil record was still very incomplete. It was especially fragmentary for land animals, since most sedimentary strata had been laid down beneath the sea.

In his discussion of volcanic rocks in the *Manual,* Lyell showed that each geological period from the Cambrian to the Newer Pliocene had been marked by volcanic activity. Adam Sedgwick had found volcanic traprocks among the Cambrian slates of Cumberland, and Sir Roderick Murchison had discovered evidence of frequent volcanic eruptions beneath the sea during the Silurian period. As a young man, Lyell had observed an abundance of traprock among the Old Red Sandstone (Devonian) strata of the Sidlaw Hills in Forfarshire. Later he had noted ancient traprocks among the coal-bearing strata of Fifeshire. They were identical in mineral composition to the modern lavas of Mount Etna in Sicily. Among the New Red Sandstone, or Triassic, strata of Devonshire, Sir Henry De La Beche had described volcanic rocks formed at the same time as the stratified rocks were laid down.[12] In 1850 Edward Forbes showed that some of the volcanic rocks of the Hebrides were of Oolitic (i.e., Jurassic) age.[13] In Greece, Theodore Virlet identified certain traprocks formed during the Cretaceous period.[14] As early as 1833, Lyell had described in southern France, Italy, and Spain volcanic rocks and extinct volcanoes belonging to each of the Tertiary periods. In the *Manual* he summarized his earlier discussion, now supplemented by later observations. The total effect was to demonstrate that every geological period had been marked by volcanic activity and each had produced volcanic rocks of similar character.

Lyell likewise showed that plutonic rocks such as granite, formed deep within the earth, had been produced at various geological periods. Granite was much more difficult to date than volcanic traprocks. When it intruded sedimentary strata, it usually altered them so much as to destroy their fossils, rendering them impossible to date paleontologically. Clearly granites had been formed

11. Meyer, "Reptiles of the Coal."

12. De La Beche, "Geology of Tor and Babbacombe Bays, Devon."

13. Forbes, "Distribution of Organic Remains in the Dorsetshire Purbecks."

14. Expédition scientifique de Morée, 1831–38, 2: pt. 2, *Géologie et minéralogie.*

throughout the Tertiary period, when such mountain ranges as the Alps and the Andes had gradually risen. Earlier geologists of the Wernerian school had thought of granite as one of the oldest rocks, formed before the deposition of the earliest sedimentary strata and before the beginning of organic life. In the 1790s James Hutton demonstrated that granite veins had intruded into sedimentary strata after the latter had been deposited. Hutton had connected the granite veins penetrating stratified rocks in Glen Tilt to a mass of granite in the hills to the north of the glen. He concluded that the masses of granite in the cores of mountains were, like the granite veins, younger than the stratified rocks with which they were associated.

When it was shown that the Alps and Pyrenees had been elevated since the Eocene, granite intrusions among the Tertiary strata of the Alps necessarily became of Tertiary age. By 1850 Swiss geologists began to think that the upper portion of the Nummulitic formation, called the *flysch,* might be of Miocene age. The talcose granite of Mont Blanc invaded the flysch, and "the question as to its age," Lyell wrote, "is not so much whether it be a secondary or tertiary granite, as whether it should be assigned to the Eocene or Miocene epoch."[15] Similarly, in South America, Charles Darwin had shown that the eastern range of the Andes in Chile consisted of Tertiary strata invaded and uplifted by great masses of granite. Thus the granite of the eastern Andes was also more recent than certain Tertiary formations.

The immense quantity of plutonic rocks formed during the Tertiary period, even late in the period, indicated that the scale and intensity of igneous activity within the earth's crust had been as great throughout the Tertiary as during earlier geological epochs. Noting that, since the beginning of the Eocene, almost the whole of Europe had been raised from beneath the sea, Lyell asked, "What amount of change of equivalent importance can be proved to have occurred in the earth's crust within an equal quantity of time anterior to the Eocene epoch?"[16] The sheer quantity of igneous activity and elevation of mountain ranges during the Tertiary showed that there had been no decline in the level of igneous activity within the earth as compared with earlier geological periods. On the contrary, the production of volcanic and plutonic rocks at every geological period revealed a steady and uniform level of igneous activity in the

15. Lyell, *Manual,* 453.
16. Ibid., 454–55.

interior of the earth, as far back in time as the geological record went. The new geological data accumulated in the twenty years since 1830 thus tended strongly to support the geological viewpoint that Lyell had presented in 1830 in the *Principles.*

On 12 January 1851, after he had seen the *Manual* through the publication process, Lyell began writing his second anniversary address to the Geological Society.[17] In his address he sought to counter the belief that "a gradual development in the scale of being, both animal and vegetable, from the earliest periods to our own time, can be deduced from palaeontological evidence."[18] He wrote the paper in less than six weeks, although he had been reading and thinking about the subject much longer.

The Fossil Record and the Species Question, 1851

At London in October 1844, the publisher John Churchill had issued an anonymous book entitled *Vestiges of the Natural History of Creation.* It set forth the theory that all living species had originated by descent from a common ancestor. The various great groups of animals and plants had developed successively one from another through the long course of geological time. The evidence for the progressive development of life on earth came from geology, especially from the geological discoveries of the preceding quarter century. The author assumed that the earth had condensed from its original nebulous or vaporous state into a crystalline rock, such as granite. The original surface of the earth, he assumed, consisted of enormous granite mountains interspersed with seas far deeper than the modern oceans.

As the granite mountains weathered and began to wear down, the author of the *Vestiges* argued, the earliest stratified rocks were deposited in the seas. They formed the "Gneiss and Mica Slate System" of the Scottish Highlands and the West of England. The great thickness of the beds demonstrated the depth of the seas in which they were formed. Furthermore, the contortions of their strata showed, the author thought, that they must have been deposited at temperatures far higher than those of modern seas, perhaps at boiling heat. The earliest appearances of life were in the limestone and slates of northern Wales (Cambrian), which contained zoophytes, sponges, crinoids, shells, and crustacea. In the overlying rocks of

17. Lyell to Mantell, 12 January 1851, Mantell MSS.
18. Lyell, *Address . . . Anniversary Meeting* (1851), 17.

the Silurian system, described so fully by Roderick Murchison in 1839, such fossils became more abundant, but the species and even the genera had changed. In the upper Ludlow rocks of the Silurian system, cartilaginous fishes began to appear. Throughout the Silurian rocks were traces of fuci, the brown alga abundant along rocky seashores. Furthermore, the fossils described by Murchison in England had also been found in North America and in various countries on the continent of Europe, showing that during Silurian times similar forms of life occurred uniformly over the surface of the earth. In the Old Red Sandstone system (Devonian), overlying the Silurian, the zoophytes, sponges, shells, and crustacea continued, but now were accompanied by a much richer variety of fishes. By 1842 Louis Agassiz had distinguished twenty genera and sixty species of fossil fish from the Old Red Sandstone. All the species differed from those of the Silurian period. As in the Silurian rocks, the Old Red Sandstone fossils were of the same forms in North America and Russia as in England.

According to the author of the *Vestiges,* the Secondary period was distinguished from the Primary by the appearance of dry land. The oldest and lowest of the Secondary rocks was the Carboniferous formation, which contained the coal beds formed from accumulations of plant material. The coal plants appeared to be predominantly ferns, including tree ferns, gigantic horsetails (*Calamites*), and club mosses (*Lepidodendron*). Fossil wood of a cone-bearing tree (*Araucaria*), with modern representatives only in New Zealand, occurred also in the Coal at Edinburgh and Newcastle. The rich vegetation of the Coal period presupposed, thought the author of the *Vestiges,* an atmosphere so rich in carbon dioxide as to prevent the existence of air-breathing land animals.

On the basis of his survey of the geological record, the author of the *Vestiges* concluded that volcanic forces appeared to have declined and that the earth's surface had gradually become more suitable for the life of higher animals. Together with such changes had occurred a parallel development of plant and animal life. Among animals, the various classes of invertebrates appeared first, followed in succession by fishes, reptiles, birds, and mammals. "That there is thus a progress of some kind, the most superficial glance at the geological history is sufficient to convince us." In addition to progress, constant extinction produced a constant change of species. New species of animals and plants must come into existence through the action of natural law, rather than as a result of individual acts of the

Creator. From work in comparative anatomy, the author showed that the various groups of vertebrate animals were formed on a common plan. In embryological development the higher animals passed through stages in which they resembled successive groups of lower animals. "Nor is man himself exempt from this law," he wrote. "His first form is that which is prominent in the animalcule. His organization gradually passes through conditions generally resembling a fish, a reptile, a bird, and the lower mammalia, before it attains its specific maturity." Furthermore, the various organ systems, such as the central nervous system and the heart, passed through stages in which they resembled those of fishes and reptiles before reaching the mammalian form. Through the whole long history of geological time, development had occurred *"from the simplest forms of being, to the next more complicated, and this through the medium of the ordinary process of generation."*[19]

During the 1830s, with the aid of the newly introduced achromatic microscopes, biologists had found that the basic unit of all plant and animal bodies was the nucleated cell and that the embryological development of all animals began from a single cell. Many of the animalcules visible in infusions were single, nucleated cells living independently. To the author of the *Vestiges,* such facts showed that life had originated on earth as single cells, or, as he called them, *germinal vesicles.* From "the whole history of the progress of animal creation as displayed by geology," the common basis of life in the cell, and the common patterns of embryological development, he deduced that *"the simplest and most primitive type . . . gave birth to the type next above it, that this again produced the next higher, and so on to the very highest."*[20] In each case the transitions were small, being only from one species to another.[21] The *Vestiges* concluded with a discussion of the evidence that the various races of mankind shared a common origin.

The publication of the *Vestiges* in the autumn of 1844 created a sensation. Despite scientific errors in the book, it was essentially accurate in its account of the fossil record. From that record the author deduced a progressive development through geological time, with which many geologists concurred, including those who vehemently attacked the book. Furthermore, the *Vestiges* pointed out ex-

19. [Chambers], *Vestiges,* 148–49, 152, 199, 205 (italics in the original).

20. Ibid., 222 (italics in the original).

21. M.J.S. Hodge has discussed the primary role of embryology in the origin of the developmental theory presented in *Vestiges.* See M.J.S. Hodge, "Chambers' *Vestiges* and *Explanations.*"

plicitly what the public had suspected for some time, namely, that the fossil record could not be reconciled with the biblical account of Creation and required a new orientation of thought. In less than a year, and even before the book was denounced by reviewers, four British and three American editions of the *Vestiges* were sold out. Excoriating though they were, the reviews did not diminish the sale.[22] Between 1844 and 1860 the *Vestiges* went through twenty editions in Britain and America and was translated twice into German. For months after its publication, at every party and scientific meeting in London, people speculated about the identity of the author. Among novelists, phrenologists, and scientists, the names of Lyell and even Prince Albert were suggested as possible authors. The true author was the Edinburgh writer and publisher Robert Chambers. Although Chambers's authorship was soon suspected, he preserved his anonymity throughout his lifetime.[23]

At the time the *Vestiges* was published, Lyell was writing his *Travels in America,* which he finished only shortly before departing for the United States in September 1845. On his return to London in June 1846, he was preoccupied with moving to a new house, arranging his collections, publishing papers on American geology, the reform of the Royal Society and of university education, and, finally, the writing of his *Second Visit to the United States* and the revision of the *Elements.* It was not until Lyell came to prepare his second anniversary address as president of the Geological Society that he confronted the issues raised by the *Vestiges.* Lyell could view the book with greater detachment than such critics as Adam Sedgwick, William Whewell, or Hugh Miller, because they shared with the author a belief in a progressive and catastrophist geology. Lyell did not accept progression in geology. He thought it was based on a misreading of geological evidence.

The theory that the fossil record showed a progressive development in both animal and plant life had been stated by Richard Owen, Adam Sedgwick, and Hugh Miller, curiously enough, in writings intended to refute the doctrine of a gradual development of living forms as set forth in the *Vestiges.* The author argued that the progressive development of animal and plant forms had occurred by the repeated transformation of one species into another by natural forces. Owen, Sedgwick, and Miller believed that the progressive de-

22. One of the most violent reviews was by Adam Sedgwick; see Sedgwick, "Natural History of Creation."

23. Millhauser, *Just before Darwin,* 116–40.

velopment of life had occurred entirely by the separate creation of new species. It represented a gradual unfolding of the Divine plan of Creation.[24]

Lyell also believed in the successive creation of new species. The difference was that he thought that new species were created individually, adapted to the requirements of a particular set of environmental conditions. As the slow but relentless forces of geological change destroyed or altered old environments, some species became extinct while others took their place. In stable environments such as the oceans, where change was slight and extremely slow, the change in species would likewise be limited and slow. Lyell saw the assemblage of species living during any one geological period as reflecting the distribution of land and sea throughout the world, the depths of the seas, the patterns of winds and currents, temperature, and rainfall. For Lyell it was environmental conditions, rather than a particular stage in the development of a Divine plan, that determined the character of the species existing at a particular period in the history of the earth. The advocates of progressive development thought that the scale and intensity of geological causes had declined steadily from the early history of the earth.[25] Miller wrote that "the signs of convulsion and catastrophe gradually lessen as we descend to the times of the Tertiary," and asked his readers to "consider how exceedingly partial and infrequent these earth-tempests have become in the recent periods."[26] He thought that the earth had gradually improved as a dwelling place for living things until it finally became a habitation fit for man.

In 1849 Adolphe Brongniart postulated an apparent development among fossil plants. During the Paleozoic, cryptogams predominated, especially in the Carboniferous. From the Triassic through the Wealden, conifers and cycads were the prevalent fossil plants. Flowering plants did not become abundant until the Tertiary.[27] Similarly at Heidelberg, Heinrich George Bronn argued from his catalogue of twenty-four thousand species of fossil animals and plants that the number and variety of species were greater in

24. Peter Bowler has shown that whereas during the 1830s the concept of progression was linked to the theory of a cooling earth, Louis Agassiz in 1842 altered the concept of biological progression to that of the unfolding of a Divine plan of Creation. See Bowler, *Fossils and Progress,* 45–65.

25. Michael Bartholomew has discussed Lyell's theory of nonprogression without seeming to appreciate fully the degree to which the concept of progressive development in the fossil record was linked to geological catastrophism. See Bartholomew, "Lyell and Evolution."

26. Miller, *Footprints,* 3d ed., 311.

27. Adolphe Brongniart, *Tableau des genres de végétaux fossiles.*

more modern formations. Higher groups of organisms had appeared successively through geological time.[28]

In the face of so many opposing authorities and a formidable array of facts, Lyell sought in his anniversary address to show that the fossil record was extremely incomplete. Moreover, the knowledge of that record so far acquired was fragmentary and haphazard. He suggested that the fossil record must be considered in relation to the nature of the rock strata in which the fossils were found. The oldest sedimentary strata then known, the Silurian, had been laid down in a deep sea. Silurian fossil plants and animals, therefore, were marine forms. They constituted a rich invertebrate fauna, with representatives of many of the same classes of invertebrates still living in modern seas. Marine fossils told nothing of the plants or animals that might have lived on land during Silurian times.

Before 1844 no fossil reptiles had been discovered among land animals earlier than the Permian. Since 1844 skeletons and footprints of fossil reptiles had been discovered in Carboniferous strata, disproving the assumption that no reptiles had lived during the Coal period. Again and again, said Lyell, the period of the first appearance of various classes of higher animals had to be pushed backward in time by the discovery of their fossil forms in earlier rock strata. Since the fossil reptiles of the Permian were as highly developed as any modern species of reptile, it was difficult to define what sort of progressive development the reptiles had undergone in the course of geological time. Only two genera of fossil mammals had been discovered among Secondary formations. Geologists could not yet be certain whether the seeming scarcity of mammals during the Secondary period was a consequence of the relative abundance of reptiles or reflected the slight knowledge of fossil land and freshwater animals. So far, geologists had discovered no land shells and only a few freshwater shells in Secondary rocks.[29]

Some five hundred species of plants were known from the Coal, but the coal plants represented only the flora of low-lying swamps. Although more than eleven thousand species of living plants were known from modern Europe, probably fewer than five hundred species could be found in the deltas of the Po and Rhone Rivers, environments similar to those in which the ancient coal plants had grown. The fossil plants found in the Coal probably represented only a portion of the plants that lived during Carboniferous times.

28. Bronn, *Index palaeontologicus.*

29. Lyell, *Address . . . Anniversary Meeting* (1851), 22.

They told nothing of plants that might have lived on Carboniferous uplands.

In the Permian, the Carboniferous plants began to disappear. They were replaced in the Triassic by an essentially new flora, followed again by another significantly new assemblage of plant species in the Liassic. In the Cretaceous, Lyell said, Adolphe Brongniart had described the beginnings of the age of flowering plants, or angiosperms. During the Cretaceous many species of flowering plants lived together with cycads and with such large reptiles as *Iguanodon, Ichthyosaurus, Plesiosaurus,* and *Pterodactyl.* Thus the atmosphere of the Cretaceous period was as capable of supporting flowering plants as it was of supporting cycads and large reptiles. As to the flora of the Tertiary period, Lyell said that the number of fossil plants obtained from Tertiary strata, though still relatively small, was rapidly increasing. The plants already known from Tertiary strata belonged to a broader range of families and classes than did the fossil plants of older formations. Lyell thought that the greater variety of plants might reflect the variety of Tertiary freshwater formations. Knowledge of these fossil plants was sufficient, Lyell thought, to show that four (possibly five) major changes in plant species had occurred since the Eocene. Nevertheless, the changes of species had not been accompanied by any "manifest elevation in the grade of organization, implying a progressive improvement in the floras which succeeded each other."[30]

Repeatedly, Lyell emphasized the incomplete and frequently misleading nature of the fossil record. When Edward Forbes and Robert MacAndrew reported in 1850 on the results of dredging the sea floor at various places around the British Isles, they found many marine invertebrates of the same classes that occurred among Silurian fossils. They found few remains of bony fishes and none of whales, porpoises, other sea mammals, or land animals.[31] The remains even of the most primitive groups of land animals were seldom found fossil. No fossil land shells were found in the Coal, although Lyell felt certain that land shells must have lived during the Carboniferous period. Lyell had hunted for land shells in the alluvial beds of the Mississippi Delta without finding any, although many species of land shells lived in the Mississippi Valley. Similarly, he had never found moths or butterflies fossil in recent deposits,

30. Ibid., 35.

31. Forbes, "British Marine Zoology by Means of the Dredge"; MacAndrew, "Distribution and Range in Depth of Mollusca."

despite the thousands of species of moths and butterflies living in Great Britain alone. Before 1846 such freshwater shells as *Planorbis* and *Lymneus* were not known from strata older than the Eocene. In that year Wilhelm Dunker discovered *Planorbis* and *Lymneus* fossils among the Wealden beds of Germany. Subsequently, Edward Forbes had found representatives of these two genera occurring with other genera of freshwater shells and the green alga *Chara* throughout the Purbeck formations (of Jurassic age).

The fossil footprints of the Connecticut Sandstone implied the presence of birds as early as the Triassic. The Stonesfield fossil mammals showed the existence of at least two genera of herbivorous marsupials in the Oolitic (Jurassic) period. Richard Owen suggested that the presence of several species of rapidly breeding herbivores would require the presence of other carnivorous species to keep down their numbers. For Lyell, the lower jawbones of the Stonesfield fossil mammals implied the existence during Oolitic times of an extensive mammalian fauna, just as the jawbones themselves implied the former existence of the skeleton of a whole animal.

Finally, as to what was and was not in the fossil record, Lyell said that the preservation of animal or plant remains as fossils was exceptional. As a rule they were destroyed, but, in the immensity of geological time, exceptional cases had accumulated until they seemed to constitute the rule. "Our knowledge, therefore, of the living creation of any given period of the past," wrote Lyell, "may be said to depend in a great degree on what we commonly call chance, and the casual discovery of some new localities rich in peculiar fossils may modify or entirely overthrow all our previous generalizations."[32]

In contradictory fashion, Lyell continued to insist on the recent origin of man. Among the many mammalian fossil bones taken from cave deposits of newer Pliocene age, he thought there were no well-authenticated instances of human bones. He was thus prepared to accept negative evidence to support the view that man had not lived at the same time as the extinct animals whose bones were found in caves. He dismissed the few reported instances of human bones found with those of cave animals as not well authenticated. Lyell wished to claim both that the fossil record did not provide evidence for progressive development and that it did indicate the recent appearance of man. Even if the development doctrine were

32. Lyell, *Address . . . Anniversary Meeting* (1851), 51–52.

true, Lyell thought that the appearance of man was an event different in kind from the appearance of a new plant or animal species. "The intellect of man and his spiritual and moral nature," Lyell said, "are the highest works of creative power known to us in the universe."[33]

On 22 February 1851, the day after the anniversary meeting of the Geological Society, the Lyells went to visit Charles Darwin at Down. With Darwin, Lyell discussed his presidential address. Darwin agreed that the fossil record was extremely incomplete. He disagreed with Lyell about the fixity of species. After discussing the occurrence of glacial boulders on the Jura Mountains in France, the parallel roads of Glen Roy, and the atheistic views of Harriet Martineau ("There is no god & Harriet is his prophet," said Darwin), they spoke of the widening of valleys and finally of species. Darwin told Lyell of his conviction that species originated by the slow transformation of preexisting species. Lyell recorded under the heading "Species": "All have come by transformation. Tho' one species of a genus dies out, the genus continues & varies indefinitely—C.D. The Fuegian savage very low compared to a civilized man."[34] Although Darwin told Lyell of his belief that species changed, he evidently did not go on to describe the mechanism by which he thought such changes were brought about.

The following week, for a few days' rest and change, the Lyells went to visit Lyell's aunt, Mrs. Heathcote, at Winchester. Henry and Katherine Lyell accompanied them, taking along their infant son, Leonard. At Winchester, Lyell made excursions to the chalk hills nearby to learn how the surface of the chalk had been shaped by denudation. On his return to London, he found that the *Manual,* out only a few weeks, was selling rapidly. Two months after publication, only six hundred of the two thousand copies were left, and they were selling at the rate of fifty copies per week. Lyell regretted that Murray had not printed a larger edition; it was too soon to prepare a new edition, yet he did not like simply to reprint the book without correction or revision.[35]

In March 1851, Lyell spent some time preparing written answers to questions sent to him by the newly appointed Royal Commission on Oxford University about the reforms needed in university education. Since his answers would be published in the commission's

33. Ibid., 59.

34. Lyell, Notebook 165, 115, Kinnordy MSS.

35. Lyell to Leonard Horner, 22 March 1851, Kinnordy MSS.

report, he prepared them carefully.[36] Late in March he received printed copies of his anniversary address, which he sent to friends and scientific colleagues.

On 4 April, Lyell delivered the Friday evening lecture at the Royal Institution, Prince Albert presiding for the occasion. The title of his talk was "On Recent and Fossil Rainprints." Drawing on his observations in America, Lyell described the first fossil rain prints he had seen in New Jersey. In 1841 William Redfield took him to a quarry in New Red Sandstone (Triassic) strata where they saw the impressions. The following summer on the mud flats of the Minas Basin in Nova Scotia, Lyell observed how raindrops made impressions on the smooth surface of the mud and how such impressions were later filled with mud brought by the incoming tide and thereby preserved. Redfield found additional fossil impressions of raindrops and hailstones at Pompton, New Jersey, and sent Lyell an account of them.[37] Similarly, Richard Brown had found impressions of raindrops in the Carboniferous strata of Cape Breton Island in Nova Scotia.[38] Impressions of raindrops showed that passing showers of rain had fallen on ancient Carboniferous and Triassic beaches, just as they fell on modern beaches. Wind and clouds, mud flats and tides were the same then as today. A small but significant detail, raindrop impressions showed that conditions on the earth's surface during the geological past were essentially like those still existing.

At Easter time, Lyell went on a six-day excursion to the Isle of Wight and the Dorsetshire coast with Edward Forbes and two other members of the Geological Survey, Henry Bristow[39] and Andrew Ramsay. They were going to examine the Purbeck formation. In 1849 Forbes began a detailed examination of the fossils of the Purbeck beds. He arrived at striking results of particular interest to Lyell.[40] At the time Bristow was collaborating with Forbes on a geological map of that part of Dorsetshire. On Monday, 21 April, Lyell traveled with Bristow and Ramsay by train to Southampton, where Forbes joined them. The four geologists took a steamboat to Yarmouth on the Isle of Wight, arriving that evening at Freshwater Gate. Somewhat in awe of Lyell, the younger geologists were pre-

36. Lyell to Frances Lyell, 29 March 1851, Kinnordy MSS.

37. Redfield, "Newly Discovered Ichthyolites"; cf. Redfield, "Fossil Rain-Marks Found in the Red Sandstone Rocks of New Jersey."

38. Lyell, "Fossil Rain-Marks."

39. Henry William Bristow (1817–89).

40. Forbes, "Distribution of Organic Remains in the Dorsetshire Purbecks."

pared to dislike him. They were pleasantly surprised. Ramsay recorded of their first day together: "I liked Lyell better; he was often anecdotal, but principally geological all day. He laughed tremendously when Bristow said his portmanteau was so heavy because it contained De La Beche's new Geological Observer."[41] Lyell had always felt slightly irritated by De La Beche.

For two days the group examined the strata beneath the chalk along the southwest shore of the Isle of Wight and the Tertiary strata on the north side of the island, particularly the series of nearly vertical beds of sand in Alum Bay. On Thursday, they returned on the steamboat to the mainland where they boarded a train that wended its way westward, through a countryside familiar to Lyell in his youth. At Swanage, the four men spent Friday examining, under Forbes's guidance, the series of Purbeck strata exposed in the cliffs on the east side of Warbarrow Bay. The beds were about 155 feet thick. Forbes had distinguished them into the Lower, Middle, and Upper Purbeck formations, each consisting of a series of freshwater strata, with brackish and marine strata interposed in places among the freshwater formations. In the Lower Purbeck there was also a well-developed "dirt-bed" containing many stumps of cycads. The dirt-bed was clearly a fossilized swamp representing an ancient land surface.

Two features of the series of Purbeck beds were remarkable. First, the change from freshwater to brackish or marine beds, or from marine to freshwater, occurred without any signs of geological disturbance. With the exception of one disturbed bed near the bottom, the whole series of Purbeck strata lay in perfect conformity to the strata beneath them. Thus the change from freshwater to marine conditions must have occurred in a very gentle manner. Second, the three successive freshwater formations of the Lower, Middle, and Upper Purbeck all contained many representatives of the same genera of freshwater shells—*Paludina, Physa, Lymneus, Planorbis, Valvata, Cyclas,* and *Unio*—but in each of the three formations, the species were different. A complete change of species had occurred among the freshwater shells in the transition from the Lower to the Middle Purbeck and again from the Middle to the Upper Purbeck. Forbes attributed the complete changes in the living forms to the long periods that must have intervened between the times when the successive freshwater formations were deposited. Despite the change

41. Geikie, *Ramsay*, 180.

in species, the Purbeck freshwater shells were similar to both Tertiary and living forms. Forbes wrote: "A most striking feature of the molluscan fauna of the Purbecks is this—so similar are those now existing, that, had we only such fossils before us, and no evidence of the infra position of the rocks in which they were found, we should be wholly unable to assign them a definite geological epoch."[42] Forbes also found between the Upper Purbeck and the overlying freshwater formation of the Wealden a similar change of species among fossil freshwater shells.

The succession of freshwater formations alternating with brackish and marine formations, all showing a multitude of analogies both with Tertiary formations and with modern lakes, estuaries, and seas, was of intense interest to Lyell. The successive changes of species, each implying the lapse of a long period of time, gave a new significance not only to the Purbeck series but also to the Wealden. Lyell thought that the Purbeck and Wealden might each have been as long as either the Oolitic or Cretaceous periods.

In the course of that day, the four geologists traced the succession of strata exposed in the cliffs from Meup's Bay on the east side of Warbarrow Bay around the coast to Kimmeridge Bay, which they reached at about five o'clock in the afternoon. They returned to Swanage in a carriage they had arranged to meet them. Of their day, trudging along the cliffs by the sea, Ramsay recorded:

> We all like Lyell much. He is anxious for instruction, and so far from affecting the big-wig, is not afraid to learn anything from any one. The notes he takes are amazing; many a one he had from me today. He is very helpless in the field without people to point things out to him; quite inexperienced and unable to see his way either physically or geologically. He could not map a mile, but understands all when explained, and speculates thereon well.[43]

Lyell had not done geological mapping since 1824, when he prepared a geological map and section of Forfarshire. Whether a man who had studied geology in the field for thirty years, in Europe from Sicily to Sweden and in North America from Nova Scotia to the mouth of the Mississippi, was really inexperienced in the field is doubtful. Lyell evidently did not parade his knowledge, and Ramsay, who was relatively young and proud of his own accomplishments

42. Forbes, "Distribution of Organic Remains in the Dorsetshire Purbecks," 81.

43. Geikie, *Ramsay,* 180–81. Lyell filled twenty-four pages with notes on Friday, 25 April. Lyell, Notebook 166, 94–118, Kinnordy MSS.

in geological mapping, did not recognize that knowledge. Lyell also seems to have been having increasing trouble with his vision. He wore glasses part of the day and needed them to see things at a distance. On 26 April, he and Forbes returned together by train to London.

On 1 May, a sunny day with occasional passing clouds, Lyell attended the opening of the Great Exhibition at the Crystal Palace in Hyde Park as one of the twenty-seven royal commissioners responsible for it. Just before noon, Queen Victoria and Prince Albert, accompanied by an escort of the Household Cavalry, arrived by carriage. The Crystal Palace gleamed in the sunshine, with the flags of all nations flying over it. After the Queen arrived, the formal ceremony began with the National Anthem, followed by prayers, speeches, and a declaration that the exhibition was open. The royal party then toured the multitude of displays of the raw materials, manufactures, arts, and crafts of many countries. Lyell joined in the general exuberant optimism pervading the crowds at the exhibition. To George Ticknor he wrote:

> How wonderfully the "Crystal Palace" has taken, and how the prejudices of those who were incapable of taking in such a "new idea" have given way before the tide of popularity and fashion. . . . We, the commissioners, are now doing our best to give it an educational turn, but my hope is that the chief good will be the admission of the million to see so much of the result of the highest civilization, such as even the aristocracy who have travelled to see palaces and museums cannot help admiring.[44]

In May and June, Lyell made several brief excursions from London with the Reverend H. M. De La Condamine to study the Blackheath pebble bed. In the Blackheath gravel, De La Condamine had found fossil freshwater shells belonging to the same extinct species as in lower Eocene beds at Woolwich. They included shells of *Cyrena tellinella* standing on end with both valves together. On the beach at Mobile, Alabama, Lyell had dug out of sandy mud living shells of *Cyrena* and *Gnathodon* in the same upright position. The advantage of the upright position was that the mollusc could protrude its siphon upward to draw in water to bathe its gills.[45] The *Cyrena* fossils of Blackheath provided yet another analogy between Eocene life and that of modern waters.

44. Lyell to Ticknor, 20 May 1851, printed in K. M. Lyell, *Life, Letters, and Journals,* 2:173–74.
45. Lyell, "Blackheath Pebble-Bed."

At the annual meeting of the British Association in July, Lyell delivered a short paper on a bed of gravel in the Red Crag near Ipswich. The upper surfaces of the pebbles in the gravel were covered with barnacles, showing that the stones still lay as they had lain when the barnacles grew on them. The bed of pebbles thus represented a sea bottom, where the barnacles grew in remarkably still water, not even the smallest and lightest pebbles having been moved or overturned.[46]

A few days after their return to London, Henry and Katherine Lyell sailed for India, taking with them their infant son. It was a painful parting. Mary was particularly close to Katherine, and Lyell to Henry. India was far away, its diseases frequently deadly, and Indian Army service dangerous. Charles and Mary accompanied Henry and Katherine to their ship, lying in the Thames. After seeing the cabins in which they would spend the long weeks of the voyage, they said farewell.

On 10 July, the Lyells, accompanied by Joanna Horner, left for France and Belgium. Lyell wanted to study the Tertiary formations of the Continent in order to compare them with the English Tertiaries. The trip also served to take their minds off Henry, Katherine, and the baby at sea. At Cassel in Flanders, a quiet and lovely place on a hill overlooking a plain with rich fields and woods extending into the distance, they spent a week enjoying its peace after the turmoil of London.[47] They rose at seven each morning to spend the whole day in the open air hunting for fossils among the local quarries. From Cassel they proceeded to Lille, where they were pleasantly rewarded with two finds—strata containing an abundance of nummulites and a museum containing excellent paintings and drawings.[48]

For the next six weeks, the Lyells and Joanna Horner traveled in Belgium, crisscrossing the country by train or, more often, in a hired carriage. For the most part, they stopped in small towns and villages, only occasionally going into cities. Lyell was intent on examining as completely as possible the Belgian Tertiary formations and collecting fossils from them. The southern half of Belgium was covered by a deposit of claylike loam that resembled the loess of the Rhine. In his geological map of Belgium, André Dumont labeled this loam "Hesbayan mud," because it covered the old province of Hesbaye.[49]

46. Lyell, "Stones Covered with Barnacles in the Red Crag."

47. Mary Lyell to Katherine Lyell, 12 July 1851, Kinnordy MSS.

48. Mary Lyell to Mrs. Leonard Horner, 19 July 1851, Kinnordy MSS.

49. Dumont, *Carte géologique de la Belgique.*

North of a line extending from Cassel across Belgium through Courtrai, Oudenaarde, and Louvain to Cologne on the Rhine, the Tertiary formations came to the surface. Despite the presence of fossil collectors in Belgium, the Belgian Tertiaries were little known. There had been no geological survey of Belgium to map their extent and relationships. The older Belgian Tertiaries lay below the London Clay; that is, they were older than the oldest Eocene formations of either the London or Paris Basins. In addition, the Chalk formation of St. Peter's Mount at Maestricht was younger than the White Chalk with flints, the uppermost Cretaceous formation of England and France. The Maestricht formation and the Tertiary formations of Belgium constituted an extensive series of strata intermediate between the better known Cretaceous and Eocene formations of England and France. They demonstrated the long interval that had elapsed between the deposition of the Chalk and the London Clay. Their fossils showed that the transition in living forms from the Cretaceous to the Eocene had been more gradual and had extended over a much longer time than many geologists had previously thought.[50]

Much of the time Mary and Joanna Horner helped Lyell search for fossils in the quarries, sandpits, brick pits, or any other opening in the earth where fossils might be found. All three seemed to thrive on the outdoor life. As their excursion extended into August, the weather turned cold and wet, the rain turning the clay soil to mud, but they continued to collect in the quarries. Mary wrote, "It was fighting against weather, however, the whole time, & we were all covered with mud & Charles soaked."[51]

Occasionally the party did go into cities. At Brussels, Lyell had a letter of introduction to Captain Henri Le Hon who owned a large collection of Tertiary fossils from the local formations.[52] Lyell spent several days examining Le Hon's fossils. They also made several expeditions to quarries in the surrounding country, but their fieldwork continued to be hampered by the weather. Brussels was pretty and gay, like a smaller Paris. The trees in the park had grown since the Lyells had last been there in 1838, and the shops were attractive. One evening they went to the theater. Mary enjoyed the magician who took bouquets of flowers, sugar plums, and toys out of a hat in

50. Lyell, "Tertiary Strata of Belgium," 366–68.

51. Mary Lyell to Katherine Lyell, 28 August–1 September 1851, Kinnordy MSS.

52. Henri Le Hon (1809–72).

great numbers and threw them into the audience. "We got a number of sugar plums, a bouquet & a penny trumpet!"[53]

At Antwerp, Norbert de Wael entertained them at his country house. From the examination of Wael's collection of fossils, Henri Nyst had discovered that many of the Antwerp species occurred also in the Suffolk Crag.[54] Wael told Lyell about excavations in progress for ponds at the zoological gardens. In piles of sand from the twenty-foot-deep excavations, Lyell found more than a half-dozen species of the most common fossil shells of the Suffolk Crag, along with many fossil vertebrae of whales, also common in the Suffolk Crag. At Maestricht, Joseph Bousquet helped Lyell identify local fossils. At Liège the geologists André Dumont and Laurent de Koninck accompanied Lyell to study the local formations. They generously gave Lyell fossils from their collections so that he might have them compared at London with English Tertiary fossils.

On the last day of August, Lyell, Mary, and Joanna arrived at Ghent, where they gave up their hired carriage. The last four days of driving through the Belgian countryside had been particularly cold and rainy. Their clothes showed the effects of so much clambering through quarries in all kinds of weather. Mary thought they would be home none too soon. Their clothes were almost completely worn out. On 6 September they arrived back in London.

Life at 11 Harley Street was busy and varied. The first Sunday afternoon after their return, the Lyells visited the Zoological Gardens in Regents Park, always a favorite amusement with them. The new orangutan was thriving, and the zoo now possessed, Mary wrote, "a beautiful jaguar & two young ones who play about like kittens."[55] Several times they went to see the Great Exhibition at the Crystal Palace, sometimes with relatives or friends who had come to London especially to see it. Dr. John Jackson of Boston, one of many Americans drawn to England by the exhibition, dined with them.[56] Mary's parents stayed with them a couple of nights on their return from a tour in Switzerland. While the Horners were with them, they went to look at a house in Highgate, which, a few weeks later, Leonard Horner decided to take. Mary was delighted to have her parents move closer to her than they had been at Hampton Wick. The new house was pleasant, with a wide view over London.

53. Mary Lyell to Katherine Lyell, 23 July–7 August 1851, Kinnordy MSS.

54. Nyst, *Fossiles des terrains tertiaires de la Belgique.*

55. Mary Lyell to Katherine Lyell, 11–18 September 1851, Kinnordy MSS.

56. Barnard Swett Jackson (1806–78); see Holmes, "Jackson."

In late September, Captain and Madame Le Hon arrived in London after a visit to the Isle of Wight. Le Hon had been studying its geology, aided by Gideon Mantell's book about the island.[57] The Lyells entertained the Le Hons and Lyell took Le Hon to meet Mantell and see his large fossil collection. On several occasions while Le Hon was in London, he and Lyell spent some time going over Lyell's collections of Belgian Tertiary fossils and discussing various points.

The first letter from Henry and Katherine Lyell, written from Capetown, arrived on 21 October, ten days earlier than expected. Mary described its arrival to Katherine: "When last night we saw that the Birkenhead had come in, Charles said 'I wonder if we shall have letters,' but I said, 'Oh, impossible.' Then this morning it appears Sophy [the cook] had been dreaming about you & was describing her dream at breakfast, when the postman arrived & John [the footman] instantly saw the letter was from you."[58] When Mary had several visitors that day, Sophy thought they came because they saw the *Birkenhead* was in. It was also little Leonard Lyell's first birthday. That evening the servants dined on roast beef and plum pudding, with wine to drink to the health of the baby.

Early in October, Lyell received the September issue of the *Quarterly Review,* which contained a long article on three of his works: the eighth edition of the *Principles,* the *Manual,* and the 1851 *Anniversary Address.* Lyell knew that the article was by Richard Owen, although, as was customary, it was unsigned.[59] In April, Lyell had suggested to John Murray that Owen be asked to write a review of the *Anniversary Address* and the *Manual.* None of his works had been noticed in the *Quarterly Review* since 1835, and the *Elements* had never been reviewed there. The Manual was enjoying a large sale; geology clearly attracted wide public interest. The public was even more intensely interested in "the question whether at successive periods in Geology new species of animals and plants have been introduced of higher & higher organization; ending with man the last in the order of creation." The latter question was the subject of the *Anniversary Address.* Lyell thought that Owen might be persuaded to review it, especially because Lyell controverted the theory of progressive development that Owen advocated "so that he would be put upon his mettle."[60] Owen's article in the *Quarterly Review* was thus no sur-

57. Mantell, *Isle of Wight.*

58. Mary Lyell to Katherine Lyell, 21 October–7 November 1851, Kinnordy MSS.

59. [R. Owen], "Art. 8."

60. Lyell to Murray, 19 April 1851, Murray MSS.

prise to Lyell. He was prepared to find that Owen had subjected the *Anniversary Address* to thoroughgoing criticism. When he read the article, however, he was taken aback by what he considered the unfairness of Owen's account of Lyell's arguments and his criticisms of them. Gideon Mantell, not an unbiased observer but possibly a perceptive one, considered Owen's article "a violent attack."[61]

Owen began with compliments for Lyell on the *Principles* and the *Manual.* He said that Lyell's leading principle, that the causes of change in the past had been similar in both kind and degree to the causes of change at work today, "seems to have gained a deeper and wider basis as the facts of the science have gone on accumulating."[62] He then sketched the contents of the *Principles,* noting some of the changes in the eighth edition, and, after a brief mention of the *Manual,* proceeded to discuss the *Anniversary Address.* Owen argued capably that the fossil record did show a successive development toward higher classes of animals. He was less certain of the successive development of plants. Too little was known of the comparative anatomy and means of reproduction of the various plant groups to say with confidence which groups were higher and which lower.

Three days after he read the article, Lyell wrote Owen a long letter, saying that he thought it "the best & most friendly course to speak out my mind openly to you & explain where I think you have not been fair."[63] Lyell explained that he had used the term "successive development" to describe the interpretation of the fossil record advocated by Sedgwick, Miller, and Owen. An effect of the *Vestiges* had been to associate the term "progressive development" with the theory of the transmutation of species, which Sedgwick, Miller, and Owen did not accept. Owen misstated Lyell's position when he said that Lyell assumed that the vertebrates were represented "by the same classes, and in the same proportion at the secondary period as at the present day."[64] Lyell said that he had always maintained that the more recent the geological period the more closely the flora and fauna approached those of the present. In the

61. Mantell, *Journal,* 275. Mantell thought that Owen's article was inspired by spite. Immediately after his return to London in September, Lyell had written to Sir Philip Egerton to recommend George Waterhouse to succeed the late Charles Konig as curator of the natural history collections at the British Museum. Lyell to Egerton, 8 September 1851 [copy], Kinnordy MSS. When Lyell wrote to Egerton he did so at Waterhouse's request, unaware that Owen was also seeking the position. Lyell to Bunbury, 28 September 1851, Kinnordy MSS.

62. [R. Owen], "Art. 8," 413.

63. Lyell to Owen, 9 October 1851, Owen MSS.

64. [R. Owen], "Art. 8," 424–25, 436.

Address he had suggested that when a group such as the reptiles was prevalent, its members had tended to fill roles in nature that in later geological periods were filled by members of other classes, particularly the mammals. Lyell wrote:

> I referred the successive changes of plants & animals to geographical & other physical conditions known & unknown, to which new sets of species were from time to time adapted & not to a continual improvement in the habitable state of the planet, or its passage from an embryo to a more mature, or from a chaotic to a more settled state. My opponents connected the supposed gradual elevation in the scale of organic life with those causes which I may call chronological or cosmological.

The predominance of reptiles during the Oolitic period, when some mammals were also present, had "no more to do with the age of the planet, than the inferior grade of the Australian [mammals] as compared to the contemporaneous mammalia of India or Africa has to do with time, or with the respective dates of the introduction of mammals into Australia, India and Africa."[65]

Lyell was annoyed most by Owen's treatment of the reptile footprints discovered in Carboniferous strata at Greensburg, Pennsylvania. Owen quoted from the *Manual* (p. 337) Lyell's statement concerning the footprints: "I was at once convinced of their genuineness." He then added, "We confess that we should have valued the conclusion more highly if it had been more deliberately arrived at."[66] Since Owen had read Lyell's *Travels in America,* had attended Lyell's lecture on the Pennsylvania fossil footprints at the Royal Institution, and had assisted Lyell in the interpretation of the slab and casts that Lyell had brought back to London from Pennsylvania, Owen knew that Lyell had spent considerable time both in Pennsylvania and later in London checking and confirming his first impression of the nature of the fossil footprints. After setting forth such facts in his letter, Lyell added, "I might rather have expected a compliment for my patience as an investigator than regrets at my undue haste." Owen had attacked with especial vigor Lyell's arguments for the presence of birds during the Triassic period. Lyell thought that Owen exaggerated greatly the number of fossil birds found in Tertiary strata and the chances that living birds might become fossilized in the modern period. He told Owen he had left no

65. Lyell to Owen, 9 October 1851, Owen MSS.
66. [R. Owen], "Art. 8," 423.

way to account for the absence of fossil bird remains "in the shell-marl of Scotch lakes which swarmed with land & waterfowl up to the moment of the drainage of the lakes."[67]

In the *Anniversary Address,* Lyell argued that much of the appearance of successive development in the fossil record was simply a reflection of the order in which the various kinds of fossils tended to be discovered. In such Tertiary strata as those of Hordwell Cliff in Hampshire, shells, corals, and fossil fish were discovered first, fossil mammals and birds only long afterward. The discovery of fossils of the various animal classes followed a similar order in the Tertiary formations of the United States and Belgium. One fossil tooth of a Miocene mammal had been found in America, but no fossil birds or mammals were yet recorded from the Belgian Tertiaries. In the review, Owen emphasized that generalizations about the fossil record should be based only on facts already acquired. On that basis, Lyell said, a Belgian or American geologist "must of course conclude that in the older tertiary era warm-blooded quadrupeds & birds, tho' they flourished elsewhere, had not yet extended their range into their countries."[68]

Owen also wrote that the recent fauna was distinguished from the Tertiary fauna by the introduction of two new genera, *Pithecus* and *Troglodytes* (apes), of which no fossil representatives had been found.[69] Lyell replied that although much more was known of living quadrumana than of fossil forms, in the tropics the Danish explorer Peter Lund had found many fossil skeletons of apes in the caves of Brazil, including one genus that he named *Proto-pithecus.* Yet Lund's research had by no means discovered all the fossil apes and monkeys buried in Brazilian caves. Other countries, such as Borneo, possessing populations of living apes and monkeys, might also have fossil forms preserved in their caves. Lyell commented that "by placing an unexplored Tertiary fauna on an equal footing with a known recent one you take unfair advantage of your opponents or in the words of my Address (p. 53) 'you throw with loaded dice.'"[70]

Lyell concluded by saying that he rejoiced to see the subject discussed and "40 pages of the Quarterly filled with original & most valuable lessons in palaeontology."[71] The whole letter is moderate

67. Lyell to Owen, 9 October 1851, Owen MSS.
68. Ibid.
69. [R. Owen] "Art. 8," 447.
70. Lyell to Owen, 9 October 1851, Owen MSS.
71. Ibid.

in tone and formed a thorough protest, not so much against Owen's argument as against certain devices he used to sustain it. Yet Lyell was truly angry when he wrote, for he considered that Owen's review seriously misrepresented his address. Lyell considered the letter sufficiently important that he had it copied before sending it. The letter, together with Owen's article and Lyell's *Anniversary Address,* shows that by 1851 the rapidly accumulating knowledge of the fossil record, extremely incomplete though it might be, was making the changes of species during the geological past the central question in geology.

On 22 October, the Lyells went to spend three days at Down with the Darwins. Charles Darwin had recently returned from taking the water cure for his chronic illness and seemed much better, but the Darwins felt keenly the recent death of their eldest daughter, Annie. Lyell discussed with Darwin how the changes of species of freshwater shells, which Forbes had discovered in the Purbeck and Wealden formations, might have occurred. The difficulty for Lyell was to imagine how a body of fresh water might remain in the same place for a period of geological time sufficiently long to permit such changes of species. He and Darwin decided that many geological changes might have occurred elsewhere, while the area of the Purbeck beds continued to be the delta of a large river. "The Wealden river may have spread over the Purbeck as the Mississippi over deltas of Red & other rivers which had previously encroached on [the] Gulf of Mexico."[72] They also discussed the structure of barnacles, on which Darwin had then been working for several years, and Owen's article in the *Quarterly Review.*

The third edition of the *Manual* was sold out. The two thousand copies had gone so rapidly in nine months that Lyell decided he must issue what would be essentially a corrected reprint but which he and Murray decided to call the fourth edition. When Lyell and Darwin talked over the new printing of the *Manual,* they decided that it had to contain some rebuttal of Owen's article. Several new fossil discoveries would counter some of Owen's arguments. Lyell decided to insert a postscript to the preface of the new edition to include an account of these discoveries. From Down he wrote to Murray, "My theory has been most grossly misrepresented in the *Q. R.* & C. Darwin with whom I am staying is most anxious that the new facts which knock Owen's line of reasoning on the head should

72. Lyell, Notebook 173, 105, Kinnordy MSS.

come out at once."[73] The new discoveries included fossils and fossil footprints from both Europe and America. Footprints of what was taken to be a fossil tortoise, found in the lowest Silurian strata of Canada, were described to the Geological Society in 1851, oddly enough by Richard Owen.[74] Two fossil mammalian teeth thought to be those of a small insectivorous mammal had been discovered in 1847 in the Triassic of Germany as had been fossil footprints in a Triassic formation in Saxony. When Owen studied the bones of fossil reptiles from Triassic strata in both Germany and England, he identified them as the remains of a large fossil frog, or batrachian. He named them *Labyrinthodon,* from the labyrinthine structure of the teeth.[75] *Labyrinthodon* appeared to be the animal that had made the Triassic footprints. Similar footprints in Carboniferous strata in Pennsylvania demonstrated that a large fossil frog had lived during the Coal period. If one air-breathing land animal had lived during the Carboniferous, so might others. It was unlikely that the first such animal encountered would be the highest. After his return from Down, Lyell began to write the postscript to the preface of the new edition of the *Manual.* He intended to go to press with it immediately.[76] In fact, he did not. A new and exciting scientific discovery intervened.

The Elgin Fossil

On 4 July 1850, Captain Lambart Brickenden, who had previously collected dinosaur bones for Gideon Mantell in England, wrote to Mantell from Elgin in Morayshire, Scotland. He had obtained a slab of local sandstone that bore footprints of a land animal, apparently a reptile.[77] He thought that the rock belonged to the Old Red Sandstone formation. The footprints were evidence either for the existence of a land reptile during the Devonian period, or that the rocks did not belong to the Old Red Sandstone.[78]

A year later, near the end of October 1851, Brickenden sent Mantell a paper on the fossil reptile's footprints. Just a week or two before, Brickenden's brother-in-law, Patrick Duff, had obtained from

73. Lyell to Murray, 24 October 1851, Murray MSS.

74. R. Owen, "Impressions on the Potsdam Sandstone"; cf. Logan, "Foot-Prints of an Animal in the Potsdam Sandstone." The tracks are now thought to be those of a large arthropod.

75. R. Owen, *Odontography,* 1:195–217.

76. Lyell to Mantell, 4 November 1851, Mantell MSS.

77. Brickenden to Mantell, 4 July 1850, Mantell MSS; cited in Benton, "Elgin Fossil Reptile."

78. Brickenden to Mantell, 16 September 1850, Mantell MSS; cited in Benton, "Elgin Fossil Reptile."

a sandstone quarry near Elgin the well preserved skeleton of a small fossil reptile, about four and a half inches long. In light of Duff's discovery of a fossil reptile in the same rock as the fossil footprints, Brickenden wished Mantell to communicate his paper on reptile footprints to the Geological Society of London.

Mantell replied immediately to ask whether Brickenden might borrow the fossil reptile skeleton and send it to him for exhibit at the Geological Society when the paper was read. Mantell also offered to make a thorough scientific investigation of the fossil. Brickenden thereupon sought to obtain it for Mantell to describe.[79] Lyell may also have written to Brickenden to urge that the new fossil be sent to Mantell for scientific study. Lyell wished to cite Mantell in the *Manual* for the fossil, which he assumed Mantell would describe.[80] In late November, Patrick Duff sent the block of sandstone containing the fossil to Dr. George Duff, a young physician living in London.[81] On 2 December, Dr. Duff allowed Mantell to examine the fossil and then took it to the Royal College of Surgeons to show it briefly to Richard Owen before bringing it back to his house.[82] The next day, Dr. Duff lent the fossil to Lyell to have it drawn for a woodcut to be included in the *Manual.* Mantell and Owen both wished to describe the fossil. On Patrick Duff's instructions, Dr. Duff gave it to Mantell, who presented a description of it to the Geological Society along with Brickenden's paper on the reptile footprints.[83]

On 3 December, Brickenden wrote to Lyell from Elgin that the new fossil reptile was found in what they considered the same division of the Devonian as that containing the fossil fish *Stagonolepis* and the reptile footprints, "but no two of the specimens were taken from the same quarry or site. The *Stagonolepis* was from Stotfield, the foot-prints from Cummingston which is on the Covesea Hills and the Fossil Lizard from the Hill of Spynie."[84] Although the three sites were separated by the Loch of Spynie, they appeared to belong to the same formation. At each of the three localities, a distinctive limestone overlaid the sandstone strata. Beneath them was a yellow

79. Mantell to Brickenden in Spokes, *Mantell,* 235.

80. Lyell to Mantell, 29 November 1851, Mantell MSS.

81. Lyell to Mantell, 6 December 1851 [misdated 5 December, but the true date is indicated by the further indication of the day, "Saturday morning"], Mantell MSS.

82. Lyell to Mantell, 1 December 1851, Brickenden MSS.

83. Duff to Mantell, 24 December 1851, 83, Folder 36, Mantell MSS; cited in Benton, "Elgin Fossil Reptile."

84. Brickenden to Lyell, 3 December 1851, Lyell MSS.

sandstone containing such characteristic Devonian fossil fishes as *Holoptychius*.

A suitable name for the fossil had to be found. On 4 December, Lyell wrote to Mantell that with Charles Bunbury's help he had settled on the name *Telerpeton elginense*, from the Greek τελε ("at a distance") and ηρπετον ("reptile"), meaning "far-off reptile." He asked whether he might publish the name as Mantell's, saying that Mantell was about to describe the fossil. "Telerpeton! Mantelli!" wrote Lyell, "very euphonious indeed, far better than Archegosaurus."[85] Evidently pleased, Mantell wrote to Brickenden:

> The reptile is evidently a batrachian—apparently related to the Salamanders and I propose naming it *Telerpeton Elginense* (from the Greek, signifying the remote or most ancient reptile), a very pretty name is it not? I hope Mr. P. Duff will be pleased with the appelation [*sic*] which as god-father, or rather name-father, I have imposed on his bantling.

On 7 December, Mantell again wrote to Brickenden to say that he had obtained the fossil from Lyell and would write a description of it to append to Brickenden's paper. He added that the fossil reptile was a primitive one that "blends characters of very distinct recent types. Its ribs are saurian—its pelvis and vertebrae and tail batrachian—it is in fact either a batrachian with saurian affinities or the reverse."[86]

Brickenden's and Mantell's papers were to be read at the Geological Society on 17 December, but for lack of time were held over to the 7 January meeting.[87] On 20 December, Richard Owen published in the *Literary Gazette* a description of the fossil to which he attached the name *Leptopleuron lacertinum*. He claimed that he described the fossil at Patrick Duff's request. In a lecture at the Ipswich Museum on 17 December, Lyell exhibited a cast of the fossil, given him by Mantell, and used Mantell's name for it. In the account of Lyell's lecture in the *Literary Gazette*, Lyell was reported as having cited Owen's opinion that the Elgin fossil was a lacertian. Lyell had said nothing of Owen's opinion, referring only to Mantell's study of the fossil. Furthermore, Lyell had corrected proofs of the report of

85. Lyell to Mantell, 4 December 1851, Mantell MSS.

86. Mantell to Brickenden, 4 December and 7 December 1851; quoted in Spokes, *Mantell*, 235–36.

87. Brickenden, "Reptilian Foot-Tracks and Remains in the Old Red or Devonian Strata of Moray"; Mantell, "*Telerpeton Elginense*."

his lecture in the *Literary Gazette,* but the editor, a friend of Owen's, altered the report to make it seem that Owen had described the fossil. On 7 January, Lyell wrote to the editor to point out the "strange anachronism" and to say that he first learned of Owen's opinion when he read the *Gazette* three days after the lecture. Lyell knew that Owen had had no opportunity to study the fossil, apart from his brief examination of it on 2 December. He thought that Owen had based his published account on Mantell's drawings, exhibited at the Geological Society on 17 December.[88] In fact, in mid-November Patrick Duff had sent Owen a copy of the notice in the *Elgin Courant* of 10 October, accompanied by drawings of the fossil.[89] Nevertheless, Owen's account published in the *Literary Gazette* on December 20 was not based on study of the actual fossil. Yet he continued to insist on his priority, and his name for the Elgin fossil, *Leptopleuron,* is accepted by modern paleontologists, even though without historical basis.

Richard Owen engaged in so much ingenious skullduggery, did outrageous things with such impudence, and was so insanely jealous of his colleagues that the wonder is that he continued to be tolerated in Victorian scientific circles throughout his lifetime. Despite his own irritation with Owen, Lyell continued to consult him on scientific questions and to communicate with him in a polite manner, while at the same time firmly blocking various of Owen's machinations.

The possibility remained that the sandstone strata at Elgin were not part of the Old Red Sandstone but younger in age. Lyell wrote to Sir Roderick Murchison and Hugh Miller, both of whom were familiar with the geology around Elgin, to ask their opinions of the age of the Elgin beds. On 26 December, Murchison replied that he and Adam Sedgwick had identified the sandstone formation of Ross and Cromarty and along both sides of the Moray Firth as upper Old Red Sandstone. Nevertheless, he advised caution in pronouncing on the age of the beds. A similar sandstone was associated with the Brora Coal, which was Secondary. Miller was similarly cautious. Although the Elgin strata were probably Old Red, he said, "their stratigraphy had not been worked out, nor had Old Red Sandstone fossils been found in them."[90] Accordingly, Lyell added a note to his postscript to the preface of the *Manual* saying that although the El-

88. Lyell's letter to the *Literary Gazette* is quoted in Spokes, *Mantell,* 238–39.

89. Benton, "Elgin Fossil Reptile."

90. Miller to Lyell, 29 December 1851, Darwin-Lyell MSS.

gin sandstone probably belonged to the Old Red Sandstone, he could not be absolutely certain of its age.[91]

In the postscript, completed in December 1851, Lyell discussed the Elgin fossil reptile along with the supposed footprints of a fossil tortoise in the lowest Silurian strata of Canada, and additional reptile fossil footprints found in lower Carboniferous strata in Pennsylvania. Henry Rogers had noted that the Pennsylvania footprints were accompanied by shrinkage cracks, such as might have been produced by the drying of mud in the sun, as well as marks like those produced by raindrops and the trickling of water on a wet sandy beach.[92] After mentioning other instances of fossil impressions of raindrops, Lyell discussed the two fossil mammalian teeth (named *Microlestes antiquus*) discovered in 1847 in the Triassic. He described the discovery in Cretaceous strata of numerous gastropods, showing that Cretaceous mollusca were much more similar to those of the Tertiary and living species than previously had been thought. On the basis of fossil plants, Adolphe Brongniart likewise included the Cretaceous with the Tertiary period in his "Age of Angiosperms." He considered the Cretaceous flora transitional between the floras of the Secondary and Tertiary periods. The existence of flowering plants, cycads, conifers, and many large land animals in the late Secondary period, Lyell thought, destroyed the idea that the earth's atmosphere was then different from the modern atmosphere.

Lyell also discussed the general features of the progressive development theory, recapitulating many of the arguments he had presented in his anniversary address to show the fragmentary, essentially random nature of geologists' knowledge of the fossil record. Lyell's theory necessarily implied that in successive geological periods, great changes would occur in the proportionate representation of various groups of plants and animals. Great changes might occur in climate and geography in the long passage of time between geological periods. In contrast, the appearance of man on earth was an event different from the appearance of any other species. At the end of the postscript Lyell wrote of man:

> Physically considered, he may form part of an indefinite series of terrestrial changes past, present, and to come: but morally and

91. Lyell, "Postscript" [to the Preface] in Lyell, *Manual,* 4th ed., ix.

92. H. D. Rogers, "Reptilian Foot-Prints in the Carboniferous Red Shale Formation in Eastern Pennsylvania."

> intellectually he may belong to another system of things—of things immaterial—a system which is not permitted to interrupt or disturb the course of the material world, or the laws which govern its changes.[93]

In late October, Lajos Kossuth, an exile from the failed movement for an independent Hungary, arrived in England.[94] He was welcomed with extraordinary enthusiasm. The mayors of various towns presented him with addresses, and he spoke before many large audiences. The English middle classes tended to be sympathetic to the Hungarian movement, but the aristocracy held aloof. His arrival was of particular interest to the Lyells, partly because they were friendly with the Hungarian Ferencz Pulszky and his wife,[95] and partly because Mary sympathized strongly with the Hungarians' desire for liberation from Austrian rule.[96] During November she was frequently involved with the Hungarian cause, writing letters of introduction for Kossuth to take with him to America when he sailed on the twelfth.

On 31 October, Lyell attended the first meeting of the Surplus Committee of the Great Exhibition, to which he had been appointed by Prince Albert. The exhibition had been such a success that after all the expenses were paid, a large sum of money remained. Prince Albert wished to use it for the education of working men. He established the Surplus Committee to determine how the funds might best be applied.

In December the Lyells went to Ipswich. On the seventeenth, Lyell delivered his lecture at the Corn Exchange to a large audience that included many working people. After attending a scientific meeting at the Ipswich Museum, they went to visit the Reverend J. S. Henslow at Hitcham, returning to London on 20 December. A few days after Christmas, the Lyells had a party for the young children of their friends, including the Pulszky and Kossuth children. "We had games & a man with a magic lantern," Mary wrote, "& then tea & coffee & cakes of all wholesome kinds at six in the dining room."[97] Leonora and Joanna Horner came to help, and twenty-two sat round the table. Thus ended the year 1851 at 11 Harley Street.

93. Lyell, "Postscript" [to the Preface] in Lyell, *Manual,* 4th ed., xxii.

94. Ferencz Lajos Kossuth (1802–94), Hungarian patriot and national leader, was minister of finance in the Hungarian government in 1848 when he chose the occasion of a dispute with Austria to proclaim the independence of Hungary.

95. Ferencz Aurel Pulszky (1814–97).

96. Mary Lyell to Katherine Lyell, 21 October–7 November 1851, Kinnordy MSS.

97. Mary Lyell to Katherine Lyell, 28 December 1851–6 January 1852, Kinnordy, MSS.

CHAPTER 11

America Revisited, and the Emerging Species Question, 1852–1853

THE SUCCESS OF Lyell's two books on America increased the demand for his geological works. The rapid sale of the *Manual of Elementary Geology* took John Murray by surprise. The sale of the *Principles* remained strong. Recognized both as an authority on America and as a scientist, Lyell had become a confidante of Prince Albert. Cabinet ministers sought his acquaintance and asked his opinion. Lyell had exercised, and would continue to exercise, a significant influence on university reform. In America, the friendly tone of Lyell's travel accounts reinforced the success of his 1845 series of Lowell lectures. Nor were Americans indifferent to Lyell's knighthood; it represented royal recognition of scientific worth. When John Amory Lowell was in London to attend the exhibition, he invited Lyell to Boston to give the Lowell lectures in the autumn of 1852. In his third series of Lowell lectures, Lyell would begin to grapple with the question of the origin of species, its interest intensified by the publication of the *Vestiges* and by new fossil discoveries.

THIRD VISIT TO AMERICA, 1852

For many months Lyell kept his plan to make a third visit to America secret, perhaps because he feared that circumstances might arise to prevent him from going. The fourth edition of the *Manual* appeared in January 1852. The postscript to the preface was also printed and sold separately, so that readers who had bought the third edition might bring their copy up to date simply by adding the postscript. Lyell also sent copies of the postscript to a long list of scientific friends to whom he had given copies of the third edition or to whom he usually sent separate copies of his scientific papers. In January he paused briefly to have his portrait painted by Henry Selous. It was to be used for Selous's large painting of the scene at the opening of the Great Exhibition when the Archbishop of Canterbury was reading the prayer.

Through the winter and spring of 1852, Lyell worked at a paper on the Belgian Tertiary formations, but with many interruptions for social and scientific engagements. On 30 January he and Mary gave an evening party with more than fifty guests, including many of Lyell's scientific friends—Chevalier Bunsen and his family, Lord Lansdowne, and others. Two weeks later, on 14 February, the Lyells dined at Lord Lansdowne's, and on 20 February Lyell attended the anniversary dinner of the Geological Society of London. The regular monthly meetings of the Geological, the Royal, and the Linnean Societies and occasional meetings of other societies consumed many of his evenings. From time to time, he also attended meetings of the Surplus Committee of the Great Exhibition. The committee was considering various plans to gather the scientific societies of London into one place, to affiliate the many Mechanics' Institutes scattered throughout the British Isles, and to create a school of industrial education or, in more modern terms, a school of technology.

On 5 March, the Lyells went to Hyde Park to take a last look at the Crystal Palace, now standing empty except for some fittings and the floor planks. It was to be sold by auction. The late winter sunshine showed the great glass and iron building to advantage. "It really looks beautiful," Mary wrote, "so very large there is something quite dreamy about the extent of it & the light colouring."

They attended a Friday evening lecture by Gideon Mantell at the Royal Institution. Although desperately ill and in great pain from a tuberculous abscess pressing against his spine, Mantell roused himself to give a spirited and entertaining lecture. To accompany his lecture on the *Iguanodon* and other fossil reptiles, he displayed "a splendid collection of bones on the table."[1] Poor Mantell knew that it would be his last performance at the Royal Institution. Perhaps many of his audience also sensed the fact. After the lecture he stood at the table for more than half an hour, explaining the bones to the eager crowd. The next morning he received a note from Lyell congratulating him on his great success.[2] To Benjamin Silliman in America, Mantell wrote that "the room was crowded to the ceiling. Herschel, Brewster, Babbage, Faraday, Murchison, even Lyell, & all our best men were there." With characteristic kindness, Michael Faraday told Mantell it was "the most successful & eloquent discourse ever delivered there," a compliment that was balm to Mantell's ravaged spirit.[3]

1. Mary Lyell to Katherine Lyell, 28 February–6 March 1852, Kinnordy MSS.

2. Mantell, *Journal,* 283.

3. Mantell to Silliman, 1 April 1852, Silliman Family Papers.

During March, Mary read aloud to Lyell in the evenings Charles Dickens's novel *David Copperfield,* which they enjoyed greatly. On the seventeenth they attended a breakfast party at Samuel Rogers's house. In the evening the prime minister, Lord Derby, gave a party at 10 Downing Street, where Lyell met many members of the new Tory government. The happiest event of the spring came on the first of April. They received a letter from Henry Lyell in India announcing that Katherine had given birth to a second son, named Francis Horner Lyell. The next morning, Mary wrote to Katherine that "Charles has just been talking to our bird & telling him he will be tied by the leg when the two boys arrive from India."[4]

On 2 April, Lyell delivered his own Friday evening lecture at the Royal Institution, discussing the Blackheath pebble bed. He suggested that it had been deposited in a delta under conditions similar to those of the delta of the Alabama River at Mobile. He related it to the series of English Eocene formations worked out by Joseph Prestwich and to the Miocene and Pliocene formations. The lecture room was crowded, but the lecture was not one of Lyell's best. Mary thought him not as animated as usual; he may have been tired.

On 5 April, the Lyells, accompanied by Joanna Horner, set off for Paris on the four o'clock train for Folkestone; they shared a compartment with Edward Forbes. At Reigate, Robert Austen came aboard, to join Forbes on a trip to Belgium to study the Belgian Tertiary formations. At Folkestone next morning, as the group was waiting for the steamboat to Boulogne, Joseph Prestwich and John Morris arrived to accompany Forbes and Austen on their expedition, a trip inspired by Lyell's study of the Belgian Tertiaries the previous summer. That evening the Lyells and Joanna reached Paris.

In 1852, Paris, under the new dictatorship of Louis Napoleon, showed many improvements, especially in the provision of sidewalks, but the political atmosphere was oppressive. The weather presented similar contrasts—bright sunshine but a piercingly cold wind. The Lyells visited Mary's two aunts and then called on various geologists. During the next two weeks Lyell saw much of the Parisian geologists, visited the Jardin des Plantes and other museums, and went on a geological excursion to Meudon with his old friend Constant Prevost. The geologists were very active, but Lyell feared that they might lose their positions, which depended on government support. Many professors at the Sorbonne, especially the professors

4. Mary Lyell to Katherine Lyell, 2–4 April 1852, Kinnordy MSS.

of history, had been removed from their chairs because of the hatred of the church for the university.

Back in Harley Street, Lyell was soon caught up in scientific and social life, especially intense in May and June during the London season. On the evening of their return on 21 April, he attended the meeting of the Geological Society and reported informally on his visit to Paris. At the next meeting of the Geological Society, on 5 May, he read the first half of his paper on Belgian Tertiary formations. After the meeting he went home, changed into evening dress, and near midnight went to the Queen's Ball at Buckingham Palace. He joined Mary, who had arrived earlier with the Sir James Clarks; she was watching a quadrille in which Queen Victoria and Prince Albert were dancing. "They all seemed to be enjoying themselves very much," Lyell observed, "& it was a very gay scene."[5] During the season, the Lyells attended other parties, including a very crowded party given by Lord Derby at 10 Downing Street. On 19 May, he read the second half of his Belgian paper to the Geological Society, the next day attended a meeting of the Surplus Committee, and on the twenty-second went with Mary to a promenade at the Zoological Gardens. So many public and social activities placed considerable demands on Lyell's time and attention, making it difficult for him to keep up with his scientific work. He tried to devote his mornings to geology and felt the need to do much reading on developments in the science. One morning Charles Darwin called to chat with Lyell over breakfast. On 15 June, Lyell attended the last meeting of the Geological Society for the year.

With meetings of the scientific societies over for the summer, the Lyells went in late June to Mildenhall in Suffolk to spend ten days with the Charles Bunburys in their pleasant old gabled mansion, built in the time of James I. Lyell took books and papers with him and continued to spend most mornings at work. The last day of June he gave a lecture on the Temple of Serapis to an audience of local people in the Mechanics' Institute. Returning to London in early July, they found a serious smallpox epidemic threatening the metropolitan area. At Highgate the whole Horner family had been vaccinated, and Mary hastened to have everyone in the Harley Street household vaccinated.

July was much quieter than the spring had been. Lyell made final revisions on his paper on the Belgian Tertiary formations. He be-

5. Lyell to Frances Lyell, 12 May 1852, Kinnordy MSS.

gan to consider revision of the *Principles* for a new edition, the eighth edition being completely sold. He also took up again a study of drift deposits and their associated freshwater formations. In 1840 he had published a paper on the Norfolk drift, but now he began to examine drift deposits at Highgate and other localities around London. On 9 July, the Lyells went into the City to book passage on the Cunard steamship *America,* to sail from Liverpool on 21 August. Their staterooms would be next to those of the governor of New Brunswick, Sir Edmund Head and his family. The Lyells were planning to visit them at Fredericton at the beginning of their American trip. Head and Lyell had known each other for many years and had many friends in common. In 1832, when they met by chance in Switzerland, Head had accompanied Lyell on a trip to study the granite veins of Valorsine. They also shared a common interest in education. In 1852 Head was concerned that King's College at Fredericton (later to become the University of New Brunswick) was not providing a kind of university education adapted to the needs of the colony.[6]

On 12 July, the Lyells celebrated their wedding anniversary with the Horners at Highgate, a particularly happy time. After weeks of warm, sunny weather, the garden at Highgate was full of flowers. "Everybody was cheerful & pleasant," Mary wrote, "& we had a famous walk across the fields to Highgate." On the eighteenth they went to Samuel Rogers's house for tea to meet Charles Dickens, whom they liked for his "nice hearty unaffected manner."[7] On 5 August they went to Sydenham in Surrey for the reopening of the Crystal Palace, reconstructed on a permanent site. A few minutes before the ceremony, Lyell was asked to make a short speech and move a vote of thanks on behalf of the political, literary, and scientific visitors to the opening. Although unprepared, he got through it well and, Mary thought, with great fluency. "He stood on a bench beside me," she wrote, "& I was only afraid he would tumble down."[8] After the ceremony at Sydenham, the Lyells went to Down, a few miles away, to spend the weekend with the Darwins.

From Down, Lyell made excursions to examine drift deposits in the neighborhood. He and Darwin had their customary long conversations on scientific subjects. Darwin described to Lyell some of the curious features of barnacles he had observed. In his effort to

6. Kerr, *Head,* 102–205.

7. Mary Lyell to Katherine Lyell, 11–19 July 1852, Kinnordy MSS.

8. Mary Lyell to Katherine Lyell, 10–18 August 1882, Kinnordy MSS.

describe and classify barnacles from all over the world, Darwin had formed a clearer idea of the range of variation within species. He found that many species created by naturalists were merely geographical variants. Distinctions drawn between closely related species were often arbitrary. The number of species that Darwin could distinguish shrank rapidly as he examined specimens from different parts of the world.

After their return to London, the Lyells began to pack for their trip to America. On 20 August they took the train to Liverpool. The voyage across the Atlantic on the *America* was not as pleasant as past voyages: the ship was crowded, much of the time there was a heavy swell, and they encountered a storm. Almost everyone was seasick, including both Lyells. Despite such inconveniences, Lyell talked a good deal with Sir Edmund Head. He learned from Head that the United States census of 1851 had produced a significant transfer of political power from the eastern states to those west of the Allegheny Mountains. During the voyage Lyell met another passenger, a Mr. Gisborne, who had manufactured a submarine telegraph cable. He was going to build an electric telegraph line from Nova Scotia to Newfoundland to connect with the telegraphic network already existing among the principal American cities. From St. John's, Newfoundland, a steamship might cross in five and a half days to Galway in Ireland, whence a message might be sent by electric telegraph to the principal European cities. Before the end of 1852, it would be possible to send a message from New Orleans to Vienna in five and a half days. On shipboard, Lyell also wrote notes on the extinction of species for one of his Lowell lectures. One night they saw a splendid display of northern lights. On the morning of the thirty-first they landed at Halifax, Nova Scotia, where John Dawson met them.

Lyell had asked Dawson to join him for a geological expedition to the fossil cliff at Joggins. During their first meeting at Pictou in the summer of 1842, Lyell had encouraged Dawson to submit papers on the geology of Nova Scotia to the Geological Society of London. He had continued to correspond with the young Nova Scotian. In the intervening ten years, Dawson had spent a second year of study at the University of Edinburgh and married a young Scotswoman. He had been appointed superintendent of education in the common schools of Nova Scotia. In this capacity, Dawson traveled all over Nova Scotia, allowing him to study the geology of the whole province.

On 1 September, Lyell and Dawson left Halifax in a hired buggy drawn by two horses. They drove to Truro at a rapid rate, changing horses every twenty miles. They had ordered their horses in advance from Halifax by the new electric telegraph line that marched across the rocky, fir-covered hills on tall tamarack poles. The telegraph delighted Lyell by its cheapness and its speed—a message of ten words cost sixpence. He seized the first opportunity to send a telegram to Mary and in the next few days sent several more.

In 1852, after the economic depression of the forties, Nova Scotia was enjoying a wave of prosperity. The telegraph line was only one of many signs of growth and change along the road from Halifax to Truro. There were new farms with stumps still in the fields and new wooden schoolhouses. The weather was beautiful. Along the roadside goldenrod was in bloom, and on the hills the maples were beginning to turn red. At one place Dawson showed Lyell where the ice in winter had pushed up boulders around the shore of a lake. He told Lyell the names of such plants as the blueberry (*Vaccinium*), famous for its edible berries, and the meadowsweet (*Spiraea*). They stopped at an Indian village of five wigwams made of birch bark. Lyell, ever curious, was about to look into a wigwam to see what the interior was like, when a curtain was suddenly dropped. Dawson warned him that the Indians were ceremonious, resenting anyone's entering their houses without an invitation.

At Truro that evening, the efficiency of the telegraph again impressed Lyell. Dawson sent a telegram to his wife at Pictou to ask about one of their children, who was ill. Although the telegraph line was busy with news and messages brought from Europe by the *America,* being transmitted to Canada and the United States, they received a reply from Mrs. Dawson while they were at tea. The telegraph, Lyell found, was also used to insure Nova Scotian sailing vessels at New York. He was equally impressed by Dawson's accomplishment of the year before in organizing at Truro a convention attended by eighty public school teachers from all over Nova Scotia.

On 3 September, Lyell and Dawson drove from Truro through the birch and maple forest on the Cobequid Hills to Amherst. They saw a schooner of forty tons being built on a hill three hundred feet above the sea and four miles from the shore. When completed, it would be hauled down to the shore by oxen during the winter, when it could be moved easily over the snow. During the next five days they measured the section of Carboniferous strata exposed in the cliffs at Joggins, on the Chignecto Basin southwest of Amherst. Extending for almost ten miles along the shore of the basin, the South

Joggins section included some fourteen thousand feet of strata, representing almost the whole series of beds of the Nova Scotia Carboniferous. Dawson and Lyell examined a portion of the section that extended about a mile along the shore east of the Joggins mines, where many erect fossil trees were exposed. Lyell wished to learn the circumstances that had favored the preservation of so many fossil trees in an erect position, in the same place, through the long period required for the accumulation of such a thick series of strata. He also sought additional evidence that the so-called *Stigmariae* fossils in the underclays beneath coal beds were the roots of the fossil tree *Sigillaria.* Though in 1846 Edward Binney demonstrated the relationship of *Stigmaria* to *Sigillaria* clearly enough, some geologists continued to question it.[9] Dawson and Lyell also thought they would study the difference between the deposits in the strata around the erect fossil trees and those forming a cast of the trunk enclosed inside the cylinder of bark converted to coal. They suspected that any fossils found in the interiors of the fossil trees might be different from those in ordinary sedimentary strata. To help them in their work, Lyell hired a laborer to dig erect fossil trees out of the cliff so that they might break them up to study their contents.

Lyell and Dawson made detailed notes on each of twenty-nine groups of beds in the series of strata at South Joggins with a total thickness of 2,819 feet.[10] The strata included underclays, beds of coal of varying thickness, beds of limestone containing multitudes of fossil *Cypris,* freshwater shells, and fish scales, and beds of sandstone. The fossils showed that the ancient coal beds at Joggins had accumulated under conditions much like those existing in the marshes of the Bay of Fundy or in the deltas of great rivers like the Mississippi. The stumps and erect fossil trees, visible at so many levels in the Joggins cliff, reminded Lyell of the trees he had seen in 1846, buried at different levels in the bank of the Mississippi at Port Hudson, Louisiana. The accumulation of such a great thickness of strata had been accompanied by slow subsidence, the conditions alternating from those of a swamp, subject to occasional flooding, to those of a freshwater lake, and back again.

One stormy day, while Lyell, Dawson, and their hired assistant were working at the base of the Joggins cliff, they were in considerable danger. A high wind blowing up the Chignecto Basin was whip-

9. Binney, "Dukinfield Sigillaria," 392.

10. Dawson, "Coal-Measures of the South Joggins," 10.

ping the trees on the top of the cliff back and forth. At any moment a tree at the brink might be loosened and, in falling down the cliff face, might loosen some of the heavy fossil trees and bring them down on the men's heads. Despite the danger, Lyell was too fascinated by the Joggins cliff to stop work. He could envision in its succession of underclays, coal beds, and other strata, and in those erect fossil trees, the succession of ancient forests, swamps, and lakes of the Carboniferous period.

After two days' work, they got their first fossil tree out of Group XV in the series of strata, a succession of underclay, coal, and shale, capped by layers of sandstone. They dug it out of the cliff to expose both trunk and roots and were even able to trace the branching roots into the underclay.[11] "In the morning's work we have proved," Lyell noted, "the undoubted connection of an ordinary fluted & leaf-scar[re]d Sigillaria with the ordinary round scar[re]d Stigmaria root & the 4 roots bifurcating."[12]

Next, they began to break up the dark sandstone cast of the interior of the trunk. They hoped to find any plants that had fallen into the trunk of the ancient hollow tree as it was filling up. After finding fragments of various plants that appeared to be remnants of ferns, *Sigillarias,* and *Calamites,* they came upon some dermal plates and several bones, one of which looked remarkably like the femur of a reptile. It certainly was not a fish bone. Although neither Dawson nor Lyell was an anatomist, they thought they could tell "a Hawk from a Handsaw." The next day, they returned to break up the remaining portion of the tree trunk and found in it what they thought might be the jaws and teeth of a *Labyrinthodon.*[13] The discovery greatly excited Lyell. Various instances of reptile footprints were known in American Carboniferous strata, but no actual fossil remains of a reptile had hitherto been found in the American coal beds to match the *Apateon* and *Archegosaurus* found in the Carboniferous of Germany. The bones were scattered about in what had been the hollow interior of the tree, mingled with pieces of wood converted into charcoal that apparently had fallen there while the tree was rotting away from above.[14] Many years later, Dawson recalled Lyell's almost childlike delight when they discovered the bones in that tree on the windblown Joggins shore:

11. Ibid., 20–22.
12. Lyell, Notebook 178, 31, Kinnordy MSS.
13. Lyell, postscript to Mary Lyell to Leonard Horner, 4–15 September 1852, Kinnordy MSS.
14. Lyell and Dawson, "Remains of a Reptile (*Dendrerpeton Acadianum,* Wyman and Owen)." The charcoal resulted from fires that raged through the Carboniferous forests.

> His thoughts ran rapidly over all the strange circumstances of the burial of the animal, its geological age, and its possible relations to reptiles and other animals, and he enlarged enthusiastically on these points, till, suddenly observing the astonishment of a man who accompanied us, he abruptly turned to me and whispered, "The man will think us mad if I run on in this way."[15]

Lyell and Dawson spent several more days together, traveling to New Brunswick and visiting the Albert mine near Hillsborough. The Albert mine yielded an unusually bituminous coal, almost like asphalt. The strata corresponded to the lowest members of the series exposed on the Joggins shore, twenty miles to the southeast across the Chignecto Basin.[16] After collecting specimens of fossil fish at Hillsborough, Lyell and Dawson parted company. As they had prearranged, Lyell met Sir Edmund Head at St. John and together they went by steamboat up the St. John River to Fredericton.

After Lyell left on his excursion with John Dawson, Mary joined Sir Edmund Head's party for the trip to Fredericton. At Windsor, Nova Scotia, they boarded a steamboat. As they passed through the Minas Basin, Mary saw from the deck Cape Blomidon, Cape Split, and other places that she and Lyell had visited ten years before. The overnight trip down the Bay of Fundy brought them early the next morning into St. John. After disembarking, they went to a hotel where, Mary said, they made themselves "a little in order." They walked about the town until a breakfast of broiled ruffed grouse was ready at the hotel. When they came out of the hotel after breakfast, an honor guard of Highlanders was drawn up with a piper playing. "It was very amusing to see the contrast of things," Mary wrote, "as we all packed into a very shabby hackney coach while five officers on horseback were escorting us."[17] After a pleasant trip up the St. John River, there was again a guard of honor at Fredericton for the governor. The whole population of the town turned out, and cannon were fired in salute.

After Lyell arrived, he and Mary spent four days with the Heads at Government House, taking drives and walks in the country. Lyell had many long talks with Head about the needs of New Brunswick. From a population of 154,000 in 1840, the province had grown to nearly 194,000 in 1851. The people were scattered in small towns

15. Dawson, *Fifty Years of Work in Canada,* 54–55.

16. Dawson, "On the Albert Mine"; cf. Egerton, "Fossil Fish from Albert Mine."

17. Mary Lyell to Katherine Lyell, 9–23 September 1852, Kinnordy MSS.

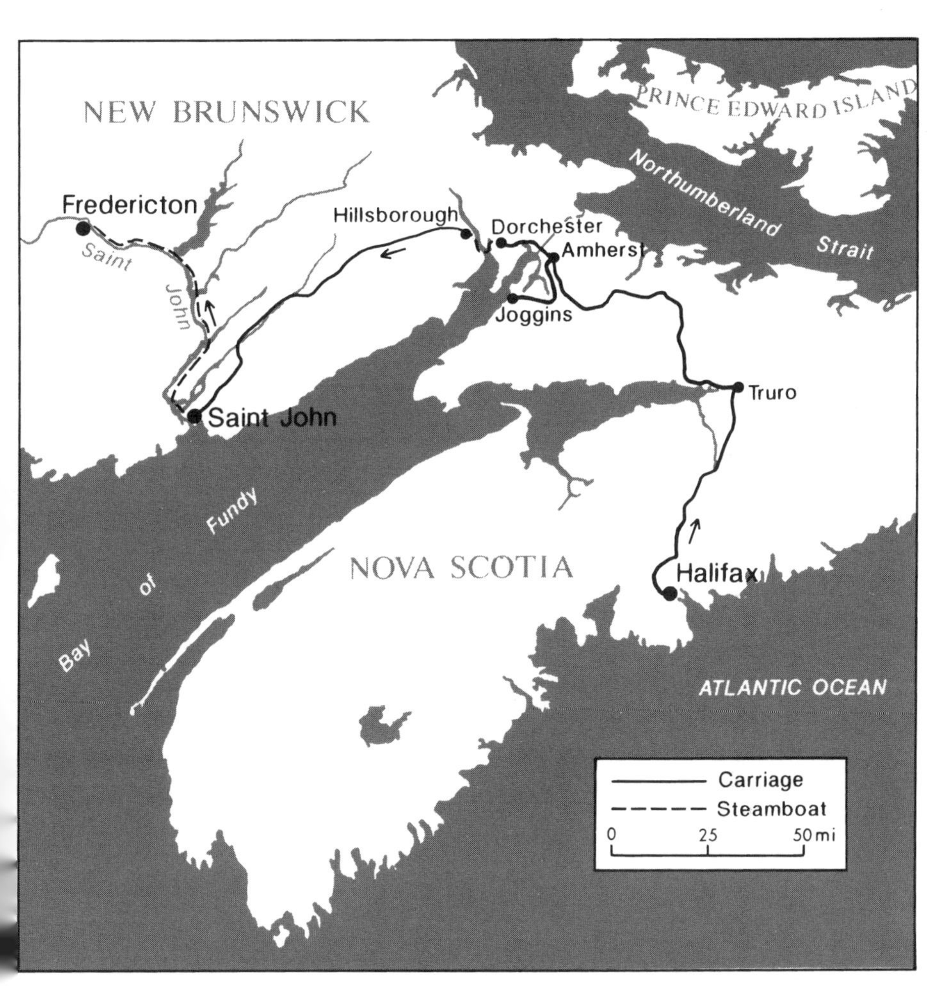

Lyell's Travels through Nova Scotia and New Brunswick, 1852. Map by Philip Heywood.

and settlements, located mostly along the shores of the Bay of Fundy and the Gulf of St. Lawrence and in the valley of the St. John River. The chief industries were fisheries, timber-cutting, and shipbuilding. Agriculture was backward. At St. John many ships were being built for the Australian trade. The introduction of the electric telegraph had provided rapid communication among the widely separated towns and villages. The governments of New Brunswick and Nova Scotia were planning to build a railroad to connect the two colonies with Canada and the United States.

Sir Edmund Head was particularly anxious to develop the educational system of New Brunswick. Like Lyell, he had presented testimony to the Oxford University Commission in favor of reform. King's College at Fredericton, under the influence of its president, an Anglican clergyman and Oxford graduate, adhered strictly to the narrow classical education of the Oxford system. Founded in 1787 as the Academy of Liberal Arts and Sciences, it was modeled on King's College at New York, whence many of the Loyalist settlers of New Brunswick had come after the American Revolution. In 1800 the academy received a charter as the College of New Brunswick and in 1828, under a new royal charter, it became King's College, New Brunswick. The charter of 1828 provided that all members of the College Council must be Anglicans in Holy Orders.[18] Although the college did not exclude students who were not Anglicans, it attracted Anglican students primarily. A more serious difficulty was that few New Brunswick students could afford to devote several years exclusively to classical studies. For a more practical form of higher education, they had to go elsewhere, usually to the United States.[19] As governor of New Brunswick, Head was also Royal Visitor to King's College. He and Lyell talked over the problems of the college, which were, Lyell wrote, "just what you might suppose the Oxford system transplanted to a colony [would be], a complete failure."[20]

Shortly after Lyell's visit to Fredericton, Head wrote an official letter to the College Council, urging them to consider seriously whether the college met the real needs for higher education in the colony.[21] Head's letter was printed and distributed throughout

18. Trueman, *University of New Brunswick,* 8–14.

19. Lyell to Leonard Horner, 12 September 1852, Kinnordy MSS; printed in K. M. Lyell, *Life, Letters, and Journals,* 2:178–84.

20. Lyell to Leonard Horner, 30 October 1852, Kinnordy MSS.

21. Kerr, *Head,* 104.

New Brunswick. The following year, the New Brunswick government appointed a commission, modeled on the Oxford University Commission, to study King's College. During his visit at Fredericton, Lyell may have drawn Head's attention to the scientific and educational qualifications of Dawson, whom he had introduced to Head when they landed at Halifax. Dawson was appointed a member of the New Brunswick commission. The report of the commission resulted in 1859 in the reorganization of King's College into the University of New Brunswick. In 1854, after Head was appointed governor-general of Canada, he was also instrumental in the appointment of Dawson as principal of McGill College at Montreal.[22]

On 16 September, the Lyells left Fredericton on the early-morning steamboat. The next day they arrived at Boston and went to George and Anna Ticknor's house. Although Boston was said to be empty, many old friends and acquaintances called to see them, William Prescott coming in from the country for the purpose.

At Boston, Lyell learned of a paper by the brothers Henry and William Rogers, in which they described two remarkable trains of erratic boulders in Berkshire County in western Massachusetts.[23] Both trains of boulders began in a depression on the crest of the Canaan ridge. The boulders appeared to have been broken off from the rock of the ridge. They extended in a southeasterly direction over a succession of valleys and ridges nearly parallel to each other for almost twenty miles. The Rogers brothers specifically rejected Lyell's explanation of the transport of erratic boulders by icebergs. Instead, they suggested that the trains of boulders had been transported by the sudden rush over North America of a great wave of water, at a speed of perhaps five miles per minute. Such a wave, they suggested, might be caused by a

> paroxysmal, or sudden and violent, disturbance of the slightly flexible crust of the earth, causing in the period of the northern drift, a partial elevation and displacement of the bed of the great frozen sea which occupies the arctic latitudes, and sending its waters, with all their ice, in a sudden inundation, over the northern lands of the two continents.[24]

22. Dawson, *Fifty Years of Work in Canada,* 92.

23. H. D. Rogers and W. B. Rogers, "Trains of Angular Erratic Blocks"; cf. Silliman, "Richmond Boulder Trains."

24. H. D. Rogers and W. B. Rogers, "Trains of Angular Erratic Blocks," 317–20.

Lyell wished to examine any phenomenon associated with such a direct challenge to his own theory. He wrote to James Hall to ask whether Hall might be able to go with him to examine the Berkshire erratic boulders. At Boston, Augustus Gould showed Lyell some remarkable fossil fruits from a deposit of brown coal at Brandon, Vermont. Lyell invited Hall to go with him also to Brandon, if he could spare the time.[25]

On 22 September, the Lyells took the new express train from Boston to New Haven, where they stopped for a day and saw only James Dwight Dana. Benjamin Silliman was out of town. From New Haven the Lyells went to Burlington, New Jersey, to visit William McIlvaine and his sisters. Lyell spent a day in Philadelphia. At the Academy of Natural Sciences he called on Joseph Leidy and David Dale Owen, the latter on a visit from Indiana. Leidy was describing the fossils collected by one of Owen's assistants in the Badlands of Nebraska.[26] Leidy and Owen showed Lyell the fossil bones of a large *Palaeotherium* from Nebraska. On 28 September, the Lyells returned to New York in order to take the train to Troy.

After spending a couple of days with the Halls, the Lyells and Hall set out together on 1 October for Brandon, Vermont, going first by rail to Amherst, Massachusetts. In Amherst that evening Edward Hitchcock, president of Amherst College, and his wife entertained them. At dinner, to Mary's amusement, the ladies were seated in a row on one side of the table and the gentlemen on the other, an arrangement she took to be an old Puritan custom.[27]

The Lyells and Hall continued their tour up the Connecticut Valley in a hired carriage, accompanied by Hitchcock and a group of Amherst students in an omnibus. At Turners Falls, they stopped to look at fossil reptile footprints in Triassic strata and then went on to Bernardston, at which point Hitchcock and the students turned back, while the Lyells and Hall boarded a train for Brandon, Vermont. During the next two days at Brandon, Lyell and Hall examined the bed of lignite, nine feet thick. Containing many fossil fruits and seeds, it proved to be of Tertiary age. Lyell collected specimens to take to England for Charles Bunbury, who was studying Tertiary fossil plants.

On 6 October they returned by railroad to Troy, New York. To Lyell's astonishment, the 102-mile trip took less than four hours,

25. Lyell to Hall, 21 September 1852, Hall MSS.

26. Leidy, "Extinct Mammalia and Chelonia."

27. Mary Lyell to Leonora Horner, 3 October 1852, Malcolm Lyell MSS.

along a line through mountainous country that yet had no tunnels. The next day Mary traveled to Boston to stay with the Ticknors, while Lyell and Hall went to study the erratic boulders in the Berkshire Hills around Pittsfield, Massachusetts. In 1845 Edward Hitchcock published the first scientific description of the Berkshire trains of erratic boulders. He had not attempted to explain them.[28] The boulders were distributed in long parallel rows, extending in nearly straight lines across ridges and valleys from their starting points on the Canaan ridge. Their direction was nearly at right angles to the lines of the ridges and bore no relation to the direction of the streams and rivers. The boulders were rounded like the glacial boulders called in Switzerland *roches moutonnés*. Near the Meeting House at Richmond, Massachusetts, Hall and Lyell measured one of the larger boulders. It was fifty-two feet long, forty feet wide, and, although partially buried, fifteen feet high.[29] The boulders rested on a deposit resembling the European "northern drift." Where the underlying rock was exposed, its surface was polished, striated, and furrowed, with the furrows running in the same direction as the trains of boulders. Lyell thought that the trains of boulders must have been transported by floating ice at a time when the Berkshire Hills stood at a much lower level, with only their highest ridges protruding above the sea. He thought their transport could not be explained by glaciation, because if glaciers had transported the boulders, the trains of boulders should have been distributed down the valleys instead of across them. In fact, the boulders had been transported by glaciers, but by continental glaciers rather than by mountain glaciers, the only ones with which Lyell was familiar.

After spending several days geologizing with Hall, Lyell rejoined Mary at the Ticknors' house in Boston. The next day the Lyells and Ticknors went to Pepperell, about forty miles from Boston, to spend three days with the Prescotts at their country house. Mary wrote:

> The house is a real old fashioned New England farm house, built of wood, at different times, first by Colonel Prescott his grandfather, who fought at Bunker Hill, then by his father Judge Prescott, whom we knew, & a part added by himself. It is furnished in the plainest way, but most comfortably (plenty of rocking chairs). It reminded me of Bonaly in former times, when one might go in or out without fear of spoiling anything. The drawing room

28. Hitchcock, "Dispersion of Blocks of Stone."
29. Lyell, "Trains of Erratic Blocks."

opens on the piazza & there are some old butter-nut trees (a kind of walnut) in front.[30]

They had a wonderful time, going for drives through the countryside and long walks in the woods, then in the full glory of their brilliant fall colors. After dinner the conversation was constant and lively. "The bursts of laughter & mirth," Mary wrote, "did one's heart good."[31] On 15 October, the Lyells and Ticknors returned to Boston. The Lyells went to the Tremont Hotel, which was to be their home for the next six weeks while Lyell delivered the Lowell lectures.

On 19 October, Lyell delivered his first Lowell lecture, taking as his subject the successive changes of the Temple of Serapis. It was essentially the lecture he had delivered at Mildenhall in July. The first eight of Lyell's Lowell lectures in the fall of 1852 were modified versions of lectures he had delivered previously at the Royal Institution in London, at Ipswich, or, in the case of the first, at Mildenhall. They treated diverse aspects of geology but were united by a common theme: that conditions in the geological past were essentially similar to those existing in the modern world and geological changes had been brought about in the past by the same kind of processes that were going on at present and at the same rates. The intensity of geological processes, Lyell argued, had not varied significantly throughout the history of the earth.

The audiences at Lyell's lectures were large, larger than had attended the Lowell lectures for several years. His success with the initial lectures made Lyell strive even harder to give of his best.[32] He revised, rewrote, and rethought each lecture before it was delivered and gave his freshest reflections on the topic. After his seventh lecture on 9 November, Mary, perhaps partial, but nonetheless a clear-sighted critic, wrote to her father, "I feel very sure these are by far the best lectures Charles has ever given, but," she added, "they fatigue him a good deal."[33] Near the end of the course, she wrote to her mother that she thought "Charles's lectures have been excellent."[34]

One of Lyell's principal concerns when he arrived at Boston was

30. Mary Lyell to Mrs. Leonard Horner, 17 October 1852, Kinnordy MSS. Bonaly was Henry Cockburn's much loved country retreat outside Edinburgh.

31. Mary Lyell to Katherine Lyell, 21 October 1852, Kinnordy MSS.

32. Lyell to Hall, 28 October 1852, Hall MSS.

33. Mary Lyell to Leonard Horner, 11–15 November 1852, Kinnordy MSS.

34. Mary Lyell to Mrs. Leonard Horner, 21 November 1852, Kinnordy MSS.

to determine the nature of the bones that he and Dawson had found in the hollow fossil tree at Joggins. Immediately after his arrival, Lyell took the jawbone and what he and Dawson thought was the humerus to Louis Agassiz to ask his opinion. Agassiz recognized that the fossil was entirely new and unknown; he thought the jawbone might be that of a fish rather than a reptile. If the skeleton were that of a fish, the supposed humerus might be the hyoid bone, as in *Lepidotis,* although it was different in form from the latter. Lyell then took the two bones, together with the rest of the fossil skeleton, to Jeffries Wyman at Cambridge. Wyman noted first the reptilian character of a number of the bones, but he was puzzled by a circular scalelike bone that might be the operculum of a fish. After comparing notes with Wyman, Agassiz told Lyell confidently that "the Joggins skeleton is a sauroidal fish of a new genus allied to Pygoterus! a carnivorous beast."

In the face of such expert opinion, Lyell felt deflated in his hope that the Joggins skeleton would prove to be a new fossil reptile. Yet he could not give up his first impression that the fossil was a reptile. He wrote to Dawson: "I still think that the interior of the tree will produce some new & unknown fossil."[35] After Wyman had examined the bones more fully, he found that they were definitely reptilian and resembled those of the batrachian *Menobranchus,* living in the Ohio River and Lake Champlain.[36]

While Wyman was searching through some of the carbonized plant remains and other debris from the interior of the Joggins fossil tree in the hope of finding additional bones, he came across another fossil. At first he thought it might be a coprolite or a shell, possibly a *Pupa.* Lyell suspected that it was a shell of the genus *Clausilia.* When they showed the specimen to Augustus Gould, Gould recognized that the whorls were reversed, or counterclockwise, a feature that occurred only in *Clausilia cylindrella* and a few other land shells. The fossil was a land shell—the first land shell to be discovered in rocks as old as the Carboniferous. Since only part of the fossil was visible, the rest being buried in the stony matrix, Gould began the delicate task of carving the fossil out of the rock. When he got it out, he thought at first that it was something quite different from what he had supposed, "some isopod, a wood louse or oniscus perhaps."[37]

35. Lyell to Dawson, 22 September 1852 [copy], Kinnordy MSS.
36. Lyell and Dawson, "Remains of a Reptile (*Dendrerpeton Acadianum,* Wyman and Owen)."
37. Lyell to Dawson, 12 November 1852 [copy], Kinnordy MSS.

At Joggins, Lyell had thrown away many fragments of the interior of the fossil tree similar to the one containing the fossil shell. Realizing now that he should have kept the entire contents of the tree, he wrote to Dawson to tell him of the find and to ask him, if possible, to go to Joggins to try to recover more of the fragments of the tree's interior, adding, "I shall tell Wyman that I pay any expenses of specimens you may send to him for our paper & I should wish much to pay for any incurred in a trip to the Joggins."[38]

When Wyman broke up fragments of the stony interior of the Joggins fossil tree, he found within one of them a series of nine small vertebrae, two with what seemed to be short ribs attached. He decided that they must be dorsal or lumbar vertebrae belonging to a small reptile, perhaps only six inches long. The new fossil was much smaller than the other reptile, whose bones indicated that it was about two and a half feet long. Thus, two species of reptile had been entombed in the ancient hollow tree. On the evening of 9 November, Lyell presented this series of remarkable fossil discoveries to his audience at the Lowell lectures.

Lyell's next lecture, on 12 November, on impressions of raindrops in rock strata, was a lecture he had given at the Royal Institution in April 1851. The next four lectures, the final ones of his 1852 series, Lyell delivered for the first time at Boston. On 16 November, he lectured on the Berkshire trains of erratic boulders, accounting for the transport of the boulders by his iceberg theory.

In his final three lectures Lyell discussed the species question. He noted that "it is a still more unexpected and wonderful result of geological investigation, the land and the sea have been inhabited at many, perhaps 30, successive periods by distinct assemblages of animals and plants—by distinct groups of species, Man belonging to the very latest of these successive assemblages." He proposed to consider in what manner "the changes in the organic world have taken place in geological times & whether the same mutations may be now in progress," and whether there had been a progressive development from the simplest organisms to the highest forms. Lyell showed that changes of species occurred gradually and successively, taking place simultaneously on land and in the sea. Land animals were not exterminated by an eruption of the sea over the land nor by any other kind of catastrophe, earthquake, or deluge. Instead, the forces tending toward the extinction of species were constantly

38. Ibid.

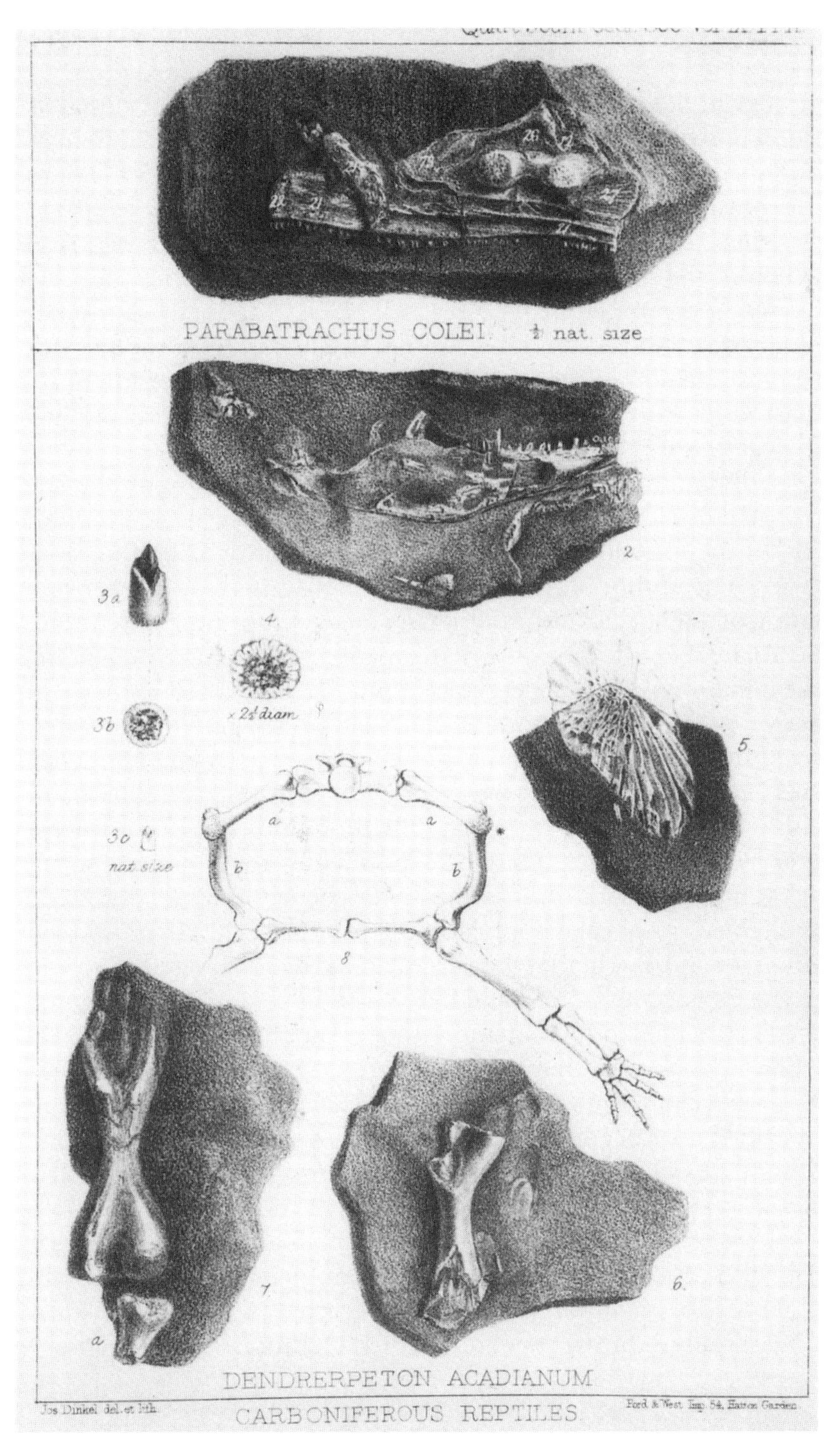

Dendrerpeton Acadianum. Fossil Reptile from Joggins, Nova Scotia. Lyell and Dawson, "On the Remains of a Reptile" (1853).

at work in the natural order. They included "changes of climate & the action of one or many species on others." Geographical changes might alter climates and permit migrations of species that had not previously been possible. The geographical distribution of species was such that each must have originated at a single center, from which it had later spread. It was difficult to determine the exact causes, or combination of causes, that brought about the extinction of any given species. As an example, Lyell cited John Collins Warren's discussion of the mystery surrounding the extinction of the mastodon, which had lived in North America in very recent geological times.[39]

Lyell then took up the question of the introduction of new species. He discussed briefly the Lamarckian theory, stressing that it solved the question of the origin of species by denying their existence. The author of the *Vestiges,* who had argued for the transmutation of one species into another, seemed to think that there should be no need for the Supreme Being to intervene in nature each time the creation of a new species was required. "But," Lyell replied, "no act of power or forethought can well be imagined, more transcending that of a finite being, than the knowing whether any given species, having given attributes physical & instinctive, shall be capable of maintaining its ground for thousands of ages in the elemental strife to which it is to be exposed in many generations."[40]

Lyell was perhaps most seriously concerned in his lecture to oppose the views of catastrophist geologists on species. He noted that when it was thought that the Pyrenees had been elevated at the end of the Cretaceous period and before the Eocene, their sudden upthrow was alleged to coincide with a great change in species. Yet, when it was shown that the Pyrenees had been elevated after the Eocene period, the same geologists asserted that the event had put an end to the Eocene plants and animals. Lyell argued that if mountain ranges were elevated suddenly, the fossils in the strata at their summits should be identical with those in the horizontal strata of the lower country at their base, but such was almost never the case. The difference between the fossils of rock formations in mountains and those of the lower country was a measure of the long period of time required for the slow and gradual elevation of mountains.

39. Lyell, MS notes, "Lect. 10 Boston November 17, 1852, Extinction of Species," Kinnordy MSS. The notes were written on 17 November; the lecture was delivered on the nineteenth. Cf. Warren, *Mastodon.*

40. Ibid., 13.

Lyell mentioned the observations of Charles Darwin and Alcide d'Orbigny on the large-scale elevation of South America in the post-Pliocene period and added:

> I cannot point out the difference of the points of view from which two excellent naturalists may see the same geological phenomena better than by alluding to the opinion of Mr. Ch.[arles] Darwin & M. Alcide d'Orbigny on the origin of the Pampean mud. According to Mr. Darwin it is an estuary & delta formation in which certain brackish water shells & the carcasses of animals were gradually buried; according to d'Orbigny it was caused by the upheaval of the Cordillera of the Andes, an event which caused the destruction of the land animals inhabiting that portion of the Continent which had previously emerged.

Orbigny asserted that in eight years in South America he had never seen the carcass of a single animal floating down any of the great rivers. By contrast, the German travelers Johann von Spix and Carl von Martius had described many animals carried down the Amazon on rafts of tangled trees. Sir Woodbine Parish described the drowning in 1828 of hundreds of thousands of animals in the Parana River. "Such events may happen at intervals which we term rare," said Lyell, "& yet recur many times every century for thousands of ages in a geological period."[41] Lyell also cited the observation that in the delta of the Alabama River near Mobile each year many cattle foundered in the mud of a great morass near the lighthouse.

In his eleventh lecture Lyell discussed how new species were introduced to replace those that had become extinct. He cited three theories for the origin of new species: (1) the Lamarckian theory, (2) the evolution of one species directly out of a preexisting species, and (3) the independent creation of new species. The last theory involved the difficulty of imagining how a new species might come into being by Divine Creation, but the first two theories really only pushed that difficulty back to the first origin of life. There was a similar difficulty in imagining how one species might evolve out of another. "The greater or less probability of all such hypotheses must be determined by a careful consideration of facts," Lyell declared. "The mystery of the genesis is as great in every theory yet suggested."[42] Although no theory was satisfactory, Lyell thought that none

41. Ibid., 20–22.

42. Lyell, MS notes, "Boston. Nov. 23d. 1852. Mode of Introduction of New Species," 5, Kinnordy MSS.

of them should be treated with religious intolerance. Each should be considered in relation to the geographical distribution and fossil history of species.

On 26 November, Lyell delivered his twelfth and final Lowell lecture on the subject "Progressive Development." It was largely a restatement of views and arguments presented in his 1851 anniversary address, but Lyell spoke with perhaps a more intense eloquence. Of the evidence in favor of progression in the fossil record, he said, "I by no means deny that there is a semblance of reality in the facts as they appear in the present state of our science. If otherwise they would not have found so many eloquent advocates." One class of organisms might take the place occupied by another in an earlier geological period. Today the flowering plants occupied many stations (ecological niches) formerly occupied by ferns during the Carboniferous period, and today mammals filled places in nature once occupied by a great variety of reptiles. Yet higher organisms were continually being discovered in geological periods earlier than they were thought to have been present. "We have fish in the Silurian strata," said Lyell, "though I remember the time when not one had been discovered. We have new reptiles in the carboniferous though eight years ago none had been detected. We have the teeth of a mammifer in the trias & the jaws & teeth of three species in the oolite." He stressed that there was no geological evidence that the earth had by gradual stages become a habitation fit for man:

> You will have learnt from some of my former lectures, that I can recognize in the geological history of the globe from the Silurian period to the present no epoch of general convulsion, when the condition of the habitable world was rudely disturbed. A disposition to regard this state of things as unstable may be derived in some degree from the conviction that the globe was not tenanted by man, nor by any rational being holding the same relation to any extinct fauna & flora, which the human race now holds to contemporary animals & plants. Had the Earth been as quiet & as beautiful as now, of what use would it have been without a rational lord of creation. It is Eve's question in Paradise Lost, of what use are the stars,
>
> "Wherefore all night long shine these
> When sleep has closed all eyes"

The Archangel is made to answer:

> "These, though unbeheld, shine not in vain
> Men think tho' men were none
> That Heaven would want spectators, God want praise
> Millions of spiritual creatures walk the earth unseen."

> The same question may be asked of thousands of beautifully coloured corals & other zoophytes, crustacea & fish which people the depths of the sea unseen by man to whom the larger part of the habitable surface of the globe is unbeheld.[43]

The mingling of the theological with the geological may now seem strange, but it was not so to Lyell nor to his audience. The continuing strength of the catastrophic view of the history of the earth in the nineteenth century was its connection with theological views of the origin of the world and of man.

On 29 November, three days after the final Lowell lecture, the Lyells left Boston by train, seen off at the station by George Ticknor and his family. Two days later they sailed from New York on the Cunard steamship *Asia,* which landed at Liverpool on 12 December. The next day, they arrived back at 11 Harley Street.

The Lowell lectures had been stimulating but severely demanding for Lyell. On the last three lectures, he had worked especially hard, his whole time taken up with them. Mary wrote to Katherine that "Charles's lectures were most successful & well attended to the last. I never heard him keep up so well all the time." The Lyells' social life at Boston had been relatively restricted, although they usually dined at the Ticknors' or the Prescotts' on Sundays. The last week at Boston, William Prescott came every day to take Mary either for a walk or for a drive in his carriage, a kindness she appreciated greatly. There had grown up between the aging, nearly blind historian and Mary, livelier and happier in middle age than she had ever been, an unusually warm friendship. "He is the most unchanging person," she told Katherine, "always the same bright look, though a cloud sometimes comes over him when he speaks of those he had lost, which he often did to me."[44] The Lyells went home to a merry Christmas in old England, but Mary remained deeply attached to her friends in New England.

43. Lyell, MS notes, "Boston. Lect. 12, 1852, Progressive Development," 5–6, Lyell MSS.
44. Mary Lyell to Katherine Lyell, 3 December 1852, Kinnordy MSS.

New Light on the Joggins Fossils

After Christmas, Lyell went to the Isle of Wight to see a series of strata that Edward Forbes had discovered in the Eocene freshwater formation on the north side of the island. The strata added a new phase to the Eocene, lengthening the geological time represented by the period. It was just the sort of discovery that Lyell anticipated would occur again and again, and in which he delighted.

On Lyell's return from the Isle of Wight, he and Mary joined the Bunburys in a large family house party at Mildenhall to bring in the new year. They enjoyed a week's holiday with beautiful weather, and Lyell made geological excursions to quarries and gravel pits in the neighborhood. At the same time, he was writing a paper on the Joggins fossil reptile to present to the Geological Society.

At Jeffries Wyman's request, Lyell asked Richard Owen to examine the fossil bones from Joggins. Owen could compare them with bones of other animals, both fossil and living, in the extensive collections at the Hunterian Museum. He confirmed Wyman's main conclusion that the Joggins fossil was a gill-bearing reptile. He offered differing interpretations of some anatomical details and at Wyman's suggestion, added his own notes to Wyman's paper.[45] Owen showed the similarity of some of the skull bones of the larger Joggins fossil reptile to those of *Archegosaurus* and *Labyrinthodon.* Wyman and Owen concurred in naming the fossil *Dendrerpeton acadianum,* or "the forest reptile of Acadia," Acadia being the Indian name for Nova Scotia. When John Quekett examined sections of the fossil bones under the microscope at the Royal College of Surgeons, he found that they possessed the same general structure as those of living batrachian reptiles, the cells being particularly like those of *Menopoma* and *Menobranchus.*

The other fossil that both Wyman and Lyell had immediately suspected was a land shell proved more difficult to interpret. Even after Augustus Gould had separated the fossil from its stony matrix, he could not locate its mouth to determine its genus. Yet Gould thought the fossil was unlike any known seashell. At London, Lyell was able to show the fossil to Gerard Paul Deshayes, then visiting from Paris. Deshayes immediately recognized the fossil as a land shell, belonging to the same family as *Helix* or *Pupa.*

On 19 January, despite a bad cold, Lyell read his and Dawson's

45. Wyman, "Notes on the Reptilian Remains"; R. Owen, "Notes on the Above-Described Fossil Remains."

paper to the Geological Society. The announcement of the first fossil reptile discovered in the Coal of North America aroused great excitement. Charles Bunbury came up from Suffolk to hear it, staying a couple of nights at 11 Harley Street. After Lyell's paper, two geologists, Lovell Reeve and John Phillips, questioned whether the supposed land shell really was a shell. Phillips thought it looked like a fish's tooth. Lyell then gave the fossil to John Quekett for microscopic examination. When the striated surface of the fossil was magnified fifty times, Quekett found that it presented the same pattern as the surface of the common English land shell *Pupa juniperi.* Also, when Quekett made thin sections of the fossil and examined them under the microscope, he found the same prismatic or hexagonal appearance and tubular structure as occurred in living mollusc shells.

On 18 March, Lyell gave a Friday evening lecture at the Royal Institution on the Joggins fossils. He presented Quekett's conclusive microscopic evidence that the enigmatic little fossil from Joggins was in fact a land shell—the first land shell discovered in the Paleozoic. The Joggins fossils demonstrated how similar the forested swamps of the Carboniferous period were to those of the modern world. Lyell showed the Royal Institution audience a part of the stem of a freshwater reed that he had collected in 1846 at the Balize. When one of the mouths of the Mississippi became blocked for a time, several years before Lyell's visit, the lack of inflow of freshwater caused sea water to invade several acres of such reeds, killing them. The dead reeds remained standing and became covered with barnacles. When the fresh water returned a few years later, the barnacles also were killed, leaving the dead reeds covered with dead barnacles. If such frail reeds could remain standing for several years when dead, it was easy to imagine, said Lyell, how *Sigillaria* trees might have stood for many years in fresh or salt water in the ancient coal swamps, while silt collected around them. Lyell calculated that the Mississippi River would require more than 2 million years and the Ganges River more than 375 thousand years to bring down the quantity of sediment represented by the Nova Scotia coal beds.[46] The Royal Institution lecture was a great success. Profiting from his recent experience with the Lowell lectures at Boston, Lyell had attained greater poise and fluency as a speaker. Not once pausing to clear his throat or allowing his voice to drop to a whisper, he spoke

46. Lyell, "Fossil Reptilian Remains . . . Coal-Measures of Nova Scotia."

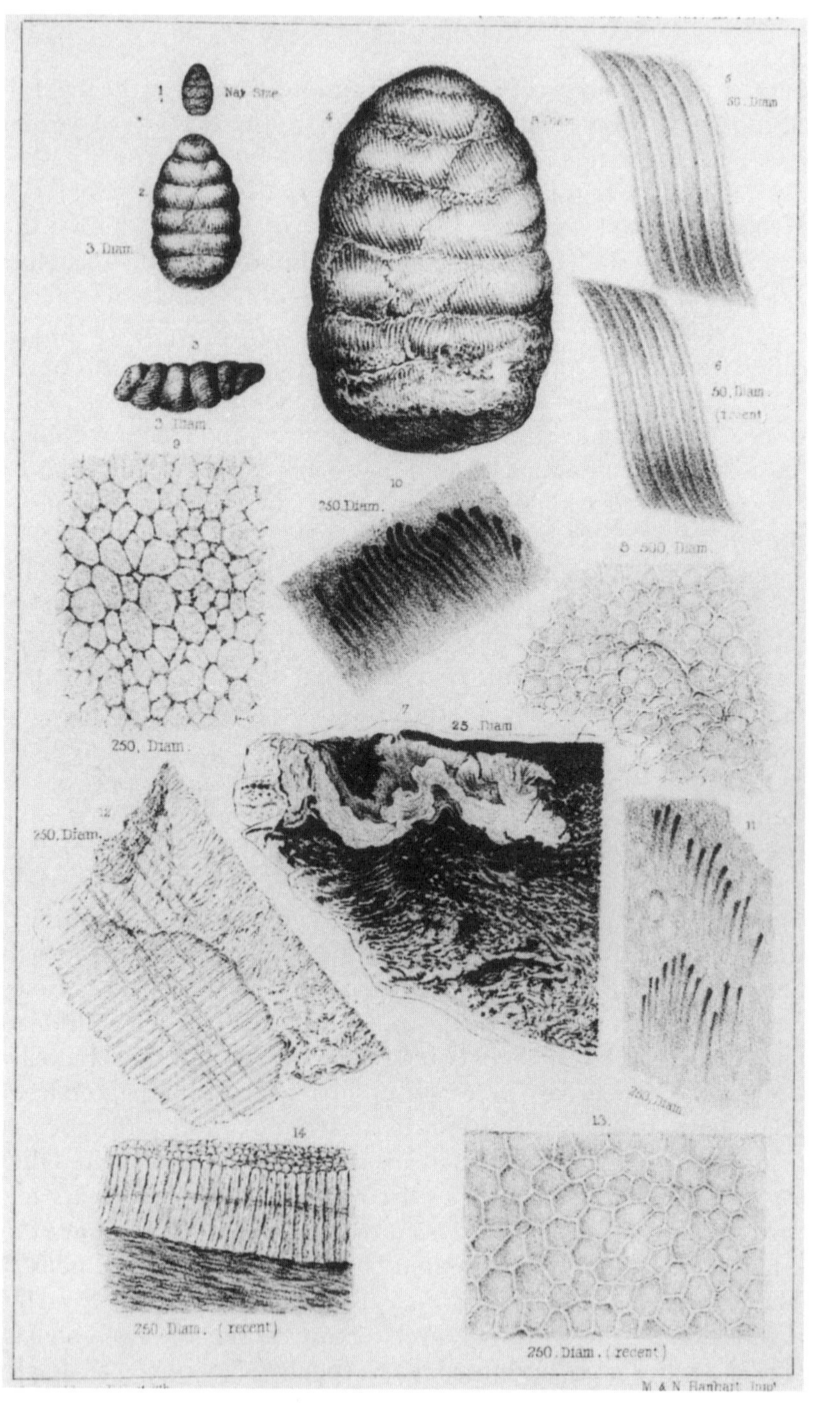

Carboniferous Land Shell. Joggins, Nova Scotia. Lyell and Dawson, "On the Remains of a Reptile" (1853).

clearly and distinctly to the end. The audience listened in rapt attention.

The day after the Royal Institution lecture, Lyell went to dine at the Duke of Somerset's house to meet William Gladstone, chancellor of the exchequer in the new government. They discussed education and the United States. Gladstone and Lyell had both been members of the commission for the 1851 exhibition, and both were interested in the educational schemes beginning to take shape with funds derived from the exhibition. They also talked about geology, "on which," Lyell wrote, "I gave him an elementary lecture, especially touching coal fields, as new to him as it would have been to any eminent Oxford scholar one or two centuries ago."[47] Gladstone listened so intently and asked such intelligent questions that with the time and inclination, Lyell thought he would soon learn the science.

During the spring of 1853, Lyell remained steadily at work on the ninth edition of the *Principles,* as he had been since the beginning of the year. The edition was to include important revisions, and Murray planned to print five thousand copies. As in earlier editions, Lyell tried to incorporate the latest geological information. The ninth edition, like the seventh and eighth, reflected his successive travels in America, particularly in his account of the valley and delta of the Mississippi River and his discussion of the conditions under which beds of coal had accumulated. Frequently, Lyell referred to Charles Darwin's observations on geology and natural history.

As he had in every edition, Lyell took pains to refute Léonce Élie de Beaumont's theory of the sudden elevation of parallel mountain chains. Lyell dealt particularly with the most recent modifications to the theory. In 1852 Élie de Beaumont recognized in the Pyrenees six, and possibly seven, episodes of dislocation that had occurred at different times.[48] Nevertheless, Élie de Beaumont thought that the most recent dislocation, by which the Pyrenees were elevated to their present height, had occurred suddenly between the end of the Cretaceous and the beginning of the Tertiary. Lyell showed that geological evidence disproved the sudden elevation of the Pyrenees. The Cretaceous strata on the sides of the Pyrenees were not necessarily the last deposited during the Cretaceous; nor were the Tertiary strata at their foot the first to have been deposited during

47. Lyell to Leonard Horner, 22 March 1853, Kinnordy MSS.

48. Lyell, *Principles,* 9th ed., 163–75; cf. Élie de Beaumont, "Système des montagnes." The article was also published separately as Élie de Beaumont, *Notice sur les systèmes de montagnes.*

the Tertiary. A long period of time separated the deposition of the two formations. In other parts of Europe, Cretaceous formations younger than those of the Pyrenees, and Eocene formations older than those at the foot of those mountains, indicated that the two formations were widely separated in time.

In the ninth edition of the *Principles,* Lyell expanded his discussion of the introduction of new species. He had no solution to offer, but the space he devoted to the question suggested its growing importance in geology and in his thinking. He suggested that each year one species might become extinct and one new species might appear, without being noticed among the many thousands of species both known and unknown. Such a rate of change would imply considerable instability in the living world and, if continued for many thousands of years, would produce a profound change in living forms. Lyell looked to the observations of naturalists to provide insight into the laws governing the appearance of new species. Recent geological formations, particularly those containing the remains of living species, could provide information as to whether new species had appeared successively or all at once, and whether man appeared as early as other animals now living on earth "or whether the human species is one of the most recent of the whole."[49]

Fourth Visit to America, 1853

In late April while Lyell was finishing the *Principles,* Lord Granville called on him to say that the government wanted to send commissioners to attend the industrial exhibition scheduled to open at New York City in June. Lord Ellesmere had offered to lead the commission, and the government would like Lyell to go as the second commissioner, both as a representative of science and as a person who might be considered a friend of the United States. The British government intended the commission not only as a complimentary gesture to the United States, but also as a reporting commission on American progress in manufacturing, shipbuilding, agricultural machinery, and other industrial activities.[50] The commissioners were also to study the United States Patent Office and American systems of industrial or technical education.

It was an invitation that Lyell could not refuse, although he was

49. Lyell, *Principles,* 9th ed., 704–8.

50. "Draft of Commission. Earl of Ellesmere, Sir Charles Lyell, C. Wentwork Dilke Esq., Professor Wilson, Joseph Whitworth Esq., George Wallis Esq. May 3d. 1853," Kinnordy MSS.

seriously concerned about finishing the *Principles* before he left. He agreed to receive his travel expenses but declined payment for his time, because he wanted to limit to two months the period that he would be out of England. Thus, during May, while Lyell was making final corrections to the *Principles,* he and Mary were in a flurry of preparations for their unexpected American trip. Mary decided to take her maid, Maria. She would have to appear frequently at official functions and could not by herself take care of her own clothes and look after Lyell. She also had to order light summer dresses for the hot weather that they would encounter in New York. "Here it is so cold," she wrote to Katherine, "I have all my winter clothing still on."[51]

Lord Ellesmere invited the Lyells to sail with him in a naval frigate on 11 May, but Lyell did not wish to take the time. He appreciated too much the speed and comfort of the Cunard steamships. On 28 May he and Mary sailed from Liverpool for Boston on the *Canada.* Bearing a royal commission, they traveled in unaccustomed state. They were given the captain's cabin, a spacious room comfortably furnished with a sofa and many drawers and lockers in which to put their belongings. They dined at the captain's table along with Colonel George Hughes, a topographical engineer who had served with the United States Army in the Mexican War of 1846. Hughes was returning from a tour of various European countries to arrange for manufactured articles to be sent to the New York Exhibition. A man of wide experience, he was interested in science and had attended meetings of the British Association, where he had met Mary's father, Leonard Horner.[52] Another passenger, with whom Lyell walked the deck a good deal, was the Reverend Calvin Stowe, professor of sacred literature at Andover Theological Seminary in Andover, Massachusetts. He was the husband of Harriet Beecher Stowe, whose novel *Uncle Tom's Cabin* was then the rage. More than a million copies of the book had been sold in England, and the Lyells had read it with great interest. Other Americans were on the boat as well. Their friendliness was in sharp contrast to the Lyells' earlier visits, when there was talk of war between Great Britain and the United States. One American remarked to Lyell about the New York Exhibition: "If your Queen had come, I guess we sh[oul]d have carpetted the streets for her."[53]

51. Mary Lyell to Katherine Lyell, 29 April–3 May 1853, Kinnordy MSS.

52. Mary Lyell to Leonard Horner, 2 June 1853, Kinnordy MSS.

53. Lyell, Notebook 183, 32, Kinnordy MSS.

During the voyage Lyell worked on revisions for a new edition of the *Manual.* He also spent some time conferring with Joseph Whitworth, another commissioner on board. Whitworth, a leading manufacturer of machine tools at Manchester, was a true mechanical genius. As a young man, he had devised a method for achieving true plane surfaces for the sliding parts in machines. He had created standards for a uniform system of screw threads and invented a fine measuring instrument, or micrometer. In 1851 various of Whitworth's tools were displayed at the Great Exhibition in London. Whitworth was also interested in education. He and Lyell planned how they might study methods of industrial education in the United States.

After a brief stop at Halifax, the *Canada* entered Boston harbor early on the morning of 8 June. Mary came out on the upper deck at six o'clock to see George Ticknor putting off in a small boat in pouring rain to come on board. During the next hour or so, while the ship was being maneuvered to the wharf, the Lyells heard news of their Boston friends. Ticknor had brought a carriage for them and, when they had got their luggage through customs, took them to the Tremont Hotel. A few minutes after he left them, Anna Ticknor came to greet them. When Mary went out for a walk, she met the Prescotts, who welcomed her back. Their arrival at Boston was like a homecoming, the whole atmosphere of the city sympathetic. Soon after they landed, the weather cleared to reveal Boston in the full beauty of June, the trees and lawns a rich green and fountains playing. "Here we are in this dear old place, as happy as possible," Mary wrote to her mother. "How I wish you all knew it, and then you would enter into the delight I feel at being here once more, & so kindly welcomed. Even the very servants at the Ticknors' & Prescotts' & here in this hotel & the very washerwoman expressed such hearty welcomes."[54] The day after their arrival, Lyell went to Cambridge to see Louis Agassiz, while at the Tremont Hotel Mary received a succession of visitors—Eliots, Winthrops, Lowells, Ticknors, Prescotts, and others. Mary's maid, Maria, liked Boston a great deal, spending much of her time at the Ticknors', where the servants treated her kindly. One evening the Lyells went with George Ticknor to see *Uncle Tom's Cabin* performed at the theater.

Soon after they arrived, Lyell received a note from the president of the New York Exhibition to say that the exhibition would not

54. Mary Lyell to Mrs. Leonard Horner, 9 June 1853, Kinnordy MSS.

open until 15 July. In view of the postponement, the Lyells had to revise their plans. Lyell decided to spend the time until 15 July doing as much work as possible on his report so that he would not be delayed long after the exhibition opened. Meanwhile, Joseph Whitworth, who wished to see as much as possible of American industry, went to tour the cotton mills at Lowell, Massachusetts, and returned much impressed.

Over the next several days, all of the British commissioners, including Lyell, went to New York, where the Clarendon Hotel became their headquarters. During the next week Lyell attempted to collect information about the geological displays to be included in the exhibition. He wrote to William Redfield to ask about the economic products of the principal geological formations, telling Redfield:

> I have also to speak of coast surveys, charts & maps & geographical works of that description. Our Government also wants to know about school books & school maps, geological & geographical, subjects in which they think you may be far ahead of them as there is so much done by the State in education; and so little with us, tho' much needed.[55]

On 18 June, the Lyells left New York for Washington, so that Lyell could learn about scientific work being carried out by the United States government. Over the weekend they stopped at Philadelphia to visit Joseph Leidy. Leidy was continuing his study of vertebrate fossils from the Nebraska territory and had just been appointed professor of zoology at the University of Pennsylvania.

At Washington, in the midst of summer heat, Lyell called on Lieutenant Matthew Fontayne Maury, superintendent of the Depot of Charts and Instruments in the Navy Department and of the Naval Observatory. In 1847 Maury began to exchange his wind and current charts with mariners for their abstract logs of winds and currents observed during their voyages. In 1851, to accompany the wind and current charts, he published sailing directions. They proved particularly valuable in shortening the time of sailing voyages, especially from the east coast of the United States around Cape Horn to San Francisco and to Australia and China. Maury's directions came at a time when the number of sailing ships and the lengths of their voyages were increasing rapidly and when, on some

55. Lyell to Redfield, 15 June 1853, Redfield MSS.

routes, sail was in sharp competition with steam. At the time of Lyell's visit, Maury was about to go to Brussels to attend an international congress of countries that had agreed to collect data on winds and ocean currents according to his plan.

Lyell also visited the Coast Survey Office to learn the methods used to prepare charts of the east coast of the United States. At the Smithsonian Institution he presented the director, Joseph Henry, with a copy of the paper on the Joggins fossil reptile. The British minister, John F. Crampton, held a dinner in the Lyells' honor at his house in Georgetown.[56] Mary's old Bonn friends, the Prussian minister von Gerolt and his wife, entertained the Lyells at their house in Georgetown. The party included various government scientists and several ambassadors. After an elegant dinner, which included soft-shell crabs from the Chesapeake Bay, they sat on the piazza in the warm evening air, watching the fireflies in the dark, the gentlemen smoking cigars.

The next morning, unable to sleep because of the heat and flies at their hotel, Mary rose at half past four and, taking Maria with her, walked to the top of Capitol Hill. They enjoyed the early morning view. "I showed her the pictures of various scenes in American history by Col. Trumbull & Washington's statue," Mary wrote. "It was quite warm when we returned at six." At nine o'clock that morning, Lyell went for a drive with President Franklin Pierce, with whom he had a pleasant talk. When Lyell asked the president why there were so few parks in American cities, Pierce suggested that their absence arose from narrow views of economy and a suspicion on the part of the people that parks might be used as political favors. Scientific appointments in the United States were good, Pierce explained, because it was against the interest of any official to appoint incompetent men to such posts. Political parties tended not to interfere in scientific appointments.[57]

That evening, the Lyells went by train to Philadelphia, traveling after dark to avoid the heat. The next morning they found that the weather had broken and the air was refreshingly cool. On 24 June, they went on to Burlington, New Jersey, to stay overnight with the McIlvaines. After so many hotels, they enjoyed being in a private home, so clean and quiet, surrounded by its garden overlooking the Delaware River. Mary wrote:

56. John Fiennes Crampton (1805–86) had served in the British diplomatic service since 1826 and at Washington since 1844. He was appointed minister in 1852.

57. Lyell, Notebook 183, 90, Kinnordy MSS.

> Maria got on famously with their coloured servants, and I was quite amused at seeing her seated between two coloured women on the bank of the river, all chatting merrily together. They ate all their meals together too which I was glad to see & [the] Miss McIlvaines and I agreed it was best to leave them to their own ways.[58]

While they were at the McIlvaines', Timothy Conrad came to see Lyell. They discussed fossils and the use of marls for agriculture in New Jersey and Maryland. On 25 June, the Lyells returned to New York.

On Sunday morning despite pouring rain, the Lyells went to the Church of the Messiah. They were immediately invited into a pew, "as is always the case here," Mary noted. In the evening they went to look at the outside of the Crystal Palace; it seemed to them much smaller than the one at London. At the Clarendon Hotel the next morning, large brushes of peacocks' feathers were waved over the breakfast table to clear away "those troublesome flies."[59]

The Lyells were still marking time until the exhibition opened, and because of the heat and the flies, remaining in the city was not pleasant. Lyell wished to visit the geographer Arnold Guyot, who was preparing a series of charts for the teaching of geography in Massachusetts schools.[60] On their way to Boston, they stopped for a few days at Newport, Rhode Island. Lyell absentmindedly left his watch on the boat. When he asked at Newport how he might send word to New York to try to recover it, a Negro shoeblack suggested that he use the electric telegraph. The white watchman had forgotten to mention such a possibility.[61]

On Independence Day, the Lyells went to the Prescotts' new house by the seashore at Lynn, Massachusetts, where they were guests for several days. In the mornings Lyell worked without interruption; in the afternoons he collected shells along the beach and studied erratic boulders, including a particularly large one known as the Ship Rock. Soon it was necessary to start back for New York. Lyell went on alone to New Haven to see James Dwight Dana and Benjamin Silliman, while Mary stayed with the Ticknors. William Prescott accompanied her on the train to New York to

58. Mary Lyell to Katherine Lyell, 22–28 June 1853, Kinnordy MSS.

59. Ibid.

60. Arnold Henry Guyot (1807–84). In 1854 Guyot was appointed professor of physical geography and geology at Princeton University. Under the auspices of the Smithsonian Institution, he played a leading role in the establishment of weather observation stations throughout the United States.

61. Lyell, Notebook 183, 108, Kinnordy MSS.

spend some days with the Lyells. He attended with them the opening of the New York Industrial Exhibition.

The opening of the Industrial Exhibition on 15 July was celebrated by a great dinner at which the president of the United States and the British commissioners were honored guests. As a result of Lord Ellesmere's taking ill, Lyell was obliged to act as chairman of the British commission. In responding to a toast to the foreign commissioners, Lyell said that they were much impressed by the many labor-saving inventions in American machinery. "To such inventions, far more than to the soil, or any other cause, they ascribe the great wealth which has in so short a period accumulated in this land." It was "a most cheering sight for a foreigner to witness." Yet, Lyell added, in his travels in the United States, he had never been allowed to feel a foreigner, even though on each previous visit the newspapers had been full of exciting disputes with Great Britain that seemed to threaten the peace between the two countries. Those of us "engaged in this Industrial Exhibition," said Lyell, "may regard ourselves as members of a great Peace Association."[62]

When the exhibition was open, the Lyells and Prescott went through it to look at the various displays. Although much smaller than the one at London, the Crystal Palace was attractive.[63] Some of the exhibits were beautiful. The British silversmiths had sent fine pieces, as had the china manufacturers. Many displays had not yet arrived or were not unpacked. The geological exhibits were especially incomplete, and Lyell had long since given up hope of writing his portion of the commissioners' report. As an alternative, he asked James Hall to prepare a report when the geological part of the exhibition was finally complete.

Lyell was anxious to get home to work on a new edition of the *Manual.* Having done all that he could do for the commission, on 27 July he and Mary sailed from New York on the Cunard steamer *Arabia.* Mary's maid, Maria, remained behind. She had intended to come to America anyway, and since Anna Ticknor needed a maid, she was happy to stay at Boston. The *Arabia,* a large modern vessel, was commanded by Captain Judkins, with whom the Lyells had returned from Halifax in 1842. On 6 August, they landed at Liverpool after a voyage of nine days, seventeen hours, and twenty-five minutes, returning to cool weather and leaden skies. It was their final crossing of the Atlantic.

62. Ibid.

63. Mary Lyell to Katherine Lyell, 13 August 1853, Kinnordy MSS.

Epilogue

IN THE TWELVE YEARS since Lyell first sailed for America in 1841, he and the world had changed profoundly. In 1841, regular trans-Atlantic steamship travel was just beginning; in both Britain and the United States, railroads were few and short. Most land travel was still by stagecoach. By 1853, railways formed a steadily spreading network throughout England, on the Continent, and in the United States. Trains were becoming faster and more comfortable. In the eastern United States, railroad lines were linked by ferry and steamboat to form a connected system of steam-powered transportation that extended from Maine to Georgia. On the western rivers, steamboats were becoming larger and more luxurious. Parallel with the railroad lines marched the poles carrying the electric telegraph wires, humming with messages passing from Halifax, Nova Scotia, to New Orleans and points between.

From North America, timber, wheat, cotton, maize, rice, and sugar poured into world commerce in an ever growing stream, carried in large part by sailing ships built in North American ports. For Europe, North America provided a steadily expanding market for manufactured goods. For dispossessed or unemployed people it provided a hopeful haven. With fascination and delight, Lyell observed the rapid development of the United States and Canada. He liked the people, and, usually, after some hesitation, they decided they liked him. Whatever reservations Americans occasionally may have felt about Lyell, they had none about his wife. Mary's courtesy and intelligence, her unfailing cheerfulness in the face of any and all hardships, her sure moral sense, and her warm heart charmed everyone from Negro slave to Boston Brahmin.

In 1841 Lyell came to America as the advocate of a radical viewpoint in geology, accepted by few among the older generation of geologists. By 1853 Lyell's theories were winning increasing acceptance. The discovery of vast areas of horizontal, undisturbed Paleozoic strata thousands of feet thick in Russia and in North America revealed that tranquil conditions had prevailed for long periods of time over large areas of the earth's surface in the ancient geological past. As James Hall and other geologists studied the fossils of the New York strata, they observed that particular kinds of marine organisms were associated with particular types of sediment and water depth in ancient seas, just as they were in modern seas. The Paleozoic seas were calm rather than chaotic, populated with corals, crinoids, crustaceans, molluscs, sea urchins, and starfish, differing often in genera and species but belonging to the same classes as those of modern waters. During Paleozoic times, volcanic activity and disturbance of strata were restricted to particular localities. Through the immense interval that separated the Paleozoic period from the present, the Silurian and Devonian strata extending some four hundred miles from the Hudson Valley to the Niagara River and on into Canada have remained undisturbed, unfolded, and unintruded by igneous rocks. In contrast to the prolonged calm that had reigned over large areas of North America and Europe throughout geological history, geologists had determined by the 1850s that great mountain ranges, such as the Alps, the Pyrenees, and the Andes, had been elevated during the Tertiary period. The continuance of mountain building through the Tertiary suggested that mountain building had occurred throughout the whole course of geological time. In the *Manual of Elementary Geology*, Lyell demonstrated the presence in every geological period of volcanic activity, which he associated with mountain building. By the 1850s geologists generally accepted Lyell's concept of metamorphic rocks and with it his analysis of the structure of mountain ranges.

Lyell came to America in part to learn whether his classification of European Tertiary formations would prove applicable to the American Tertiary. It did. The strata of the Atlantic coastal plain were Eocene, Miocene, and Pliocene, each with the same proportion of extinct to living species among their fossils as those of Europe. In America, Lyell learned much more. The cypress swamps of the Atlantic coastal plain and Mississippi Delta suggested to him the conditions under which coal had accumulated during the Carboniferous period. The carbonized trees in the bluff at Port Hudson on the lower Mississippi River demonstrated to him how the fossil trees had become embedded in the strata of the great cliff at Joggins, Nova Scotia. Between 1841 and 1853 geologists gained not only an understanding of the origin of coal, but also a new view of the conditions on the earth's surface during the Coal period. The discovery in Carboniferous strata of reptile footprints and, even more, the discovery of reptile skeletons showed that air-breathing land animals had lived in the ancient forests. Previously, geologists had thought that during the Coal period the earth's atmosphere contained so much carbon dioxide as to prevent animal life on land.

In America, Lyell was confronted more directly than ever with the extinction of species and the origin of new species to replace them. He had firsthand evidence of the recent extinction of several species of large mammals—the *Megatherium* and, most especially, the mastodon. It was a question he dealt with in 1852 in his Lowell lectures and one with which he would continue to wrestle. In April 1856, Charles Darwin would reveal to Lyell his theory of the origin of species by natural selection. Over the years, Lyell had readily accepted the evidence for an ever-older earth, a concept not always easy for others to embrace. For the next ten years Lyell would grapple with the implications of natural selection, especially for the origin of man.

Great as Lyell's scientific accomplishments were, his contributions to his country were also significant. As an adviser to Prince Albert, he played an important role in the reform of English universities. The Prince's education in American affairs from reading Lyell's accounts of the country and talking with him may have been of critical importance at the time of the Trent incident in the early stages of the American Civil War. On his deathbed in December 1861, Prince Albert softened Great Britain's note of protest to the United States to avoid a war that would have been disastrous for both countries.

During the years of the Civil War, when British opinion was intensely hostile to the United States, Sir Charles and Lady Lyell remained steadfastly loyal to America. In London they befriended the American ambassador, Charles Francis Adams, and his family when they were shunned by much of London society. On social occasions great and small, Sir Charles and Lady Lyell defended the Union. They felt keenly the agony of the conflict, for they had friends on both sides with whom they mourned the loss of sons and relatives.

During his trans-Atlantic years, Lyell was concerned primarily with paleontology, the correlation of American with European formations, and the imaginative reconstruction of the environments in which fossil organisms had lived. Although he would not make another trip to the United States after 1853, he continued to make long, strenuous geological field expeditions to Madeira and the Canary Islands, to Sicily, and through Germany and the Swiss and Austrian Alps. In Great Britain he traveled regularly to examine new findings by members of the Geological Survey, such as Andrew Ramsay's discovery of what he thought were signs of glaciation in the Permian period. Through the 1850s Lyell gave his attention mainly to volcanic geology, until in 1858 he utterly destroyed the theory of craters of elevation and, with it, the central pillar of geological catastrophism. He then turned his attention to the geological origin of man, a question that would haunt his thoughts to the end of his life.

Bibliography

A Note on Sources

Sources for the period of Charles Lyell's American travels and writings are extraordinarily extensive. In addition to the two accounts that he wrote, *Travels in North America, Canada, and Nova Scotia* and *A Second Visit to the United States,* each in two volumes, Lyell incorporated much of his findings on American geology into *A Manual of Elementary Geology* and into the seventh and subsequent editions of the *Principles of Geology.* At a time when the written letter was the main means of communication, he and Mary Lyell regularly wrote to members of their families. Their letters from America were especially long and detailed. Many of their travel letters, together with diaries, travel journals, and Lyell's scientific notebooks, are at Kinnordy House in Scotland, the seat of Lord Lyell of Kinnordy. The series of 266 notebooks that Lyell kept from the beginning of his scientific career in 1824 until the end of his life are a particularly valuable source. He recorded not only his scientific observations and reflections but also his impressions of the places he went and the people he met.

The other principal body of Lyell papers is in the University of Edinburgh Library, which holds more than two thousand letters sent to Lyell, most from scientists. They include letters from Americans and Canadians, who frequently sent him information and sometimes scientific specimens. One of Lyell's closest friends was Charles Darwin, whom he met in the autumn of 1836 just after Darwin's return from the voyage of H.M.S. *Beagle,* and with whom he corresponded regularly thereafter. A long series of Darwin's letters to Lyell (along with letters from a number of other scientists to Lyell) is in the American Philosophical Society Library in Philadelphia. By contrast, most of Lyell's many—and frequently long—letters to Darwin are missing. Their loss is unfortunate, because the few surviving letters, or copies of fragments of letters, show that in writing to Darwin, Lyell discussed scientific questions more fully and frankly than he did in his letters to anyone else. Lyell's letters to other individuals are usually among the papers of the person concerned, insofar as they survive. The more important series are those to Charles Babbage, James Hall, T. H. Huxley, Gideon Mantell, Roderick Murchison, John Murray, and William Whewell, to be found among these individuals' respective papers. Lyell's letters to Benjamin Silliman are scattered in various private collections.

Among printed sources, *Life, Letters, and Journals of Sir Charles Lyell, Bt.*, edited by Katherine M. Lyell, his sister-in-law, contains little on Lyell's four visits to America, because Lyell's own published accounts of his first two visits were so thorough. In addition to his travel writings, Lyell published more than thirty scientific papers on American geology, many of which include significant details about the places he visited.

Although Lyell took pains to acknowledge the hospitality and aid he received from Americans, frequently he did little more than name the individuals involved, and sometimes he left them anonymous. In seeking information about the many persons whom Lyell met in the United States and Canada, I used such standard biographical sources as the *Dictionary of American Biography*, the *Dictionary of National Biography*, and the *Dictionary of Scientific Biography*. George P. Merrill's *Contributions to the History of American Geology* was a valuable source of information about American geologists. Occasionally, I had to resort to regional biographical collections or local histories to learn something of more obscure individuals.

Manuscript Sources

Agassiz MSS. Harvard University, Houghton Library, Cambridge, Mass. The papers and correspondence to Louis Agassiz.

Babbage MSS. British Library, London. The papers and correspondence (1806–71) of Charles Babbage.

Brickenden MSS. British Museum (Natural History), General Library, London. The papers of Lambart Brickenden.

Darwin-Lyell MSS. American Philosophical Society, Philadelphia. Includes most of the surviving letters (1837–74) written by Charles Darwin to Sir Charles Lyell, plus many letters to Lyell from other scientists, particularly geologists.

De La Beche MSS. National Museum of Wales, Department of Geology, Cardiff. The papers and correspondence to Sir Henry De La Beche.

Everett Papers. Massachusetts Historical Society, Boston. The papers of Edward Everett.

Gibbes Papers. Library of Congress, Washington, D.C. The papers of Lewis R. Gibbes.

Hall MSS. New York State Library, Manuscripts and History Section, Albany. The papers and correspondence to James Hall.

Herschel MSS. University of Texas, Humanities Research Center, Austin. The papers and correspondence of Sir John F. W. Herschel.

Huxley MSS. University of London, Imperial College of Science and Technology, London. The papers of Thomas Henry Huxley.

Kinnordy MSS. Kinnordy House, Kirriemuir, Angus, Scotland. The letters, journals, and notebooks of Sir Charles Lyell, Lady Lyell, and other members of the family. Private collection of Lord Lyell of Kinnordy.

Lyell MSS. Edinburgh University Library, Edinburgh, Scotland. The papers and correspondence (1824–74) to Sir Charles Lyell.

Malcolm Lyell MSS. Dinton, Wiltshire. The Lyell family papers and letters. Private collection of Malcolm Lyell, Esq.

Mantell MSS. Alexander Turnbull Library, Wellington, New Zealand. The papers and correspondence (1801–52) of Gideon Algernon Mantell.

Morton MSS. The Library Company of Philadelphia, Philadelphia. The papers and correspondence of Samuel George Morton, M.D.

Murchison MSS. Geological Society of London Library, London. The papers and correspondence of Sir Roderick Impey Murchison.

Murray MSS. John Murray (Publishers), London. Includes Lyell's letters to John Murray. Private collection of John Murray.

Owen MSS. British Museum (Natural History), General Library, London. The papers and correspondence of Sir Richard Owen.

Peel MSS. British Library, London. The papers and correspondence of Sir Robert Peel.

Prescott MSS. Massachusetts Historical Society Library, Boston. The papers and correspondence of William Hickling Prescott.

Ravenel Collections. Charleston Museum, Charleston, S.C. The scientific collections and personal papers of Edmund Ravenel, M.D.

Redfield MSS. Yale University Library, New Haven, Conn. The papers and correspondence to William C. Redfield.

Royal Society MSS. Library of the Royal Society of London. Letters.

Silliman Family Papers. Yale University Library, New Haven, Conn. Includes correspondence to Benjamin Silliman.

Ticknor Papers. Dartmouth College Library, Hanover, N.H.

United States Military Academy Archives. United States Military Academy, West Point, N.Y.

University of Pennsylvania MSS. University of Pennsylvania Library, Philadelphia. Letters.

Wheatland MSS. Essex Institute, Salem, Mass. The correspondence to Henry Wheatland, M.D.

Whewell MSS. Trinity College Library, Cambridge, England. The papers and correspondence of Reverend William Whewell.

Windsor MSS. Royal Library, Windsor, England. Papers and letters. Property of Her Majesty the Queen.

Published Works, Primary and Secondary

Albritton, Claude C. *The Abyss of Time: Changing Conceptions of the Earth's Antiquity after the Sixteenth Century.* San Francisco, Calif.: Freeman, Cooper & Co., 1980.

Aldrich, Michele Alexis LaClergue. "New York Natural History Survey, 1836–1845." Ph.D. diss., University of Texas at Austin, 1974.

Alwyne, S. J., marquis of Northampton. "[Presidential address, 1848]." *Proceedings of the Royal Society of London* 5 (1851): 761–67.

Archiac de Saint Simon, Étienne Jules Adolphe Desmier (vicomte d'). "Essais sur la coordination des terrains tertiaires du nord de la France, de la Belgique et de l'Angleterre." *Bulletin de la Société géologique de France* 10 (1839): 168–225.

———. "Etudes sur la formation cretacée des vorsants sud-ouest, nord et nord-ouest du plateau centrale de la France." *Annales des sciences géologiques* 2 (1843): 121–43, 169–91.

———. *Histoire des progrès de la géologie de 1834 à 1859.* 9 vols. Paris: Société géologique de France, 1847–60.

Archiac de Saint Simon, Étienne Jules Adolphe Desmier (vicomte d'), and Edouard de Verneuil. "On the Fossils of the Older Deposits in the Rhenish Provinces, Preceded by a General Survey of the Fauna of the Palaeozoic Rocks, and Followed by a Tabular List of the Organic Remains of the Devonian System in Europe." *Transactions of the Geological Society of London,* 2d ser., 6 (1842): 303–410.

Association of American Geologists and Naturalists. *Proceedings and Transactions.* Boston: Gould, Kendall & Lincoln, 1843.

———. *Reports . . . .* Boston: Gould, Kendall & Lincoln, 1843.

Audubon, John James. *The Birds of America from Original Drawings.* 4 vols. London: By the author, 1827–38.

Audubon, John James, and John Bachman. *The Viviparous Quadrupeds of North America.* 3 vols. New York: J. J. Audubon, 1845–48.

Babbage, Charles. "Observations on the Temple of Serapis, at Pozzuoli, Near Naples, with Remarks on Certain Causes Which May Produce Geological Cycles of Great Extent." *Quarterly Journal of the Geological Society of London* 3 (1847): 186–217.

Barrande, Joachim. *Notice préliminaire sur le système silurien et les trilobites de Bohème.* Leipzig: By the author, 1846.

Bartholomew, Michael J. "Lyell and Evolution: An Account of Lyell's Response to the Prospect of an Evolutionary Ancestry for Man." *British Journal for the History of Science* 6 (1973): 261–303.

Bartram, William. *Travels through North and South Carolina, Georgia, East and West Florida . . . Containing an Account of the Soil and Natural Productions of Those Regions.* Philadelphia: James & Johnson, 1791.

Bayfield, H. W. "Notes on the Geology of the North Coast of the St. Lawrence." *Proceedings of the Geological Society of London* 2 (1833–38): 4–5.

———. "Notes on the Geology of the North Coast of the St. Lawrence." *Transactions of the Geological Society of London,* 2d ser., 5 (1837): 89–102.

———. "Outlines of the Geology of Lake Superior." *Transactions of the Literary and Historical Society of Quebec* 1 (1829): 1–43.

———. "Remarks on Coral Animals in the Gulf of St. Lawrence." *Transactions of the Literary and Historical Society of Quebec* 2 (1831): 107.

Benton, Michael J. "Progressionism in the 1850s: Lyell, Owen, Mantell and the Elgin Fossil Reptile *Leptopleuron* (Telerpeton)." *Archives of Natural History* 11 (1982): 123–36.

Bigelow, Jacob. *Florula Bostoniensis: A Collection of Plants of Boston and Its Vicinity* . . . . 2d ed. Boston: Cummings, Hilliard & Co., 1824.

———. "Some Account of the White Mountains of New Hampshire." *New England Journal of Medicine and Surgery* 5 (1816): 321–38.

Binney, E. W. "Description of the Dukinfield Sigillaria." *Quarterly Journal of the Geological Society of London* 2 (1846): 390–93.

Bonney, T. G. *Annals of the Philosophical Club of the Royal Society.* London: Macmillan & Co., 1919.

Boué, Ami. "Ueber die Nummuliten-Ablagerungen." *Berichte über die Mittheilungen von Freunden der Naturwissenschaften in Wien* 3 (1847): 446–69.

Bowler, Peter J. *Fossils and Progress: Paleontology and the Idea of Progressive Evolution in the Nineteenth Century.* New York: Science History Publications, 1976.

Brickenden, Lambart. "Notice of the Discovery of Reptilian Foot-Tracks and Remains in the Old Red or Devonian Strata of Moray." *Quarterly Journal of the Geological Society of London* 8 (1852): 97–100.

Brongniart, Adolphe. *Tableau des genres de végétaux fossiles considérés sous le point de vue de leur classification botanique et de leur distribution géologique.* Paris: L. Martinet, 1849.

Brongniart, Alexandre. *Histoire naturelle des crustacés fossiles, sous les rapports zoologiques et géologiques. Savoir: Les Trilobites.* Paris: F. G. Levrault, 1822.

———. "Sur les caractères zoologiques des formations, avec l'application de ces caractères à la détermination de quelques terrains de craie." *Annales des Mines* 6 (1821): 537–72.

Bronn, Heinrich Georg. *Index palaeontologicus, oder Uebersicht der bis jetzt bekannten fossilen Organismen* . . . . 2 vols. in 3. Stuttgart: E. Schweizerbart, 1848–49.

Brown, Bangrell W. "The First Hundred Years of Geology in Mississippi." *Southern Quarterly* 13 (1975): 295–302.

Brown, Richard. "Description of an Upright Lepidodendron with Stigmaria Roots, in the Roof of the Sydney Main Coal, in the Island of Cape Breton." *Quarterly Journal of the Geological Society of London* 4 (1848): 46–50.

———. "Description of Erect Sigillariae with Conical Tap Roots, Found in the Roof of the Sydney Main Coal, in the Island of Cape Breton." *Quarterly Journal of the Geological Society of London* 4 (1848): 354–60.

Browne, Janet. *Charles Darwin*. Vol. 1, *Voyaging*. New York: Alfred A. Knopf, 1995.

Buckland, William. *Geology and Mineralogy Considered with Reference to Natural Theology*. 2 vols. London: William Pickering, 1837.

Buckley, S. B. "On the Zeuglodon Remains of Alabama." *American Journal of Science*, 2d ser., 2 (1846): 125–29.

Bunbury, Charles James Fox (Sir). *Life, Letters and Journals of Sir Charles J. F. Bunbury*. Edited by Frances Joanna Bunbury. 3 vols. [London: Privately Printed, 1894].

Campbell, Thomas. *The Pleasures of Hope; in Two Parts: With Other Poems*. Edinburgh: Mundell & Son, 1799.

Carpenter, W. M. "Account of the Bitumenization of Wood in the Human Era, in a Letter to Prof. Silliman. . . ." *American Journal of Science* 36 (1839): 118–24.

———. "Geological Survey of Louisiana." *American Journal of Science* 42 (1842): 390–91.

———. "Interesting Fossils Found in Louisiana." *American Journal of Science* 34 (1838): 201–3.

———. "Miscellaneous Notices in Opelousas, Attakopas, etc." *American Journal of Science* 35 (1839): 344–46.

———. "Remarks on Some Fossil Bones Recently Brought to New Orleans from Tennessee and from Texas." *American Journal of Science*, 2d ser., 1 (1846): 244–54.

Carson, James Petigru. *Life, Letters and Speeches of James Louis Petigru*. Washington, D.C.: W. H. Lowdermilk & Co., 1920.

[Chambers, Robert]. *Vestiges of the Natural History of Creation*. London: J. Churchill, 1844.

Charlevoix, Pierre François Xavier de. *Histoire et description générale de la Nouvelle France: avec le journal historique d'un voyage fait par ordre du roi dans l'Amerique septentrionnale . . . .* 6 vols. Paris, 1744.

Clark, John M. *James Hall of Albany, Geologist and Palaeontologist, 1811–1898*. Albany, N.Y.: Privately printed, 1923.

Clark, Willis G. *History of Education in Alabama, 1702–1889*. Washington, D.C.: Government Printing Office, 1889.

Cock, R. S. "William M. Carpenter, a Pioneer Scientist of Louisiana." *Tulane Graduates Magazine* (1914): 1–8.

Conrad, Timothy A. *American Marine Conchology; or, Descriptions and Coloured Figures of the Shells of the Atlantic Coast of North America*. Philadelphia: By the author, 1831.

———. "Descriptions of New Tertiary Fossils from the Southern States." *Journal of the Academy of Natural Sciences of Philadelphia* 7 (1834): 130–57.

———. "New Fossil Shells from N. Carolina." *American Journal of Science* 39 (1840): 387–88.

———. "New Species of Fossil Shells in the Medial Tertiary Deposits of

Calvert Cliffs, Maryland." *Proceedings of the Academy of Natural Sciences of Philadelphia* 1 (1843): 28–33.

———. "Observations on the Tertiary and More Recent Formations of a Portion of the Southern States." *Proceedings of the Academy of Natural Sciences of Philadelphia* 7 (1834): 116–29.

———. "Observations on the Tertiary Strata of the Atlantic Coast." *American Journal of Science* 28 (1835): 104–11, 280–82.

———. "On the Geology and Organic Remains of a Part of the Peninsula of Maryland." *Journal of the Academy of Natural Sciences of Philadelphia* 6 (1830): 205–30.

———. "On the Silurian System, with a Table of the Strata and Characteristic Fossils." *American Journal of Science* 38 (1840): 86–93.

Conybeare, William D., and William Phillips. *Outlines of the Geology of England and Wales.* London: William Phillips, 1822.

Culin, Stewart. "The Dickeson Collection of American Antiquities." *Bulletin of the Free Museum of Science and Art* (Pennsylvania University) 2 (1900): 113–33.

Cuvier, Georges, and Alexandre Brongniart. *Essai sur la géographie minéralogique des environs de Paris, avec une carte géognostique, et des coupes de terrain.* Paris: Baudouin, 1811.

Dana, James Dwight. *Geology.* Vol. 10 in United States Exploring Expedition (1838–1842) *United States Exploring Expedition . . . .* New York: G. P. Putnam, 1849.

Darwin, Charles Robert. *The Correspondence of Charles Darwin.* Edited by Frederick Burkhardt and Sydney Smith. 10 vols. to date. Cambridge: Cambridge University Press, 1985–.

———. *Journal of Researches into the Natural History and Geology of the Countries Visited during the Voyage of H.M.S.* Beagle. London: J. Murray, 1845.

———. "On the Connexion of Certain Volcanic Phenomena in South America; and on the Formation of Mountain Chains and Volcanos, as the Effect of the Same Power by Which Continents Are Elevated." *Transactions of the Geological Society of London,* 2d ser., pt. 3, 10 (1840): 601–31.

———. *The Life and Letters of Charles Darwin.* Edited by Francis Darwin. 3 vols. London: John Murray, 1888.

———. *The Structure and Distribution of Coral Reefs.* London: Smith, Elder, 1842.

Daubeny, Charles. *Description of Active and Extinct Volcanoes.* London: W. Phillips; Oxford: Joseph Parker, 1826.

Davis, Jefferson. *The Papers of Jefferson Davis.* Edited by Haskell M. Monroe Jr. and James T. McIntosh. 8 vols. Baton Rouge: Louisiana State University Press, 1971–95.

Dawson, John William (Sir). *Acadian Geology: An Account of the Geological Structure and Mineral Resources of Nova Scotia, and Portions of the Neigh-*

bouring Provinces of British America. Edinburgh: Oliver and Boyd; London: Simpkin, Marshall & Co., 1855.

———. *Fifty Years of Work in Canada, Scientific and Educational.* Edited by Rankine Dawson. London: Ballantyne, Hanson & Co., 1901.

———. "On the Albert Mine, Hillsborough, New Brunswick." *Quarterly Journal of the Geological Society of London* 9 (1853): 107–14.

———. "On the Coal Measures of the South Joggins, Nova Scotia." *Quarterly Journal of the Geological Society of London* 9 (1853): 1–42.

Dean, Dennis R. "Graham Island, Charles Lyell, and the Craters of Elevation Controversy." *Isis* 71 (1980): 571–88.

De La Beche, Henry Thomas. "On the Geology of Tor and Babbacombe Bays, Devon." *Proceedings of the Geological Society of London* [1827] 1 (1829–33): 31–331; 2 (1833–38): 161–70.

———. *Report on the Geology of Cornwall, Devon, and West Somerset.* London: Longman, Orme, Brown, Green and Longmans, 1839.

———. *Researches in Theoretical Geology.* 1834. Reprint, with a preface and notes by Edward Hitchcock. New York: F. J. Huntington & Co., 1837.

Dick, Everett. *William Miller and the Advent Crisis, 1831–1844.* Berrien Springs, Mich.: Andrews University Press, 1994.

Dickens, Charles. *American Notes for General Circulation.* 4th ed. 2 vols. London: Chapman & Hall, 1842.

Dickerson [*sic*], M. W. "On the Geology of the Natchez Bluffs." In American Association of Geologists and Naturalists, *Abstract of the Proceedings of the Sixth Annual Meeting,* pp. 77–79. New Haven, Conn., 1845.

Dott, Robert H. "Lyell in America—His Lectures, Field Work and Mutual Influences 1841–1852." *Earth Sciences History* 15 (1996): 101–140.

Duffy, John. *The Tulane University Medical Center: One Hundred and Fifty Years of Medical Education.* Baton Rouge: Louisiana State University Press, 1984.

———, ed. *The Rudolph Matas History of Medicine in Louisiana.* 2 vols. Baton Rouge: Louisiana State University Press, 1958.

Dumont, André. *Carte géologique de la Belgique et des contrées voisines représentant les terrains qui se trouvent au-dessous du Limon Hesbayen et du Sable Campinien.* Brussels, 1849.

Eaton, Amos. *A Geological and Agricultural Survey of the District Adjoining the Erie Canal in the State of New York.* Albany, N.Y., 1824.

———. *Geological Text-Book for Aiding the Study of North American Geology.* 2d ed. Albany, N.Y.: Websters & Skinners, 1832.

———. *An Index to the Geology of the Northern States.* Albany, N.Y., 1818.

Egerton, P. de M. G. (Sir). "Note on the Fossil Fish from Albert Mine." *Quarterly Journal of the Geological Society of London* 9 (1853): 115.

Élie de Beaumont, Jean Baptiste Armand Louis Léonce. *Notice sur les systèmes de montagnes.* Paris: P. Bertrand, 1852.

———. "Recherches sur la structure et l'origine du Mont Etna." *Annales des Mines* 9 (1836): 175–216, 575–630; 10 (1836): 351–70, 507–76.

———. "Sur les émanations volcaniques et metallifères." *Bulletin de la Société géologique de France,* 2d ser., 4 (1846): 1249–333.

———. "Sur quelques points de la question des cratères de soulèvement." *Bulletin de la Société géologique de France* 4 (1833–34): 225–91.

———. "Système des montagnes." Vol. 12 of *Dictionnaire universel d'histoire naturelle . . . ,* edited by Charles d'Orbigny. Paris: Renard, Martinet et cie, 1849.

Elliott, Stephen. *Sketch of the Botany of South Carolina and Georgia.* 2 vols. Charleston, S.C.: J. R. Schenck, 1821–24.

Emblen, D. L. *Peter Mark Roget: The Word and the Man.* New York: Thomas Y. Crowell, 1970.

Emmons, Ebenezer. *Comprising the Survey of the Second Geological District.* Pt. 2 of *Geology of New York,* by New York (State) Natural History Survey. Albany, N.Y.: W. & A. White & J. Visscher, 1842.

Expédition scientifique de Morée, 1831–1838. *Expédition scientifique de Morée . . . .* Vol. 2, pt. 2, *Géologie et minéralogie,* by Émile Le Puillon de Boblaye and Théodore Virlet d'Aoust. Paris: F. G. Levrault, 1833.

Faraday, M., and C. Lyell. *Report . . . on the Subject of the Explosion at the Haswell Collieries and on the Means of Preventing Similar Accidents.* London: Her Majesty's Stationery Office, 1844.

Faujas-de-St.-Fond, cit. (Barthelemy). *Histoire naturelle de la montagne de Saint Pierre de Maestricht.* 2 vols. Paris: H. J. Jansen, 1799.

Fergusson, Charles Bruce. *"The Old King Is Back": Amos "King" Seaman and His Diary.* Bulletin 23. Halifax: Public Archives of Nova Scotia, 1972.

Finch, John. "Geological Essay on the Tertiary Formations in America." *American Journal of Science* 7 (1824): 31–43.

Fisher, George P. *Life of Benjamin Silliman, M.D., LL.D., Late Professor of Chemistry, Mineralogy, and Geology at Yale College.* 2 vols. New York: Charles Scribner & Co., 1866.

Forbes, Edward. "On the Succession of Strata and Distribution of Organic Remains in the Dorsetshire Purbecks." British Association for the Advancement of Science, London. *Reports* 20, pt. 2 (1850): 79–81.

———. "Report on the Investigation of British Marine Zoology by Means of the Dredge. Part I. The Infralittoral Distribution of Marine Invertebrata on the Southern, Western and Northern Coasts of Great Britain." British Association for the Advancement of Science, London. *Reports* 20, pt. 2 (1850): 192–263.

Fossier, A. E. "History of Medical Education in New Orleans from Its Birth to the Civil War." *Annals of Medical History,* n.s., 6 (1934): 320–52, 427–47.

Fourier, [J. B. J.] "Extrait d'un memoire sur le refroidissement seculaire du globe terrestre." *Annales de Chemie et de Physique* 13 (1820): 418–28.

Fulton, John F., and Elizabeth H. Thomson. *Benjamin Silliman 1779–1864: Pathfinder in American Science.* New York: Henry Schuman, 1947.

Geikie, Archibald (Sir). *Memoir of Sir Andrew Crombie Ramsay.* London: Macmillan & Co., 1895.

Gerstner, Patsy A. *Henry Darwin Rogers, 1808–1866, American Geologist.* Tuscaloosa: University of Alabama Press, 1994.

———. "The Influence of Samuel George Morton on American Geology." In *Beyond History of Science: Essays in Honor of Robert E. Schofield,* edited by Elizabeth Garber, 126–36. Bethlehem, Pa.: Lehigh University Press, 1990.

———. "Vertebrate Palaeontology, an Early Nineteenth Century Transatlantic Science." *Journal of the History of Biology* 3 (1970): 137–48.

Gesner, Abraham. *Remarks on the Geology and Mineralogy of Nova Scotia.* Halifax: Gossipand Coade, 1836.

Gillispie, Charles Coulston. *Genesis and Geology: A Study in the Relations of Scientific Thought, Natural Theology, and Social Opinion in Great Britain, 1790–1850.* Cambridge, Mass.: Harvard University Press, 1951.

Goldfuss, Georg August. "Apateon pedestris, aus der Steinkohlen formation von Münsterappel." *Palaeontographica* 1 (1851): 153–54.

———. *Beitrage zur vorweltlichen Fauna des Steinkohlengebirges.* Bonn: Henry & Cohen, 1847.

———. "Über das alteste der mit Bestimmt heit erkannten Reptilien, einen Krokodilier und einige neue fossile Fische aus der Steinkohlen-Formation." *Neues Jahrbuch für Mineralogie, Geognosie, Geologie und Petrefakten-kunde* 15 (1847): 400–404.

Goodrich, Samuel Griswold. *The Tales of Peter Parley about America.* Boston: S. G. Goodrich, 1827.

Gould, Stephen Jay. *Time's Arrow, Time's Cycle: Myth and Metaphor in the Discovery of Geological Time.* Cambridge, Mass.: Harvard University Press, 1987.

Gray, Asa. "Obituary Notice of William Oakes." *American Journal of Science,* 2d ser., 7 (1849): 138–42.

Great Britain. *Memoirs of the Geological Survey of Great Britain and of the Museum of Economic Geology in London.* London: Her Majesty's Stationery Office, 1846.

Grove, William Robert. *The Correlation of Physical Forces.* London: Managers of the London Institution, 1846.

Hall, James. *Comprising the Survey of the Fourth Geological District.* Pt. 4 of *Geology of New York,* by New York (State) Natural History Survey. Albany, N.Y.: Carroll & Cook, 1843.

Hall, Margaret Hunter. *The Aristocratic Journey, Being the Outspoken Letters of Mrs. Basil Hall Written during a Fourteen Month's Sojourn in America 1827–1828.* New York: G. P. Putnam's Sons, 1931.

Hall, Marie Boas. *All Scientists Now: The Royal Society in the Nineteenth Century*. Cambridge: Cambridge University Press, 1984.

Hamilton, William Thomas. *The Duties of Masters and Slaves Respectively: Or, Domestic Servitude as Sanctioned by the Bible: A Discourse, Delivered in the Government-Street Church, Mobile, Ala. . . . on Sunday night, December 15, 1844.* Mobile, Ala.: F. A. Brooks, 1845.

Harlan, Richard. "Description of the Remains of the 'Basilosaurus,' a Large Fossil Marine Animal, Recently Discovered in the Horizontal Limestone of Alabama." *Transactions of the Geological Society of Pennsylvania* 1 (1835): 348–57.

———. "Notice of Fossil Bones Found in the Tertiary Formation of the State of Louisiana." *Transactions of the American Philosophical Society,* 2d ser., 4 (1834): 397–403.

———. "On the Discovery of the Remains of the Basilosaurus or Zeuglodon." *Transactions of the Geological Society of London,* 2d ser., 6 (1841): 67–68.

Hayes, J. L. "Probable Influence of Icebergs upon Drift." *American Journal of Science* 45 (1843): 316–27.

Heilprin, Angelo. *Contributions to the Tertiary Geology and Palaeontology of the United States.* Philadelphia: By the author, 1884.

Hennig, Helen Kohn. *Great South Carolinians from Colonial Days to the Confederate War.* 2 vols. Chapel Hill: University of North Carolina Press, 1940.

Herrick, Francis Hobart. *Audubon the Naturalist: A History of His Life and Time.* 2 vols. New York: D. Appleton & Co., 1917.

Hitchcock, Edward. "Description of a Singular Case of Dispersion of Blocks of Stone Connected with Drift in Berkshire County, Massachusetts." *American Journal of Science* 49 (1845): 258–65.

———. *Final Report on the Geology of Massachusetts.* 2 vols. Amherst, Mass.: J. S. & C. Adams; Northampton, Mass.: J. H. Butler, 1841.

Hodge, James T. "Observations on the Secondary and Tertiary Formations of the Southern Atlantic States. With an appendix by T. A. Conrad." In *Reports . . . ,* by the Association of American Geologists and Naturalists, 94–111. Boston: Gould, Kendall & Lincoln, 1843.

Hodge, M. J. S. "The Universal Gestation of Nature: Chambers' *Vestiges* and *Explanations.*" *Journal of the History of Biology* 5 (1972): 127–51.

Hodgson, William B. *Memoir on the Megatherium and Other Extinct Gigantic Quadrupeds of the Coast of Georgia, with Observations on Its Geologic Features.* New York: Bartlett & Welford, 1846.

Holbrook, John Edwards. *North American Herpetology: Or a Description of the Reptiles Inhabiting the United States.* 5 vols. Philadelphia: J. Dobson, 1842.

Holmes, Oliver Wendell. "John Barnard Swett Jackson." *Boston Medical and Surgical Journal* 100 (1879): 63–66.

Horsman, Reginald. *Josiah Nott of Mobile: Southerner, Physician, and Racial Theorist.* Baton Rouge: Louisiana State University Press, 1987.

Irving, John B. *A Day on Cooper River.* 3d ed. Edited by Louisa Cheves Stoney. Notes by Samuel Gaillard Stoney. Columbia, S.C.: R. L. Bryan Co., 1969.

Jackson, J. B. S. "On the Dentition of the Mastodon." *Proceedings of the Boston Society of Natural History* 2 (1845–48): 140–41.

———. "On the Fossil Bones of Mastodon." *Proceedings of the Boston Society of Natural History* 2 (1845–48): 60–62.

Kellogg, Remington. *A Review of the Archaeoceti.* Washington, D.C.: Carnegie Institution of Washington, 1936.

Kemble, Frances Ann. *Journal of Residence on a Georgia Plantation in 1838–1839.* Edited by John A. Scott. New York: Alfred A. Knopf, 1961.

Kerr, D. G. G. *Sir Edmund Head, a Scholarly Governor.* Toronto: University of Toronto Press, 1954.

King, Alfred T. "Description of Fossil Foot Marks, Supposed to Be Referable to the Classes Birds, Reptilia and Mammalia, Found in the Carboniferous Series, in Westmoreland County, Pennsylvania." *Proceedings of the Academy of Natural Sciences of Philadelphia* 2 (1844–45): 175–80.

Kirtland, J. P. "Observations on the Sexual Characters of the Animals Belonging to Lamarck's Family of Naiades." *American Journal of Science* 26 (1834): 117–20.

Koch, Albert Carl. *Description of the* Hydrargos Sillimani *(Koch): A Gigantic Fossil Reptile or Sea Serpent: Lately Discovered by the Author in the State of Alabama, March 1845* . . . . New York: [By the author], 1845.

———. *Description of the Missourium, or Missouri Leviathan . . . ; Also Comparisons of the Whale, Crocodile and Missourium with the Leviathan, as Described in the 41st Chapter of the Book of Job.* Louisville, Ky.: Prentice & Weissinger, 1841.

———. *Journey through Part of the United States of North America in the Years 1844 to 1846.* Translated and edited by Ernst A. Stadler. Carbondale, Ill.: Southern Illinois University Press, 1972.

———. "Remains of the Mastodon of Missouri." *American Journal of Science* 37 (1839): 191–92.

Kohlstedt, Sally Gregory. *The Formation of the American Scientific Community: The American Association for the Advancement of Science, 1848–60.* Urbana: University of Illinois Press, 1976.

LaBorde, M. *History of the South Carolina College.* Columbia, S.C.: Peter B. Glass, 1859.

Laudan, Rachel. *From Mineralogy to Geology: The Foundations of Science, 1650–1830.* Chicago: University of Chicago Press, 1987.

Lea, Isaac. *Contributions to Geology.* Philadelphia: Cary, Lea & Blanchard, 1833.

———. *Observations on the Genus* Unio, *Together with Descriptions of New Genera and Species* . . . . 13 vols. Philadelphia: By the author, 1832–74.

LeConte, John. "Experiments Illustrating the Seat of Volition in the Alligator." *New York Journal of Medicine and the Collateral Sciences* 5 (1845): 335–47.

Legget, Robert F. "Thomas Roy (?–1842): An Early Engineer-Geologist." *Geoscience Canada* 3 (1976): 126–27.

Leidy, Joseph. "Description of the Remains of Extinct Mammalia and Chelonia, from Nebraska Territory." In *Report of a Geological Survey*, edited by David Dale Owen, 539–72. Philadelphia: Lippincott, Grampo & Co., 1852.

Logan, William. "On the Character of the Beds of Clay Lying Immediately Below the Coal Seams of South Wales, and on the Occurrence of Coal Boulders in the Pennant Grit of That District." *Proceedings of the Geological Society of London* 3 (1838–42): 275–77.

———. "On the Coal-Fields of Pennsylvania and Nova Scotia." *Proceedings of the Geological Society of London* 3 (1838–42): 707–12.

———. "On the Occurrence of a Track and Foot-Prints of an Animal in the Potsdam Sandstone of Lower Canada." *Quarterly Journal of the Geological Society of London* 7 (1851): 247–50.

Lonsdale, William. "Notes on the Age of the Limestones of South Devonshire." *Transactions of the Geological Society of London,* 2d ser., 5 (1840): 721–38.

———. "On the Age of the Limestones of South Devon." *Proceedings of the Geological Society of London* 3 (1838–42): 281–86.

Lurie, Edward. *Louis Agassiz, a Life in Science.* Chicago: University of Chicago Press, 1960.

Lyell, Charles (Sir). *Address Delivered at the Anniversary Meeting of the Geological Society of London, on the 21st of February, 1851.* London: Richard Taylor, 1851. Also published in *Edinburgh New Philosophical Journal* 51 (1851): 1–31, 213–26.

———. "Anniversary Address of the President." *Quarterly Journal of the Geological Society of London* 6 (1850): xxvii–lxvi.

———. "Coal Field of Tuscaloosa, Alabama, Being an Extract of a Letter to Prof. Silliman, from Charles Lyell, Esq., dated Mobile, Alabama, Feb. 19th, 1846." *American Journal of Science,* 2d ser., 1 (1846): 371–76. Also published in *Quarterly Journal of the Geological Society of London* 2 (1846): 278–82.

———. *Eight Lectures on Geology Delivered at the Broadway Tabernacle in the City of New York.* New York: Greeley & McElrath, Tribune Office, 1842.

———. *Elements of Geology.* London: John Murray, 1838.

———. *Elements of Geology.* 2d ed., 2 vols. London: John Murray, 1841.

———. *A Manual of Elementary Geology.* London: John Murray, 1851.

———. *A Manual of Elementary Geology.* 4th ed. London: John Murray, 1852.

———. "A Memoir on the Recession of the Falls of Niagara." *Proceedings of the Geological Society of London* 3 (1838–42): 595–602.

———. "Notes on Some Recent Foot-Prints on Red Mud in Nova Scotia, Collected by W. B. Webster of Kentville." *Quarterly Journal of the Geological Society of London* 5 (1849): 344.

———. "Notes on the Cretaceous Strata of New Jersey, and Other Parts of the United States Bordering the Atlantic." *Quarterly Journal of the Geological Society of London* 1 (1845): 55–60.

———. "Notes on the Silurian Strata in the Neighbourhood of Christiania, in Norway." *Proceedings of the Geological Society of London* 3 (1838–42): 465–67.

———. "Notice on the Coal-Fields of Alabama; Being an Extract from a Letter to the President from Charles Lyell, Esq., F.R.S., Dated Tuscaloosa, Alabama, 15th February 1846." *Quarterly Journal of the Geological Society of London* 2 (1846): 278–82. Also published in *American Journal of Science,* 2d ser., 1 (1846): 371–76.

———. "Observations on the Loamy Deposit Called 'Loess' of the Basin of the Rhine." *Edinburgh New Philosophical Journal* 17 (1834): 110–22.

———. "Observations on the White Limestone and Other Eocene or Older Tertiary Formations of Virginia, South Carolina and Georgia." *Quarterly Journal of the Geological Society of London* 1 (1845): 429–42.

———. "The Occurrence of Graptolites in the Slate of Galloway in Scotland." *Proceedings of the Geological Society of London* 3 (1838–42): 28–29.

———. "On Ancient Sea Cliffs and Needles in the Chalk of the Valley of the Seine in Normandy." *Reports of the British Association for the Advancement of Science* 11 (1841): 111–13.

———. "On Certain Trains of Erratic Blocks on the Western Borders of Massachusetts, United States." *Notices of the Proceedings of the Royal Institution of Great Britain* 2 (1854–58): 86–97.

———. "On Craters of Denudation, with Observations on the Structure and Growth of Volcanic Cones." *Quarterly Journal of the Geological Society of London* 6 (1850): 207–34.

———. "On Foot-Marks Discovered in the Coal-Measures of Pennsylvania." *Quarterly Journal of the Geological Society of London* 2 (1846): 417–20.

———. "On Fossil Rain-Marks of the Recent, Triassic and Carboniferous Periods." *Quarterly Journal of the Geological Society of London* 7 (1851): 238–47.

———. "On Impressions of Rain-Drops in Ancient and Modern Strata." *Notices of the Proceedings of the Royal Institution of Great Britain* 1 (1851–54): 50–53.

———. "On the Age of the Newest Lava Current of Auvergne, with Remarks on Some Tertiary Fossils of That Country." *Quarterly Journal of the Geological Society of London* 2 (1846): 75–80.

———. "On the Age of the Volcanoes of Auvergne as Determined by the Remains of Successive Groups of Land Quadrupeds." *Edinburgh New Philosophical Journal* 43 (1847): 50–54.

———. "On the Alleged Coexistence of Man and the Megatherium." *American Journal of Science,* 2d ser., 3 (1847): 267–69.

———. "On the Blackheath Pebble-Bed, and on Certain Phaenomena in the Geology of the Neighbourhood of London." *Notices of the Proceedings of the Royal Institution of Great Britain* 1 (1851–54): 164–67.

———. [On the Carboniferous and Older Rocks of Pennsylvania] "A Letter Addressed to Dr. Fitton, . . . Boston the 15th October, 1841." *Proceedings of the Geological Society of London* 3 (1838–42): 554–58.

———. "On the Coal-Formation of Nova Scotia, and on the Age and Relative Position of the Gypsum and Accompanying Marine Limestones." *Proceedings of the Geological Society of London* 4 (1842–45): 184–86.

———. "On the Delta and Alluvial Deposits of the Mississippi and Other Points in the Geology of North America, Observed in the Years 1845, 1846." *Reports of the British Association for the Advancement of Science* 17, pt. 2 (1847): 117–19. Also published in *American Journal of Science,* 2d ser., 3 (1847): 34–39.

———. "On the Discovery of Some Fossil Reptilian Remains, and a Land Shell in the Interior of an Erect Fossil-Tree in the Coal-Measures of Nova Scotia, with Remarks on the Origin of Coal-Fields, and the Time Required for Their Formation." *Notices of the Proceedings of the Royal Institution of Great Britain* 1 (1851–54): 281–88. Also published in *American Journal of Science,* 2d ser., 16 (1853): 33–41.

———. "On the Evidence of Fossil Footprints of a Quadruped Allied to the Cheirotherium in the Coal Strata of Pennsylvania." *American Journal of Science,* 2d ser., 2 (1846): 25–29.

———. "On the Faluns of the Loire, and a Comparison of Their Fossils with Those of the Newer Tertiary Strata in the Cotentin, and on the Relative Age of the Faluns and Crag of Suffolk." *Proceedings of the Geological Society of London* 3 (1838–42): 437–44.

———. "On the Fossil Foot-Prints of Birds and Impressions of Rain-Drops in the Valley of the Connecticut." *Proceedings of the Geological Society of London* 3 (1838–42): 793–96. Also published in *American Journal of Science* 45 (1843): 394–97.

———. "On the Freshwater Fossil Fishes of Mundesley, as Determined by M. Agassiz." *Proceedings of the Geological Society of London* 3 (1838–42): 362–63.

———. "On the Geological Position of the *Mastodon giganteum* and Associated Fossil Remains at Bigbone Lick, Kentucky, and Other Localities in the United States and Canada." *Proceedings of the Geological Society of London* 4 (1842–45): 36–39. Also published in *American Journal of Science* 46 (1844): 320–23.

———. "On the Miocene Tertiary Strata of Maryland, Virginia, and of North and South Carolina." *Quarterly Journal of the Geological Society of London* 1 (1845): 413–29.

———. "On the Newer Deposits of the Southern States of North America." *Quarterly Journal of the Geological Society of London* 2 (1846): 405–10.
———. "On the Newer Deposits of the Southern States of North America." *Quarterly Journal of the Geological Society of London* 4 (1848): 10–16.
———. "On the Occurrence of a Stratum of Stones Covered with Barnacles in the Red Crag at Wherstead, Near Ipswich." *Reports of the British Association for the Advancement of Science* 21, pt. 2 (1851): 65–66.
———. "On the Primitive Inhabitants of Scandinavia." *Reports of the British Association for the Advancement of Science* 17, pt. 2 (1847): 31–32.
———. "On the Probable Age and Origin of a Bed of Plumbago and Anthracite Occurring in Mica-Schist Near Worcester, Massachusetts." *Quarterly Journal of the Geological Society of London* 1 (1845): 199–206. Also published in *American Journal of Science* 47 (1844): 214–15.
———. "On the Relative Age and Position of the So-Called Nummulite Limestone of Alabama." *American Journal of Science,* 2d ser., 4 (1847): 186–91. Also in *Quarterly Journal of the Geological Society of London* 4 (1848): 10–16.
———. "On the Ridges, Elevated Beaches, Inland Cliffs, and Boulder Formations of the Canadian Lakes and Valley of St. Lawrence." *Proceedings of the Geological Society of London* 4 (1842–45): 19–22.
———. "On the Structure and Probable Age of the Coal-Field of the James River, near Richmond, Virginia." *Quarterly Journal of the Geological Society of London* 3 (1847): 261–80.
———. "On the Tertiary Formations and Their Connection with the Chalk in Virginia and Other Parts of the United States." *Proceedings of the Geological Society of London* 3 (1838–42): 735–42.
———. "On the Tertiary Strata of Belgium and French Flanders." *Quarterly Journal of the Geological Society of London* 8 (1852): 277–371.
———. "On the Tertiary Strata of the Island of Martha's Vineyard in Massachusetts." *Proceedings of the Geological Society of London* 4 (1842–45): 31–33. Also published in *American Journal of Science* 46 (1844): 318–20.
———. "On the Upright Fossil-Trees Found at Different Levels in the Coal Strata of Cumberland, Nova Scotia." *Proceedings of the Geological Society of London* 4 (1842–45): 176–78. Also published in *American Journal of Science* 45 (1843): 353–59.
———. [Portion of a Letter to Benjamin Silliman, Jr.] *American Journal of Science,* 2d ser., 1 (1846): 313–15.
———. *Principles of Geology: Being an Attempt to Explain the Former Changes of the Earth's Surface, by Reference to Causes Now in Operation.* 3 vols. London: John Murray, 1830–33.
———. *Principles of Geology* . . . . 2d ed. 3 vols. London: John Murray, 1832–33.
———. *Principles of Geology* . . . . 3d ed. 4 vols. London: John Murray, 1834.
———. *Principles of Geology* . . . . 4th ed. 4 vols. London: John Murray, 1835.

———. *Principles of Geology*. . . . 5th ed. 4 vols. London: John Murray, 1837.
———. *Principles of Geology*. . . . 6th ed. 3 vols. London: John Murray, 1840.
———. *Principles of Geology*. . . . 7th ed. London: John Murray, 1847.
———. *Principles of Geology*. . . . 8th ed. London: John Murray, 1850.
———. *Principles of Geology*. . . . 9th ed. London: John Murray, 1853.
———. "Remarks on Some Fossil and Recent Shells, Collected by Capt. Bayfield in Canada." *Transactions of the Geological Society of London,* 2d ser., 6 (1842): 135–42.
———. "Remarks on Some Fossil and Recent Shells Collected by Capt. Bayfield, R.N., in Canada." *Proceedings of the Geological Society of London* 3 (1838–42): 119–21.
———. *A Second Visit to the United States.* 2 vols. London: John Murray, 1849.
———. "Some Remarks on the Silurian Strata between Aymestry and Wenlock." *Proceedings of the Geological Society of London* 3 (1838–42): 463–65.
———. "State of the Universities." *Quarterly Review* 36 (1827): 216–68.
———. *Travels in North America, Canada, and Nova Scotia.* 2 vols. London: John Murray, 1845.
Lyell, Charles (Sir), and J. W. Dawson. "On the Remains of a Reptile (*Dendrerpeton Acadianum,* Wyman and Owen) and of a Land Shell Discovered in the Interior of an Erect Fossil Tree in the Coal Measures of Nova Scotia." *Quarterly Journal of the Geological Society of London* 9 (1853): 58–63.
Lyell, Katherine Murray. *Life, Letters, and Journals of Sir Charles Lyell, Bt.,* 2 vols. London: John Murray, 1881.
Lyons, Henry. *The Royal Society 1660–1940: A History of Its Administration under Its Charters.* Cambridge: Cambridge University Press, 1944.
MacAndrew, Robert. "Notes on the Distribution and Range in Depth of Mollusca and Other Marine Animals Observed on the Coasts of Spain, Portugal, Barbary, Malta, and Southern Italy in 1849." *Reports of the British Association for the Advancement of Science* 20, pt. 2 (1850): 264–304.
Maclure, William. "Observations on the Geology of the United States, Explanatory of a Geological Map." *Transactions of the American Philosophical Society* 6 (1809): 411–28.
Mantell, Gideon Algernon. "Description of the *Telerpeton Elginense,* a Fossil Reptile Recently Discovered in the Old Red Sandstone of Moray; with Observations on Supposed Fossil Ova of Batrachians in the Lower Devonian Strata of Forfarshire." *Transactions of the American Philosophical Society* 8 (1852): 100–109.
———. *Geological Excursions Round the Isle of Wight and Along the Adjacent Coast of Dorsetshire.* London: H. G. Bohn, 1847.
———. *The Journal of Gideon Mantell, Surgeon and Geologist Covering the Years 1818–1852.* Edited by E. Cecil Curwen. London: Oxford University Press, 1940.
Martin, Theodore. *The Life of His Royal Highness the Prince Consort.* 4th ed. 5 vols. London: Smith, Elder & Co., 1877–80.

Martineau, Harriet. *Society in America.* 3 vols. London: Saunders & Otley, 1837.

Mather, William W. *Comprising the Geology of the First Geological District.* Pt. 1 of *Geology of New York,* by New York (State) Natural History Survey. Albany, N.Y.: Carroll & Cook, 1843.

McAllister, Ethel M. *Amos Eaton: Scientist and Educator.* Philadelphia: University of Pennsylvania Press, 1941.

Meigs, Charles D. *A Memoir of Samuel George Morton, M.D.* Philadelphia: T. K. & P. G. Collins, 1851.

Merrill, George P. *Contributions to the History of American Geology.* Washington, D.C.: U.S. National Museum, 1906.

Meyer, Hermann von. "The Reptiles of the Coal Formation." *Quarterly Journal of the Geological Society of London* 4, pt. 2 (1848): 51–56.

Michaux, François André. *The North American Sylva, or Description of the Forest Trees of the United States, Canada and Nova Scotia. . . .* 3 vols. Philadelphia: T. Dobson, 1817–19.

Miller, Hugh. *The Footprints of the Creator: Or, the Asterolepis of Stromness.* [2d ed.] London: Johnstone & Hunter, 1849.

———. *The Footprints of the Creator: Or, the Asterolepis of Stromness.* 3d ed. With a Memoir of the Author by Louis Agassiz. Boston: Gould & Lincoln, 1866.

Millhauser, Milton. *Just before Darwin.* Middletown, Conn.: Wesleyan University Press, 1959.

[Milman, Henry]. Review of *A Second Visit to the United States, in the Years 1845–6,* by Sir Charles Lyell. *Quarterly Review* 85 (1849): 183–224.

Mitchell, Betty L. *Edmund Ruffin, a Biography.* Bloomington: Indiana University Press, 1981.

Montor, Alexis François Artand de. *Histoire de Dante Alighieri.* Paris: A. LeClerc, 1841.

Morton, Samuel George. "Description of the Fossil Shells Which Characterize the Atlantic Secondary Formation of New Jersey and Delaware; Including Four New Species." *Journal of the Academy of Natural Sciences of Philadelphia* 6 (1829): 72–100.

———. "Descriptions of Two New Species of Fossil Shells of the Genera Scophitis and Crepidula: With Some Observations on the Ferruginous Sand, Plastic Clay and Upper Marine Formations of the United States." *Journal of the Academy of Natural Sciences of Philadelphia* 6 (1829): 107–29.

———. "Geological Observations on the Secondary, Tertiary and Alluvial Formations of the Atlantic Coast of the United States of America. Arranged from the Notes of Lardner Vanuxem." *Journal of the Academy of Natural Sciences of Philadelphia* 6 (1829): 59–71.

———. *Synopsis of the Organic Remains of the Cretaceous Group of the United States. . . .* Philadelphia: Kay & Biddle, 1834.

———. "Synopsis of the Organic Remains of the Ferruginous Sand Formation of the United States with Geological Remarks." *American Journal of Science* 17 (1830): 274–95; 18 (1830): 243–50.

Moultrie, John. *The Dream of Life; Lays of the English Church and Other Poems.* London: W. Pickering, 1843.

Murchison, Roderick I. "First Sketch of Some of the Principal Results of a Second Geological Survey of Russia, in a Letter to M. Fischer." *Philosophical Magazine,* 3d ser., 19 (1841): 417–22.

———. "On the Geological Structure of the Alps, Apennines and Carpathians, More Especially to Prove a Transition from Secondary to Tertiary Rocks, and the Development of Eocene Deposits in Southern Europe." *Quarterly Journal of the Geological Society of London* 5 (1849): 157–312.

———. "On the Silurian System of Rocks." *Philosophical Magazine,* 3d ser., 7 (1835): 46–52.

———. "The 'Permian System' as Applied to Germany, with Collateral Observations on Similar Deposits in Other Countries; Showing That the Rothe-Todte-Liegende, Kupfer-Schiefer, Zechstein, and the Lower Portion of the Bunter-Sandstein, Form One Group, and Constitute the Upper Member of the Palaeozoic Rocks." British Association for the Advancement of Science, London. *Reports* 13 (1843): 52–54.

———. *Silurian System: Founded on Geological Researches in the Counties of Salop, Hereford, Radnor, Montgomery, Caermarthen, Brecon, Pembroke, Monmouth, Gloucester, Worcester, and Stafford; with Descriptions of the Coalfields and Overlying Formations.* 2 vols. London: John Murray, 1839.

New Brunswick Geological Survey. *Report on the Geological Survey of the Province of New Brunswick, [1838]–1842,* by Abraham Gesner. 5 vols. Saint John, New Brunswick: H. Chubb, 1839–43.

New York (State) Natural History Survey. *First Annual Report.* Albany, N.Y., 1837.

———. *Second Annual Report.* Albany, N.Y., 1838.

———. *Third Annual Report.* Albany, N.Y., 1839.

———. *Fourth Annual Report.* Albany, N.Y., 1840.

———. *Geology of New York.* 4 vols. Albany, N.Y., 1842–43.

Nilsson, [Sven]. "On the Primitive Inhabitants of Scandinavia." *Reports of the British Association for the Advancement of Science* 17 (1847): 31–32.

Nott, Josiah Clark. "The Mulatto a Hybrid—Probable Extermination of the Two Races if the Whites and Blacks Are Allowed to Intermarry." *American Journal of Medical Science* 6 (1843): 252–56.

———. *Two Lectures on the Natural History of the Caucasian and Negro Races.* Mobile, Ala.: Printed by Dade & Thompson, 1844.

———. "Unity of the Human Race.—A Letter Addressed to the Editor, on the Unity of the Human Race." *Southern Quarterly Review* 9 (1846): 1–57.

Nyst, Henri. *Description des coquilles et des polypiers fossiles des terrains tertiaires de le Belgique.* Brussels: Académie Royale des Sciences, 1845.

"Obituary Notice of Asahel Clapp." *Boston Medical and Surgical Journal* 68 (1863): 44–45.

Orbigny, Charles Dessalines d', ed. *Dictionnaire universel d'histoire naturelle* . . . . 13 vols. Paris: Renard, Martinet et cie, 1849.

Ord, [George]. "Obituary Notice of William McIlvaine." *Proceedings of the American Philosophical Society* 6 (1859): 101–104.

Owen, David Dale. "On Palm Trees, Found in Posey County, Indiana." *American Journal of Science* 45 (1843): 336–57.

———. "On the Geology of the Western States of North America." *Quarterly Journal of the Geological Society of London* 2 (1846): 433–47.

———. *Report of Geological Survey of Wisconsin, Iowa and Minnesota; and Incidentally of a Portion of Nebraska Territory.* Philadelphia: Lippincott, Grampo & Co., 1852.

Owen, [Richard]. "Art. 8 . . . ." *Quarterly Review* 89 (1851): 412–51.

———. "Description of the Impressions on the Potsdam Sandstone Discovered by Mr. Logan in Lower Canada." *Quarterly Journal of the Geological Society of London* 7 (1851): 250–52.

———. "Notes on the Above-Described Fossil Remains." *Quarterly Journal of the Geological Society of London* 9 (1853): 66–67.

———. "Observations on the Basilosaurus of Dr. Harlan (Zeuglodon Cetoides, Owen)." *Transactions of the Geological Society of London,* 2d ser., 6 (1841): 69–79.

———. *Odontography; or a Treatise on the Comparative Anatomy of the Teeth; Their Physiological Relations, Mode of Development, and Microscopic Structure in the Vertebrate Animals.* 2 vols. London: H. Baillière, 1840–45.

———. *On the Nature of Limbs; a Discourse Delivered on Friday, February 9, at an Evening Meeting of the Royal Institution of Great Britain.* London: J. Van Voorst, 1849.

Percival, James Gates. *Report on the Geology of the State of Connecticut.* New Haven, Conn.: Osborn & Baldwin, 1842.

Philips, Ulrich Bonnell. *Life and Labor in the Old South.* Boston: Little, Brown & Co., 1929.

Phillips, John. *Figures and Descriptions of the Palaeozoic Fossils of Cornwall, Devon and West Somerset; Observed in the Course of the Ordnance Geological Survey of That District.* London: Longman, Brown, Green & Longman's, 1841.

———. *Illustrations of the Geology of Yorkshire; or, a Description of the Strata and Organic Remains: Accompanied by a Geological Map, Sections and Plates of the Fossil Plants and Animals.* Vol. 2, *Mountain Limestone District.* London: John Murray, 1835–36.

Pound, Reginald. *Albert, a Biography of the Prince Consort.* New York: Simon & Schuster, 1973.

Ramsay, Andrew C. "On the Denudation of South Wales and the Adjacent Counties of England." In *Memoirs of the Geological Survey of Great Britain,*

England and Wales. Vol. 1. London: Her Majesty's Stationery Office, 1846–72.

Redfield, William C. "Observations on the Hurricanes and Storms of the West Indies and the Coast of the U. States." *American Journal of Science* 25 (1834): 114–21.

———. "On Newly Discovered Ichthyolites in the New Red Sandstone of New Jersey." *Proceedings of the Geological Society of London* 11 (1844): 23.

———. "On the Fossil Rain-Marks Found in the Red Sandstone Rocks of New Jersey and the Connecticut Valley and Their Authentic Character." American Association for the Advancement of Science. *Proceedings* 4 (1851): 72–75.

———. "Remarks on the Prevailing Storms of the Atlantic Coast of the North American States." *American Journal of Science* 20 (1831): 17–51.

Riess, Karlem. *John Leonard Riddell: Scientist-Inventor; Melter and Refiner of the New Orleans Mint, 1839–1848; Postmaster of New Orleans, 1859–1862.* New Orleans: Louisiana Heritage Press, 1977.

Ritter, Karl. *Die Erdkunde im Verhaltnis zur Natur und zur Geschichte des Menschen, oder allgemeine, vergleichende Geographie, als sichere Grundlage des Studiums und Unterrichts in physikalischen und historischen Wissenschaften.* 3 pts. in 2 vols. Berlin: G. Reimer, 1817–18.

———. *Die Erdkunde im Verhaltnis zur Natur und zur Geschichte des Menschen; oder allgemeine, vergleichende Geographie als sichere Grundlage des Studiums und Unterrichts in physikalischen und historischen Wissenschaften.* 2d ed. 18 pts. in 20 vols. Berlin: G. Reimer, 1832–59.

Rogers, Henry Darwin. "An Inquiry into the Origin of the Appalachian Coal Strata, Bituminous and Anthracite." In *Reports,* edited by the Association of American Geologists and Naturalists, 433–79. Boston: Gould, Kendall & Lincoln, 1843.

———. "On the Position and Character of the Reptilian Foot-Prints in the Carboniferous Red Shale Formation in Eastern Pennsylvania." *Proceedings of the American Association for the Advancement of Science* (1850): 250–51.

Rogers, Henry D., and William B. Rogers. "An Account of Two Remarkable Trains of Angular Erratic Blocks, in Berkshire, Massachusetts, with an Attempt at an Explanation of the Phenomena." *Boston Journal of Natural History* 5 (1845–47): 310–30.

Rogers, William B., and Henry D. Rogers. "Contributions to the Geology of the Tertiary Formations of Virginia." *Transactions of the American Philosophical Society,* 2d ser., 5 (1837): 319–41.

Roy, Thomas. "On the Ancient State of the North American Continent." *Proceedings of the Geological Society of London* 2 (1838–42): 537–38.

Rudwick, Martin J. S. "A Critique of Uniformitarian Geology: A Letter from W. D. Conybeare to Charles Lyell, 1841." *Proceedings of the American Philosophical Society* 111 (1967): 272–87.

———. *The Great Devonian Controversy: The Shaping of Scientific Knowledge among Gentlemanly Specialists*. Chicago: University of Chicago Press, 1985.

Ruffin, Edmund. *An Essay on Calcareous Manures*. Petersburg, Va.: J. W. Campbell, 1832.

———. *Report of the Commencement and Progress of the Agricultural Survey of South Carolina for 1843*. Columbia, S.C.: A. H. Pemberton, 1843.

Say, Thomas. "An Account of Some of the Fossil Shells of Maryland." *Journal of the Academy of Natural Sciences of Philadelphia* 4 (1824): 124–55.

Schneer, Cecil J. "Ebenezer Emmons and the Foundations of American Geology." *Isis* 60 (1969): 439–50.

———. "The Great Taconic Controversy." *Isis* 69 (1978): 173–91.

Secord, James A. *Controversy in Victorian Geology: The Cambrian-Silurian Dispute*. Princeton, N.J.: Princeton University Press, 1986.

———. "The Geological Survey of Great Britain as a Research School, 1839–1855." *History of Science* 24 (1986): 223–75.

Sedgwick, Adam. *A Discourse on the Studies of the University of Cambridge*. 5th ed. London: J. W. Parker, 1850.

———. "Introduction to the General Structure of the Cambrian Mountains; with a Description of the Great Dislocations by Which They Have Been Separated from the Neighbouring Carboniferous Chains." *Transactions of the Geological Society of London*, 2d ser., 4 (1835): 47–68.

[———.] "Natural History of Creation." *Edinburgh Review* 82 (1845): 1–85.

———. "A Synopsis of the English Series of Stratified Rocks Inferior to the Old Red Sandstone—with an Attempt to Determine the Successive Natural Groups and Formations." *Proceedings of the Geological Society of London* 2 (1833–38): 675–85.

Sedgwick, Adam, and Roderick I. Murchison. "On the Distribution and Classification of the Older or Palaeozoic Deposits of the North of Germany and Belgium, and Their Comparison with Formations of the Same Age in the British Isles." *Transactions of the Geological Society of London*, 2d ser., 6 (1840): 221–410.

———. "On the Physical Structure of Devonshire, and on the Subdivisions and Geological Relations of Its Old Stratified Deposits." *Proceedings of the Geological Society of London* 2 (1833–38): 556–63.

Sellers, James B. *History of the University of Alabama*. University, Ala.: University of Alabama Press, 1953.

Silliman, Robert H. "Agassiz vs. Lyell: Authority in Assessment of the Diluvium-Drift Problem by North American Geologists with Particular Reference to Edward Hitchcock." *Earth Sciences History* 13 (1994): 180–86.

———. "The Hamlet Affair: Charles Lyell and the North Americans." *Isis* 86 (1995): 541–61.

———. "The Richmond Boulder Trains: *verae causae* in 19th-Century American Geology." *Earth Sciences History* 10 (1991): 60–72.

Skinner, Hubert C. "Charles Lyell in Louisiana." *Tulane Studies in Geology and Paleontology* 12 (1976): 243–48.

Smith, Crosbie. "William Hopkins and the Shaping of Dynamical Geology: 1830–1860." *British Journal of the History of Science* 22 (1989): 27–52.

Smith, James. "On Recent Depressions in the Land." *Quarterly Journal of the Geological Society of London* 3 (1847): 234–40.

Spokes, Sidney. *Gideon Algernon Mantell, LL.D., F.R.C.S., F.R.S., Surgeon and Geologist.* London: John Bale, Sons & Danielsson, 1927.

Stoney, Samuel Gaillard. *Plantations of the Carolina Low Country.* Edited by Albert Simons and Samuel Lapham. Charleston, S.C.: Carolina Art Association, 1938.

Sydnor, Charles S. *Gentleman of the Old Natchez Region: Benjamin L. C. Wailes.* Durham, N.C.: Duke University Press, 1938.

Tillyard, Alfred Isaac. *A History of University Reform from 1800 A.D. to the Present Time with Suggestions Towards a Complete Scheme for the University of Cambridge.* Cambridge: W. Heffer & Sons, 1913.

Trollope, Frances. *Domestic Manners of the Americans.* 2 vols. London: Whittaker, Treacher & Co., 1832.

Trueman, A. W. *Canada's University of New Brunswick: Its History and Its Development.* New York: Newcomen Society in North America, 1952.

United States Exploring Expedition (1838–1842). *United States Exploring Expedition . . . Under the Command of Charles Wilkes, U.S.N.* 19 vols. in 21. Philadelphia: C. Sherman, 1844–74.

Van Rensselaer, Jeremiah. *Lectures on Geology; Being Outlines of the Science, Delivered in the New York Athenaeum.* New York: E. Bliss & E. White, 1825.

Vanuxem, Lardner. *Comprising the Survey of the Third Geological District.* Pt. 3 of *Geology of New York,* by New York (State) Natural History Survey. Albany, N.Y.: W. & A. White & J. Visscher, 1842.

———. "Remarks on the Characters and Classification of Certain American Rock Formations." *American Journal of Science* 16 (1829): 254–56.

Wailes, B.L.C. "On the Geology of Mississippi." In *Abstract of the Proceedings of the Sixth Annual Meeting of the Association of American Geologists and Naturalists,* 80–81. New Haven, Conn., 1845.

Ward, W. R. *Victorian Oxford.* London: Frank Cass & Co., 1965.

Warren, John Collins. *Description of a Skeleton of the Mastodon Giganteus of North America.* Boston: J. Wilson & Son, 1852.

Weld, Charles Richard. *A History of the Royal Society, with Memoirs of the Presidents.* London: J. W. Parker, 1848.

Wheeler, Harry Edgar. *Timothy Abbott Conrad, with Particular Reference to His Work in Alabama One Hundred Years Ago.* Ithaca, N.Y.: Paleontological Research Institution, 1935.

Whewell, William. *Of Liberal Education in General, and with Particular Reference to the Leading Studies of the University of Cambridge.* 2d ed. London: J. W. Parker, 1850.

———. *On the Principles of English University Education.* London: J. W. Parker, 1837.

———. *On the Principles of English University Education.* 2d ed. London: J. W. Parker, 1838.

———. Review of *Principles of Geology* [vol. 2], by Charles Lyell. *Quarterly Review* 47 (1832): 103–32.

Wightman, Orrin Sage, and Margaret Davis Cate. *Early Days of Coastal Georgia.* St. Simon's Island, Ga.: Fort Frederica Association, 1955.

Wilson, Leonard G. "Brixham Cave and Sir Charles Lyell's . . . *The Antiquity of Man:* The Roots of Hugh Falconer's Attack on Lyell." *Archives of Natural History* 23 (1996): 79–97.

———. *Charles Lyell, the Years to 1841: The Revolution in Geology.* New Haven, Conn.: Yale University Press, 1972.

———. "The Emergence of Geology as a Science in the United States." *Journal of World History* 10 (1967): 416–37.

———. "Geology on the Eve of Charles Lyell's First Visit to America, 1841." *Proceedings of the American Philosophical Society* 124 (1980): 168–202.

———. "The Gorilla and the Question of Human Origins: The Brain Controversy." *Journal of the History of Medicine and Allied Sciences* 51 (1996): 184–207.

———. "Lyell on the Geological Similarity of North America and Europe." *Earth Science History* 1 (1982): 45–47.

———, ed. *Benjamin Silliman and His Circle: Studies on the Influence of Benjamin Silliman on Science in America.* New York: Science History Publications, 1979.

Winstanley, Denys Arthur. *Early Victorian Cambridge.* Cambridge: Cambridge University Press, 1940.

Wolfe, Suzanne Rau. *The University of Alabama, a Pictorial History.* University, Ala.: University of Alabama Press, 1983.

Wylly, Charles S. *Annals and . . . Statistics of Glynn County, Georgia.* Brunswick, Ga.: Charles S. Wylly, 1897.

Wyman, Jeffries. "Notes on the Reptilian Remains." *Quarterly Journal of the Geological Society of London* 9 (1853): 64–66.

———. "On the Fossil Skeleton Recently Exhibited in New York, as That of a Sea-Serpent, Under the Name of Hydrarchos Sillimani." *Proceedings of the Boston Society of Natural History* 2 (1845–48): 65–68.

Index

The main entry for Charles Lyell has been limited to references not readily located under other entries. In entries that refer to Lyell, the initials "CL" are used. During the period covered by this volume, Mary Lyell participated so extensively in her husband's activities that separate entries for her would be redundant.